Lecture Notes in Mathematics

Edited by A. Dold and B. Eckmann

704

Computing Methods in Applied Sciences and Engineering, 1977, I

Third International Symposium
December 5–9, 1977

IRIA LABORIA
Institut de Recherche d'Informatique et d'Automatique

Edited by R. Glowinski and J. L. Lions

Springer-Verlag
Berlin Heidelberg New York 1979

Editors
R. Glowinski
J. L. Lions

IRIA LABORIA
Domaine de Voluceau
B.P. 105
Rocquencourt
F-78150 Le Chesnay

AMS Subject Classifications (1970) 65-02, 65K05, 65LXX, 65MXX, 65NXX, 65P05, 76-04, 93E10, 93E25

ISBN 3-540-09123-8 Springer-Verlag Berlin Heidelberg New York
ISBN 0-387-09123-8 Springer-Verlag New York Heidelberg Berlin

Printed in Germany

Printing and binding: Beltz Offsetdruck, Hemsbach/Bergstr.
2141/3140-543210

INTRODUCTION

This book contains part of the lectures which were presented during the Third International Symposium on Computing Methods in Applied Sciences and Engineering, December 5 - 9, 1977, organized by IRIA-LABORIA under the sponsorship of Association Française pour la Cybernêtique Economique et Technique (AFCET), Groupe pour l'Avancement des Mêthodes Numêriques de l'Ingênieur (GAMNI), International Federation for Information Processing (IFIP WG7 - 2).

More than 400 scientists and engineers from many countries attended this meeting.

The organizers wish to express their gratitude to Mr. A. DANZIN, Director of IRIA and address their thanks to each session chairperson who directed very interesting discussions and also to all speakers.

Sincere gratitude is also expressed to the IRIA Public Relations Office whose help contributed greatly to the success of this Symposium.

The remainder of these proceedings are published as Lecture Notes in Physics, volume 91

R. GLOWINSKI J.L. LIONS

IRIA-LABORIA Institut de Recherche d'Informatique et d'Automatique
IRIA Research Laboratory

I N T R O D U C T I O N

Le présent volume rassemble une partie des travaux présentés au Troisième Colloque International sur les "Méthodes de Calcul Scientifique et Technique" organisé par l'IRIA-LABORIA du 5 au 9 Décembre 1977, sous le patronage de l'Association Française pour la Cybernétique Economique et Technique (AFCET), le Groupe pour l'Avancement des Méthodes Numériques de l'Ingénieur (GAMNI) et l'International Federation for Information Processing (IFIP WG7 - 2).

Ce Colloque a réuni à Versailles près de 400 chercheurs et ingénieurs de toutes nationalités.

Les organisateurs remercient Monsieur A. DANZIN, Directeur de l'IRIA et les divers Présidents de séance qui ont animé d'intéressantes discussions ainsi que tous les conférenciers qui ont pris part à ce Colloque.

Nos remerciements vont également au Service des Relations Extérieures de l'IRIA dont l'aide a joué un rôle essentiel dans l'organisation de cette rencontre.

L'autre partie de ce Colloque est publiée sous Lecture Notes in Physics, volume 91.

R. GLOWINSKI J.L. LIONS

IRIA-LABORIA Institut de Recherche d'Informatique et d'Automatique

IRIA Research Laboratory

Table des Matières
Table of Contents

GENERALITIES
GENERALITES

NUMERICAL ALGEBRA AND OPTIMIZATION
ALGEBRE NUMERIQUE ET OPTIMISATION

FINITE ELEMENTS
ELEMENTS FINIS

TIME DEPENDANT PROBLEMS
PROBLEMES D'EVOLUTION

NON-LINEAR PROBLEMS, BIFURCATION
PROBLEMES NON LINEAIRES, BIFURCATION

HOMOGENIZATION
HOMOGENEISATION

MEDICAL APPLICATIONS
APPLICATIONS MEDICALES

GENERALITIES

GENERALITES

SUR CERTAINS ASPECTS QUALITATIFS DE LA THEORIE DES EQUATIONS AUX DERIVEES PARTIELLES

R. THOM

Alors que la théorie qualitative des équations différentielles ordinaires, vieille déjà d'un siècle, connait actuellement un développement impressionnant (rappelons à cet égard, après les noms de H. Poincaré, de G. Birkhoff, celui de S. Smale...), rien de semblable n'existe pour les Equations aux Dérivées Partielles. Il est intéressant d'en comprendre la raison, et d'indiquer une voie possible par où une telle théorie pourrait entrevoir ses tout premiers développements.

En théorie des Equations Différentielles Ordinaires, on se préoccupe avant tout de fournir une description totale ("phase portrait") de l'espace de phase, qui est une variété différentiable M. L'intégration d'un champ de vecteurs sur M définit sur M un feuilletage de dimension un, avec singularités. Les propriétés topologiques globales de M s'expriment alors par des relations entre le nombre des singularités, leurs configurations relatives, la décomposition de M en bassins d'attracteurs ... etc ; la théorie de la stabilité structurelle permet de préciser (dans une certaine mesure) la nature des singularités et des attracteurs "génériques". Or, dans les Equations aux Dérivées Partielles, on recherche les applications F d'une variété source S dans une variété but B, satisfaisant à certaines conditions liant les dérivées partielles de F (donc en fait, on exprime que le jet d'ordre r de F, $j^r(F)$, se trouve dans une certaine sous-variété G de l'espace des jets $G \subset J^r(S, B)$). L'espace à considérer est donc un espace fonctionnel $L(S, B)$ d'applications de S dans B qui, si le problème de Cauchy est bien posé, sera lui-même le siège d'un "flot" comme notre variété M de tout-à-l'heure. On aperçoit la différence entre les deux situations : M est de dimension finie, alors que $L(S, B)$ est de dimension infinie. De plus, si B est un espace vectoriel, comme dans le cas linéaire, la topologie de $L(S, B)$ est triviale. Et si B est une variété à topologie non triviale, il est en général difficile de calculer la topologie (l'homotopie) de $L(S, B)$. Les seules tentatives faites en ce sens qu'on puisse citer relèvent de la théorie de Morse, pour les problèmes variationnels sur les espaces de courbes. Les généralisations aux dimensions supérieures se heurtent à de grandes difficultés (qu'elles soient de caractère technique ou fondamental).

Peut-être pourrait-on revenir à une situation de dimension finie si l'on s'inspirait d'idées récemment développées dans la théorie physique des milieux ordonnés. Considérons par exemple un cristal liquide tel qu'un nématique orienté. Il s'agit d'un fluide dont les molécules, allongées comme des aiguilles, s'orientent parallèlement entre elles du fait de leurs interactions. En tout point (régulier) du milieu, on peut considérer le sous-groupe d'isotropie G_x^s des isomorphismes locaux qui conservent la structure ; si G désigne le groupe total des isomorphismes locaux en x, le quotient G/G_s^x paramétrise la totalité de toutes les structures locales de type mématique orienté. Dans le cas présent il s'agit de l'espace homogène $SO(3)/SO(2)$ qui n'est autre que la 2-sphère S^2. Si le nématique était non orienté (les molécules peuvent pivoter sur elles-mêmes de 180°), cet espace Y des structures locales serait le plan projectif $\mathbb{PR}(2)$, quotient de S^2 par le groupe Z_2 des transformations antipodales.

Cela étant, un milieu ordonné présente en général des défauts ; il s'agit de sous-ensembles fermés K où la structure régulière n'est plus présente, et où la structure locale définie au voisinage de K ne peut être prolongée continuement sur K. On s'intéresse particulièrement aux défauts structurellement stables, qui persistent homéomorphiquement à eux-mêmes dans une petite perturbation des conditions extéricures.

En général, l'ensemble K des défauts a une structure stratifiée ; il est constitué par la réunion disjointe de sous-variétés de dimension décroissante (singularités régulières, singularités de singularités... etc) ; en un point k d'une strate de dimension maximum p, on prend un $(n-p)$ plan transverse à la strate, et dans ce plan une sphère s de dimension $n-p-1$, de petit rayon. Les physiciens Kleman et Toulouse ont alors énoncé le principe :
<u>Si un défaut</u> K <u>est stable dans</u> R^n, <u>l'application</u> $g : s \rightarrow Y$ <u>qui associe à tout point de</u> s <u>la structure locale en ce point est une application non homotope à zéro</u>.

<u>Exemples d'applications du principe</u>.

Pour un nématique orienté, $Y = S^2$, il ne peut y avoir de défaut en forme de courbe. Car alors s serait de dimension un, et toute application de S^1 dans S^2 est homotope à une constante. Les seuls défauts stables sont des points, car alors s est de dimension deux. Par contre, pour un nématique non orienté, il peut y avoir des courbes de défauts correspondant à l'application du cercle s^1 sur la droite projective de $PR(2)$.

Nous montrerons tout-à-l'heure pourquoi le principe de Kleman-Toulouse peut tomber en défaut (si l'on peut dire !). Revenons aux Equations aux Dérivées Partielles. La donnée d'une solution $u : S \to B$ définit en général sur S une structure analogue à celle d'un milieu ordonné.
Prenons l'exemple le plus simple d'une fonction scalaire u sur S ; en chaque point $\in S$, on prend le quotient du groupe des isomorphismes locaux (ici, difféomorphismes locaux) par le sous-groupe qui laisse invariant la fonction u. En se restreignant éventuellement à un ordre fini, on définit l'espace Y des structures locales comme espace homogène ; à chaque type de singularité de u est associé un certain espace Y ; ces singularités peuvent être considérées comme des "défauts" d'une structure ordonnée. La stabilité des défauts correspond alors à la stabilité structurelle des singularités. Dans certains cas, le principe de Kleman-Toulouse s'identifie au principe de transversalité. Par exemple si l'on veut trouver les singularités d'une un-forme différentiable θ, $\theta = \sum_{i=1}^{n} a_i \, dx_i$, il convient d'écrire que l'application section $s : x \to a_i(x)$ est transversale sur l'origine $a_i = 0$. A cette seule condition l'ensemble singulier sera structurellement stable. Mais alors l'ensemble consiste de points isolés, et l'application g du principe de Kleman-Toulouse est l'application identique de S^{n-1} sur elle-même. Dans tous les cas où les défauts sont définis "transversalement", l'application g est non seulement non homotope à zéro, mais fournit un générateur du groupe d'homotopie $\pi(Y)$.

Mais le lemme de transversalité (qui dit que "presque toutes" les applications sont transversales) peut lui-même tomber en défaut lorsque les applications sont soumises à des conditions différentielles. Il convient à cet égard de faire une étude systématique des <u>jets permis</u> i.e. compatibles avec les équations aux dérivées partielles données au départ. L'existence de singularités exclues est un puissant moyen d'analyse, en théorie elliptique notamment (principe du maximum).

De là résulte que le lemme de transversalité n'est plus valable en général, et qu'on peut avoir de manière rigide des singularités très dégénérées (une fonction holomorphe sur une variété compacte est constante ...).
Il peut cependant arriver que le lemme de transversalité retrouve (à l'intérieur d'une catégorie plus restreinte), sa validité. En effet, les seuls opérateurs différentiels réellement intéressants sont ceux qui permettent un relativement grand nombre de solutions locales, et un pseudo-groupe d'équivalences locales G ; on peut alors former l'espace des structures locales G/G_u associé à une solution locale u . Tel est le cas des fonctions holomorphes sur une variété de Stein. En voici un autre exemple :

Comparons les problèmes : Trouver les singularités génériques

a) d'une 1-forme différentiable (sans condition) θ

b) d'une 1-forme intégrable (satisfaisant à la condition de Frobenius $\theta \wedge d\theta = 0$).

c) d'une 1-forme fermée $d\theta = 0$.

A priori, on a les inclusions des solutions (a)$\supset$(b)$\supset$c.
(a) permet la transversalité ; (b) ne la permet pas. On sait en effet (phénomène de Kupka) qu'une 1-forme intégrable dans R^n a génériquement des singularités de codimension deux : la (n-2)- suspension d'un foyer (dans la terminologie de Poincaré). En effet, si la singularité foyer dans un plan R^2 était isolée dans R^n, elle se transformerait dans une 2-section parallèle ne passant par le point singulier en une singularité centre, ce qu'exclut la continuité des valeurs propres au point singulier. Par contre, la condition c) plus forte permet à nouveau la transversalité, parce qu'alors $\theta = df$ et on peut varier f de manière à rendre df transversale sur la section nulle du fibré des covecteurs.

L'effet des conditions d'intégrabilité doit être pris en compte lorsqu'on veut appliquer le principe de Kleman-Toulouse. Donnons-en un exemple emprunté à la physique des cristaux liquides ; il s'agit des smectiques, qui sont des nématiques dans lesquelles les molécules s'organisent parallèlement en couches séparées par des surfaces parallèles auxquelles la direction des molécules est normale.
La variété des structures locales est donc $Y = S^2$, normales orientées aux surfaces. Le principe de Kleman-Toulouse d'applique au champ des directions, pour lequel il n'y a pas de condition d'intégrabilité.
Ceci entraîne qu'il ne peut y avoir de surfaces "ondes de choc" (parois) limitant deux domaines de l'espace où la direction des normales présenterait une discontinuité ; or les normales à une surface dans R^3 enveloppent en général une surface "focale" ; si cette surface existait, il y aurait certainement focalisation des normales au voisinage de l'enveloppe, donc choc, ce qui est exclu. Les seules surfaces possibles de la structure sont celles dont la surface focale est réduite à deux courbes : ce sont des cyclides de Dupin, dont les normales rencontrent deux coniques focales (Friedel). En un point d'une telle conique, le principe de Kleman-Toulouse est en défaut. En effet la sphère S^1 transverse à la conique n'a aucune application essentielle dans la sphère S^2 . En fait la singularité n'en est pas une au point de vue topologique, car les surfaces s'empilent les une sur les autres comme des chapeaux de clowns, l'ensemble de leurs points coniques constituant la conique.

Dans la réaction de Zhabotinski, on observe aussi des singularités de codimension deux ; il s'agit là en principe d'un phénomène décrit par une équation dite de réaction-diffusion $u_t = F(u) + K \Delta u$, où la cinétique $u_1 = F(u)$ présente un cycle limite. Il y a alors dans l'espace produit du plan par le temps des singularités de codimension deux qui sont soit du type "centre", analogues aux surfaces empilées des smectiques, soit du type "foyer". La propagation du front d'onde (rendu visible par un réactif coloré) est régie par un processus de type Hamilton-Jacobi.

Parfois, la variété des structures locales peut être spécifiée par des modes de propagation propres à l'équation considérée, en particulier s'il y a des caractéristiques : sous-variétés de S dans lesquelles la propagation se fait de façon autonome ; en ce cas, en complétant la caractéristique à partir d'un élément de contact, on peut - à partir des données initiales ou au bord - construire une fonction V sur la fibre Y, potentiel qui doit être minimisé pour définir la solution. Un schéma de ce genre est connu pour l'équation de Riemann $u_t = f(u)_x$ (Théorème de Peter Lax). La présence de défauts dans tout milieu naturel soumis à des contraintes au bord est pratiquement inévitable, et toute théorie raisonnable d'Equations aux Dérivées Partielles se doit de les identifier et les répertorier.

Référence

G. Toulouse, M. Kleman, Principles of a classification of defects in ordered Media, J. Phys. Letters, 37, L 149, 1976.

LES PROBLEMES MAL POSES ET LES PROBLEMES MATHEMATIQUES DE TRAITEMENT AUTOMATIQUE DE RESULTATS D'EXPERIENCES

A.N. TIKHONOV
Institut de Mathématiques Appliquées
de l'Académie des Sciences de l'URSS

I. Une particularité type de la façon de poser un problème mathématique classique est de supposer que toutes les données initiales sont connues exactement et que tous les calculs se font aussi exactement. Ce point de vue a un caractère hypnotique mais il a quand même un rôle important dans le développement de la mathématique et de ses multiples applications.

Le développement des moyens de calcul électronique a considérablement élargi les possibilités des applications de la mathématique. Actuellement la mathématique est omniprésente dans tous les domaines de l'activité humaine. Les nouvelles applications ont soulevé des problèmes mathématiques nouveaux et ont montré que de nombreux problèmes pratiques ne rentrent pas dans le cadre de la conception citée ci-dessus ; cette conception n'est donc pas universelle. Faisant l'analyse de différents problèmes mathématiques, HADAMARD (1929), a introduit la notion de "problème bien posé". Si le problème consiste à trouver la solution $Z = R(u)$, où "u" est la donnée initiale, Z la solution et R la relation fonctionnelle entre u et Z , alors ce problème s'appelle problème bien posé si la solution de ce problème :

1°) existe
2°) est unique
3°) est stable par rapport aux perturbations.

Il est évident que l'on prend une classe de données d'entrée possibles $\mathcal{U}$ ainsi qu'une classe $\mathcal{Z}$ de solutions possibles et que dans $\mathcal{Z}$ et $\mathcal{U}$ on définit la notion de distance ρ_u et ρ_z (mesures de variations), cela pour pouvoir définir la notion de stabilité.

Le point de vue de HADAMARD, largement suivi par les mathématiciens, consiste en ce que les problèmes mathématiques essentiels satisfassent les règles énoncées ci-dessus et que les problèmes mal posés (problèmes n'obéissant pas à une des règles ci-dessus au moins), ne puissent pas être rencontrés dans les applications physiques ou techniques. Ce point de vue se fonde en particulier sur le fait que si 3°) n'est pas vérifié, alors des variations, même très petites sur $\tilde{u}$, ou des données initiales approximatives u , peuvent entraîner des variations pour les solutions correspondantes $\tilde{Z} = R(\tilde{u})$ ou $Z = R(u)$, si grandes que la solution approchée de tels problèmes n'a pas de sens. Cependant, dans cette argumentation il y a une proposition cachée : la solution approchée $\tilde{Z}$ est prise comme la solution exacte de l'équation fonctionnelle $\tilde{Z} = R(\tilde{u})$. Cette argumentation établit une règle générale : pour les problèmes instables, on ne peut pas prendre comme solution approchée une solution exacte donnée par R appliqué à $\tilde{u}$. Ainsi la première question qui se pose est : comment poser le problème des solutions approchées des problèmes instables ? Cela peut être appliqué non seulement aux problèmes instables, mais aussi aux problèmes mal définis.

Si $\rho_u(u,\tilde{u}) \leq \delta$ entraîne $\sup \rho_Z(Z,\tilde{Z}) = \varepsilon(\delta)$ alors $\varepsilon(\delta)$ sera appelé module de définition du problème. Si ε_o est la précision demandée sur la solution, δ_o est une précision donnée pour δ et si $\varepsilon(\delta_o) > \varepsilon_o$, alors le problème est analogue à un problème instable.

2. Pour définir un objet à partir de l'observation de ses caractéristiques, il est normal d'obtenir le plus possible d'observations (sur cet objet) ; cela doit en général donner la possibilité d'augmenter l'exactitude de la définition. Cependant cela conduit souvent à des problèmes mathématiques indéterminés qui n'ont pas de solution.

Un exemple très simple est la vérification de l'hypothèse de l'existence d'une relation linéaire entre les valeurs observées x et y et la définition de cette relation à partir de certaines observations.

La relation est de la forme :

$$ax_i + by_i + c = 0$$

où les (x_i, y_i), i=1, 2, ...n sont les observations. (On rencontre ce genre de problème en biologie, technologie chimique, etc..). En principe le système cité ci-dessus n'a pas de solution classique lorsque $n > 3$. Il faut alors introduire la notion de solution généralisée qui dans ce cas là est donnée par la méthode des moindres carrés dûe à GAUSS et LAGRANGE.

Cependant, la méthode des moindres carrés peut donner des solutions instables. On prend par exemple le système suivant :

$$x + 7y = 5$$

$$\sqrt{2}x + \sqrt{98}y = \sqrt{50}$$

et l'on regarde le comportement des systèmes approchés en résolvant par ordinateur, ce qui en quelque sorte modèlise les systèmes approchés correspondant au traitement des observations.

On introduit donc l'information dans le calculateur et l'on effectue les calculs avec la précision 100, 300, et 500 décimales; on obtient :

$$x_{100} = 0,\dots \qquad x_{300} = 1,6\dots , \qquad x_{500} = 5,\dots ,$$

ce qui montre l'instabilité de la méthode classique ainsi que de la méthode des moindres carrés. On peut montrer cette instabilité sans passer par le calcul sur ordinateur.

Dans de nombreux problèmes de l'automatisation du traitement des observations (en tant qu'une des étapes), on rencontre souvent des systèmes algébriques d'ordre élevé, pouvant présenter une pareille instabilité.

Les solutions généralisées définies comme stables peuvent être construites par la méthode de régularisation (solutions normales).

3. Le problème de l'automatisation du traitement des observations scientifiques est très actuel, ne serait-ce qu'à cause du niveau de l'automatisation des expériences, (par exemple en physique). Les observations et leur traitement sont en fait les composantes d'un même problème.

L'automatisation de l'expérience et les moyens d'enregistrement des résultats permettent d'obtenir en un laps de temps très court une quantité d'informations assez importante (des dizaines ou des centaines de milliers de dessins, d'oscillogrammes ou des données de détecteurs (capteurs), etc...). Pour interprèter cette information, il faut un traitement mathématique. Souvent ce traitement doit être exécuté pratiquement en même temps que le déroulement de l'expérience ou avec seulement un léger décalage dans le temps. Un tel traitement qui demande le traitement d'une grande quantité d'informations ne peut être exécuté qu'à l'aide d'un ordinateur.

Ce qui veut dire que l'utilisation de l'ordinateur doit être une composante nécessaire dans l'expérience de l'étude d'un objet ou d'un phénomène.

Souvent avant de commencer le traitement des observations, on fait une classification (sélection) primaire des résultats enregistrés (photos, vues d'un film, etc..) en éliminant les informations inutiles. On agit ainsi lors du traitement des photographies de la trace des particules nucléaires s'entrechoquant dans les chambres à bulle, ou lors d'autres expériences.

Pour que le calculateur puisse traiter ces données d'expériences, il faut les lui "fournir". Pour cela, il faut "récupérer" les informations des appareils enregistreurs, les digitaliser, et les stocker dans les mémoires de l'ordinateur,("archiver" ces informations). La "récupération" de ces informations et leur digitalisation se fait souvent pendant l'expérience même.

La plupart des expériences se compose des étapes essentielles suivantes de traitement mathématique des résultats d'observation :

<u>1ère étape</u> : traitement primaire. Cette étape peut comprendre une normalisation des données de l'observation, une préparation en vue d'un système de calcul bien défini, un traitement statistique, un filtrage, etc.. Son but est d'obtenir une sortie (une courbe expérimentale) et on notera ce résultat par u.

<u>2ème étape</u> : analyse du dispositif, c'est-à-dire la construction de l'opérateur A correspondant au modèle du dispositif à l'aide duquel on établit les relations fonctionnelles entre les caractéristiques quantitatives du modèle de l'objet (du phénomène) étudié et la sortie (courbe de sortie)

$$Az = u \tag{1}$$

<u>3ème étape</u> : interprétation des résultats obtenus dans la première étape. On définit donc dans cette étape, les caractéristiques du modèle de l'objet (ou phénomène) $\tilde{z}$ selon les résultats de sortie u et selon l'opérateur A, adaptées au modèle du dispositif.

Souvent le traitement des résultats d'expériences est utilisé de façon limitée, comme un traitement mathématique primaire en vue d'obtenir une sortie. Mais nous, par contre, nous considérerons le traitement mathématique des données dans un sens plus vaste, prévoyant l'utilisation de ces trois étapes y compris l'interprétation physique. Un tel traitement va être appelé traitement mathématique complet. L'ensemble des algorithmes du traitement complet (de la 1ère à la 3ème étape) et sa mise en oeuvre sera appelé système automatisé du traitement mathématique complet des résultats d'expériences ou, plus brièvement, système de traitement.

4. On va étudier plus en détail le problème de l'interprétation des résultats des observations. Ce problème consiste à reconnaître l'allure de l'objet (du phénomène) étudié selon les valeurs approchées de sortie des observations $\tilde{u}$. On note δ l'estimation de la précision sur $\tilde{u}$. Il est évident qu'il s'agit du modèle approché de l'objet (du phénomène) Z défini par une certaine classe hypothétique $Z = \{z\}$ de modèles approchés. Le modèle individuel z est défini par l'appartenance à la classe Z choisie et par certaines caractéristiques qui sont des paramètres numériques, des fonctions etc.. .
Soit u une valeur théorique du résultat des observations correspondant au modèle hypothétique , où $u=Az$. Ici l'opérateur A définit une relation de cause à effet entre z et u.

On appellera z un correspondant (correspondant formel) de $\tilde{u}$ du modèle si $\rho_u(Az,\tilde{u}) \leq \delta$ où ρ_u est une distance dans l'espace $\mathcal{U}$ de valeurs admissibles (si bien que cet espace pourra être considéré comme normé) et on pourra écrire :

$$\|Az - \tilde{u}\| \leq \delta$$

La notion de correspondance formelle est un principe de sélection dans les limites de la précision donnée des observations.

On appelle Z_δ l'ensemble de tous les modèles z qui peuvent être formellement des correspondants de $\tilde{u}^\delta$. Si Z_δ est vide (il n'y a pas de modèle correspondant aux résultats des observations) cela signifie que les modèles de Z ont une structure si simplifiée que les correspondants de $\tilde{u}$ n'existent pas pour la précision δ. Dans ce cas là, il faut élargir la classe Z en prenant si possible une suite entière de classes hypothétiques croissantes de mocèles Z_1 , Z_2, ... Z_n jusqu'à ce qu'on obtienne une correspondance.

Si Z_δ n'est pas vide et contient des modèles assez différents, alors la notion de correspondance formelle ne peut pas être une méthode d'interprétation dans Z . Cela signifie que la classe de modèles hypothétiques Z ne correspond pas à la précision des observations et que les modèles individuels z ont une structure si sensible que ces modèles assez différents sont pratiquement équivalents entre eux (pour la précision δ).

Pour créer la stabilité des méthodes d'interprétation (reconnaissance de l'objet (du phénomène)), il est nécessaire d'établir un meilleur principe de sélection du modèle z dans l'ensemble des modèles étudiés. Cette sélection peut être considérée comme un principe de sélection d'un modèle avec une complexité minimale correspondant à une précision des observations. La notion de complexité d'un modèle peut être formulée à l'aide des fonctionnelles de complexité $\Omega(z)$ - continues, non négatives et pour lesquelles l'ensemble $Z_c = \{z:\Omega(z) \leq c\}$ est compact. On appellera une solution normale du problème un élément $\bar{Z}^{\delta}$ tel que

$$\Omega(\bar{Z}^{\delta})) \leq \Omega(Z^{\delta}) \qquad z^{\delta} \in Z^{\delta} .$$

L'unicité d'un tel élément dépend du choix de $\Omega(z)$. On peut énoncer des conditions assez générales suffisantes pour l'unicité de $\bar{z}^{\delta}$. On peut prouver que Z^{δ} est stable par rapport aux petites variations $\tilde{u}$, δ et aussi par rapport aux variations de l'opérateur A. On peut aussi montrer que la solution normale Z^{δ} est la solution exacte d'une équation $A\bar{z} = \bar{u}$, si $\rho_u(\bar{u},\tilde{u}^{\delta}) \to 0$. Cette affirmation signifie que $\tilde{Z}^{\delta}$ est une solution stable généralisée (régularisée) de l'équation $Az = \tilde{u}^{\delta}$.

Si la caractéristique des modèles $z \in Z$ est une fonction $Z(s)$, $(a \leq s \leq b)$, alors la fonctionnelle de complexité $\Omega(z)$ peut être choisie par exemple comme

$$\Omega(Z) = \int_a^b (Z'^2 + Z^2)ds \qquad \text{(I)}$$

Si la classe Z des modèles hypothétiques se compose des fonctions constantes par morceaux $z(s)$ $(a \leq s \leq b)$, alors dans certains cas on peut prendre comme fonctionnelle de complexité le nombre n d'intervalles de la fonction $z(s)$ (si la fonction est univalente). Les solutions généralisées stables permettent de construire des algorithmes stables pour les systèmes automatisés du traitement des résultats d'expériences.

L'exactitude de l'interprétation sera considérée dans le sens de l'exactitude de la définition de la solution généralisée (normale) dans la classe donnée Z , dépendant du choix des individuels $\tilde{u}^{\delta}$.

5. Les systèmes de traitement automatisé sont construits sur le principe de modèles hiérarchisés.

La structure du système de traitement dont nous allons parler, est une structure type pour un grand ensemble d'expériences physiques et son système moniteur est inchangé lors des divers remplissages des modules. C'est pour cela que le système décrit ici peut être appelé, pour de nombreuses raisons, système à plusieurs buts.

Soulignons les problèmes typiques apparaissant lors du traitement de nombreuses classes d'expériences physiques et les étapes essentielles du traitement de ces problèmes en vue d'obtenir des résultats.

Comme déjà signalé ci-dessus, l'une des étapes du traitement du résultat de mesures est un traitement primaire. Son but est d'obtenir les résultats de sortie des expériences ("courbes de sortie"). C'est pour cela que le premier module du système de traitement doit être un module de traitement primaire. L'action du système lors du traitement primaire sera appelée travail du système dans le "cadre de traitement primaire". La première partie de l'interprétation est une analyse du dispositif. Pour résoudre le problème de l'interprétation, il est recommandé d'utiliser les modèles du dispositif auquel correspondent les différentes approches de l'opérateur A. Si l'on utilise les modèles de dispositif dans la classe $\{A\}$, alors le but de l'analyse est de trouver le modèle individuel A dans la classe (c'est-à-dire trouver l'opérateur correspondant), qui se rapproche du dispositif réel et une estimation de son écart par rapport à ce dispositif.

Si chaque modèle A de la classe est défini par un nombre fini de paramètres, alors le but de l'analyse du dispositif est :

a) trouver la valeur des paramètres permettant d'approcher le plus le modèle du dispositif de la classe $\{A\}$ du dispositif réel,

b) estimer (apprécier), cette approximation.

Dans de nombreux cas l'analyse du dispositif se fait à l'aide de :

a) montage d'expériences particulières,

b) simulation mathématique (numérique).

Le deuxième module du système de traitement doit être un module d'analyse du dispositif fonctionnant dans le "cadre de l'analyse du dispositif".

Le trosième module du système sera un module "interprétation". Le but essentiel du travail dans ce cadre est la définition des caractéristiques du phénomène dans la classe des modèles considérés.

On a aussi besoin d'une appréciation de l'influence que peuvent avoir sur l'interprétation des erreurs sur les données d'entrée, c'est-à-dire l'analyse de la précision de l'interprétation. Cette précision se détermine par une simulation mathématique : le modèle du phénomène étudié étant choisi - à l'aide du modèle du dispositif - (appelé plus tard "modèle quasi-réel"), on calcule des données de sortie exactes de "l'expérience modèle", on introduit ensuite dans ces données un bruit de niveau typique et les résultats ainsi obtenus sont ensuite soumis au traitement mathématique du système étudié.

La réalisation d'une telle étape sera appelée expérience quasi-réelle. L'expérience quasi-réelle permet de prévoir une expérience réelle. De la même façon, elle permet d'estimer le niveau admissible des erreurs sur les données expérimentales pour une précision d'interprétation donnée, et de choisir les paramètres optimaux dirigeant le dispositif.

Le quatrième module du système, qui est la simulation numérique, est fait pour accomplir cette tâche.

Les dispositifs expérimentaux ont souvent des paramètres directeurs (principaux), par le choix desquels on peut changer les résultats du traitement des données expérimentales. C'est la raison pour laquelle le système doit fonctionner sous le régime des "paramètres directeurs". Le but essentiel du système dans ce régime est de définir la valeur de ces paramètres directeurs, qui permettent, à partir des éléments caractéristiques des structures types étudiées, d'obtenir la meilleure solution. En conclusion, le système doit travailler dans le cadre suivant :

a) "traitement primaire"
b) "analyse du dispositif"
c) "interprétation"
d) "analyse de la précision"
e) "paramètres directeurs"
f) "simulation numérique".

Le système satisfaisant les exigences énumérées ci-dessus est un système à plusieurs buts, ce qui signifie que lors du traitement des résultats correspondants - correspondant au dispositif concret - il peut être utilisé pour la solution de différents problèmes répondant aux régimes décrits et donner des réponses à un certain nombre de questions posées par les chercheurs et, de plus, il peut être facilement appliqué à de nouveaux fonctionnements. On donne ensuite la description du système concret EOS (destiné au traitement des expériences sur les plasmas - (diagnostic du plasma)), satisfaisant toutes les exigences énumérées et, par conséquent, ayant plusieurs buts. Sa structure logique et son système moniteur sont tels qu'ils contiennenet les éléments essentiels nécessaires pour le traitement d'une grande classe d'expériences. En conséquence, le terme "à plusieurs buts", a, pour ce système un autre sens encore, c'est-à-dire la possibilité d'utiliser celui-ci pour traiter d'autres expériences en initialisant de façon analogue les modules correspondants.

Pour la commodité d'utilisation du système, il est bon d'avoir un langage spécial de commande. Un tel langage est créé pour le système décrit ci-dessus. Il permet, en se basant sur l'information correspondant à l'expérience physique, de faire entrer le système dans le mode de travail correspondant. Ce langage est un langage orienté, de niveau plus élevé que les langages ordinaires universels utilisés pour les algorithmes. L'existence d'un tel langage permet de simplifier l'utilisation du système et n'exige pas de l'expérimentateur une préparation spéciale, dans le domaine des mathématiques et de la programmation. En utilisant le système avec ce langage, on peut se limiter à la connaissance du contenu physique de l'information. La définition, à l'aide de ce langage, de l'information pour "mettre en marche" le système n'est qu'une transcription de l'information d'entrée.

La "translation" de ce langage génère, suite aux phrases choisies, une séquence définie d'accomplissement des opérations dans les modules essentiels ainsi que dans les sous-modules. Avec un tel langage pour mettre en marche le système, il est nécessaire d'entrer dans le calculateur une petite information sur les données initiales (provenant des archives ou non), ainsi que les données de base correspondant au mode de travail qui nous intéresse, les données de bases relatives aux classes de modèles de dispositifs étudiés, et les données de base du phénomène qui ont déjà été archivées).

6. Comme exemple, citons le résultat du traitement des expériences définissant les sections (atomiques) (γ,n), c'est-à-dire les sections de sortie des neutrons de l'atome d'une matière soumise à un bombardement de rayons γ .

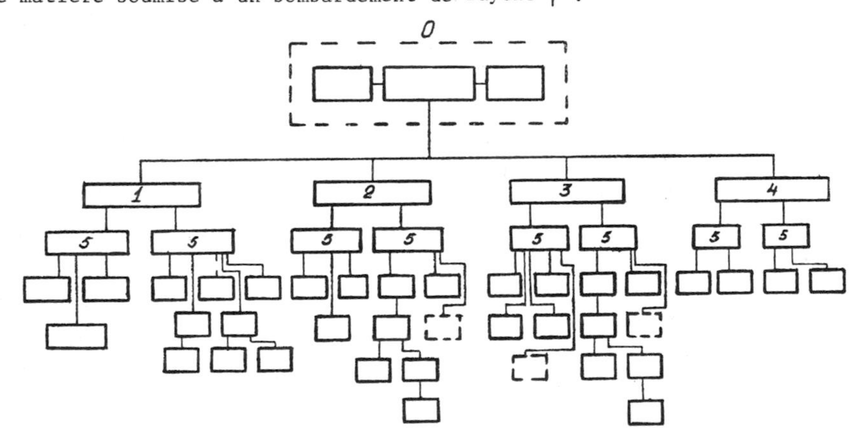

Fig.1 : Schéma descriptif pour un système de traitement d'expériences (système à plusieurs buts)

0	module directeur
1,2,3,4	modules essentiels du traitement primaire, de l'analyse du dispositif, de l'interprétation et de la simulation numérique
5	sous-modules (d'ordre supérieur)

Ce problème a été étudié en automatisation complète à l'Université de Moscou en 1967 et fonctionne encore actuellement. Cette automatisation comprend le traitement primaire et l'interprétation. Sans nous arrêter à la description de l'expérience, remarquons que l'interprétation conduit à la résolution d'une équation intégrale de première espèce pour les sections $\sigma(E)$ (E = énergie de Kant de γ). Sa résolution est faite par un algorithme stable basé sur la méthode de régularisation dans la classe des solutions "lisses". La fonctionnelle de complexité est :

$$\int_{E_1}^{E_2} \left(\frac{d\sigma}{dE}\right)^2 dE$$

elle donne sur les sections $\sigma(E)$ correspondant aux observations une classe compacte Σ_c. On se limite à l'exemple (Ca^{40}), en montrant les résultats de l'interaction (γ,n) Cf. Figure 2 ci-dessous.

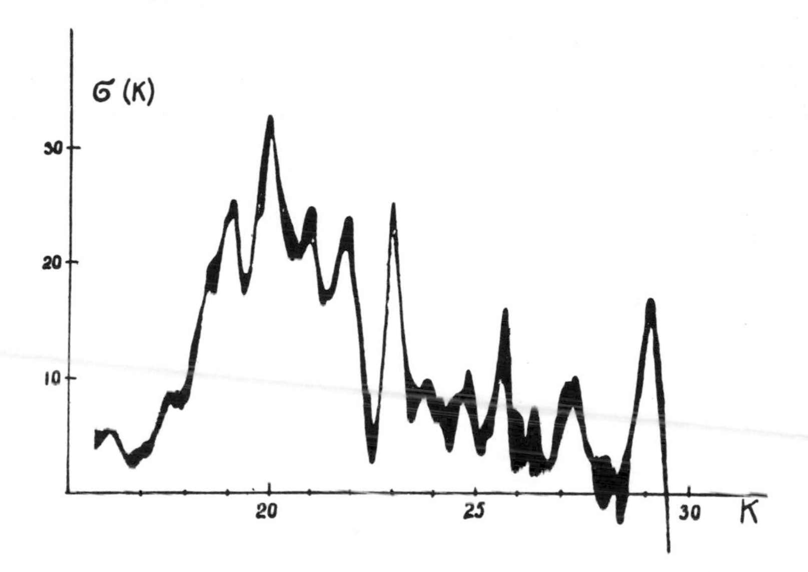

Fig. 2 : Section de l'interaction (γ,n)

Après avoir obtenu la solution régularisée, en vue d'apprécier la précision, on l'a soumise au traitement quasi-réel. Selon la section obtenue $\sigma(E)$ et à l'aide de l'équation intégrale, on a calculé le second membre avec la précision de l'ordinateur. On a ensuite ajouté à ce second membre différentes réalisations de bruits éventuels (1%) et les résultats quasi-réels ainsi obtenus de l'expérience mathématique ont été soumis au traitement. L'épaisseur du tracé sur la Figure 2 est obtenu à l'aide de plusieurs réalisations différentes de bruit (10) et donne une estimation approximative de la précision de la définition de $\sigma(E)$.

Comme deuxième exemple de système automatisé de traitement des observations, on prend le système EOS à plusieurs buts fonctionnant à l'I.P.M. (Institut de Mathématiques Appliquées) de l'U.R.S.S. depuis 1976. Le système est principalement orienté sur le problème du diagnostic des paramètres physiques - densité et température d'objets transparents ayant une structure cylindrique (par exemple faisceau de plasma). Ce système a une structure hiérarchisée de modules et a des sous-modules d'interprétation basés sur différentes classes hypothétiques de modèles comparables. Il contient un algorithme de résolution de problème inverse dans la classe des courbes lisses, avec une fonctionnelle de complexité du type (I) ; il comprend également un module fondé sur la régularisation à l'aide de classes de modèles paramétriques K_n qui se compliquent successivement. On a réalisé dans ce système le traitement primaire d'enregistrements photographiques obtenus à l'aide de techniques interographiques et de techniques d'obscurogrammes.

Sur la figure 3, on donne les résultats de l'interprétation pour l'expérience modèle définissant la densité d'un tube de verre creux. La densité pouvant être connue auparavant, nous pouvons ainsi apprécier visuellement les erreurs obtenues pendant le fonctionnement du système : dispositif expérimental et système automatisé.

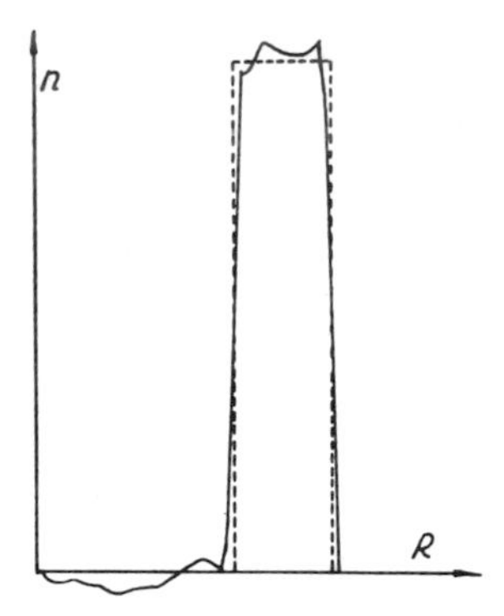

Fig. 3 : Densité d'un tube de verre creux.

L'enregistrement photographique pour ce problème a été effectué à l'aide d'un interferomètre. L'interprétation a été faite dans la classe des modèles comparables (lisses), avec la même fonctionnelle de complexité citée dans l'exemple précédent. Sur la Figure 4, on donne les résultats définissant la densité et la température d'une formation cylindrique de plasma à l'aide d'enregistrements photographiques (par camera spéciale). L'enregistrement s'effectue à l'aide de deux filtres différents, dont le choix optimal se fait à l'aide de l'expérience mathématique.

Le dessin ci-après montre les deux courbes d'obscurcissement correspondant aux deux filtres et les résultats définissant la densité et la température ainsi que les résultats du traitement quasi-réel.

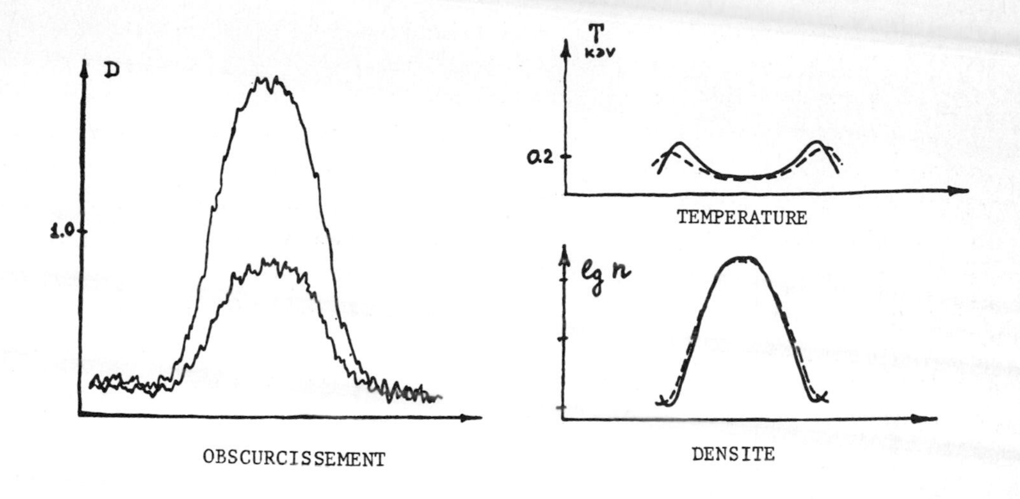

Fig. 4 : Densité et température d'une section d'un faisceau de plasma.

Sur la figure 5, on donne des résultats de l'expérience mathématique (en traitement quasi-réel) de détermination de la densité d'un cordon cylindrique, d'un modèle quasi-réel paramétrique à 4 degrés, pour des niveaux de bruits 4%, 8%, 16%. Ici, l'interprétation se fait dans les classes de modèles paramétriques K_n qui se compliquent successivement 1er, 2ème, 3ème, etc.. degré (classes K_1, K_2, K_3, K_4).

Pour les niveaux de bruit cités ci-dessus, on a inscrit les erreurs de l'observation δ_o à l'aide du modèle quasi-réel du dispositif. En calculant $\min \| Az-\tilde{u} \| = \delta_n$ $(z \in K_n)$, où $\tilde{u}$ sont des courbes de sortie quasi-réelles pour les niveaux de bruit de 4%, 8%, 16%, on a établi les résultats suivants :

- pour le niveau de bruit de 16%, la précision de toutes les classes de modèle est comparable à la précision δ_o des observations (16%), de telle sorte que seule la densité moyenne du cordon donnée par le modèle K_1 a une valeur significative à ce niveau de bruit.

- pour le niveau de bruit à 4%, les classes de modèles K_1, K_2, K_3 ne sont pas comparables aux observations, mais la classe K_4 (à 4 degrés) est comparable aux observations. Ainsi, avec cette précision des observations, on peut constater la réalité de la chute de la densité sur l'axe du cordon. Les modèles de K_1 avec un bruit de 16% et de K_3 avec 4%, sont définis de manière polyvalente ; en choisissant en plus le critère de sélection, on peut définir la précision des différentes caractéristiques du modèle. Par exemple, avec un bruit de 16% on peut estimer la densité moyenne maximale et minimale comparable avec les observations et pour le bruit de 4%, on peut estimer la profondeur de la chute de la densité au centre de l'objet étudié.

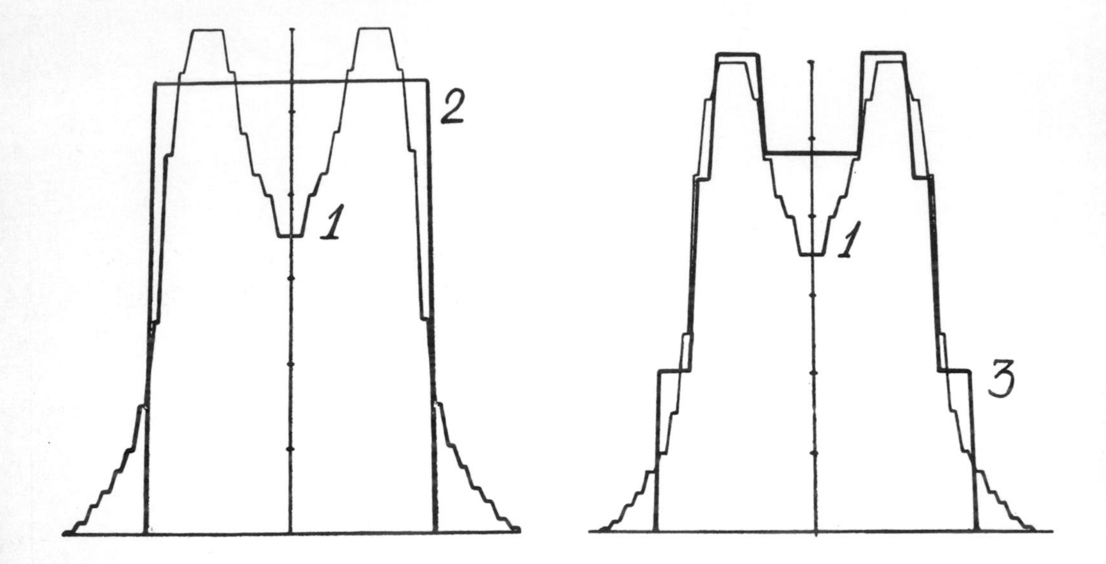

Fig. 5 : Définition de la densité dans l'expérience quasi-réelle
1. Modèle quasi-réel
2. Modèle à 1 degré
3. Modèle à 4 degrés.

Remarquons pour conclure que la réalisation de l'expérience quasi-réelle du modèle mathématique du dispositif, avec le traitement automatisé, permet de considérer l'aspect quantitatif de tout le système et d'estimer de façon optimale les paramètres du dispositif permettant d'obtenir les plus grandes possibilités de solutions.

REFERENCES

[1] TIKHONOV A.N., ARSENIN B.Ya : Les Méthodes de Résolution de problèmes mal posés Naouka, Moscou (1974)

[2] TIKHONOV A.N., SHEVCHENKO V.G., GALKIN V.Ya, GORYACHIEV B.I., ZAIKIN P.N. & al. Proceedings of IFIP Congress, 2 (1968), pp. 1549-1551.

[3] TIKHONOV A.N., ARSENIN B.Ya, DOUMOVA A.A., MITROFANOV V.B., PERGAMENT A.KH, PERGAMENT M.I. Sur les systèmes à plusierus buts du traitement des résultats d'expériences. Reprint I.P.M. de l'Académie des Sciences de l'U.R.S.S., N° 142, Moscou 1976.

NUMERICAL ALGEBRA AND OPTIMIZATION

ALGEBRE NUMERIQUE ET OPTIMISATION

ITERATIVE METHODS IN NUMERICAL SOLUTION OF DIFFERENTIAL EQUATIONS

V.P. Il'in, Yu. A. Kuznetsov

Novosibirsk

1. Introduction

The paper contains a survey of iterative methods for solving systems of linear equations which arise in discretization of boundary-value problems for differential equations. Section 2 is concerned with algebraic aspects of some classes of iterative algorithms for solving linear equations in subspaces. In Section 3 the emphasis is made on "composite" methods based on the use of "false" domains, triangulation of curvilinear domains, topologically equivalent to rectangles and on application of a sequence of auxiliary nets.

2. Iterative methods in subspaces

Let us consider the system of linear algebraic equations

$$Au = f \qquad (1.1)$$

with a real symmetric and positive semidefinite matrix A of the order n and the vector $f \in U_A$, where U_A is subspace of n dimensional real space E_n with the scalar product $(u,v) = \sum_{i=1}^{n} u_i v_i$ and the norm $\|v\| = (v, v)^{1/2}$, with U_A belonging to the image A. We introduce a symmetric positive definite, in space U_A, matrix D_o, matrix $D = AD_o A$, a nonsingular matrix H and space $U = HU_A$.

In what follows we assume that space U is invariant with respect to HA, and introduce the scalar products $(\xi,\eta)_{D_o} = (D_o\xi, \eta)$, $(u, v)_D = (Du, v)$ and the norms $\|\xi\|_{D_o} = (\xi,\xi)_{D_o}^{1/2}$, $\|v\|_D = (u,u)_D^{1/2}$, U_A and U correspondingly. It is assumed that HA is a symmetric positive definite matrix in U with respect to the scalar product $(,)_D$. The latter means $(HA\,u,v)_D = (u, HA\,v)_D$ for all $u,v \in U$ and $(HAu,u)_D > 0$ for all nontrivial $u \in U$.

From the above assumptions it follows that space U is a linear span of some D orthonormal system of eigenvectors $\Psi_1,\Psi_2,\ldots,\Psi_s$

of HA corresponding to its positive eigenvalues $m = \lambda_1 < \lambda_2 < \dots$ $\dots < \lambda_s = M$ (without loss of generality we assume all $\lambda_1, \dots, \lambda_s$ to be different):

To solve (1.1) we apply the iterative method

$$u^{k+1} = u^k - \alpha_k H(Au^k - f), \tag{1.2}$$

$$u^o \in U, \quad k = 0,1,\dots,$$

where

$$\alpha_k \equiv \alpha = \frac{2}{M + m}, \tag{1.3}$$

or

$$\alpha_k = \frac{(AH\xi^k, \xi^k)_{D_o}}{\|\xi^k\|_D^2}, \quad \xi^k = Au^k - f. \tag{1.4}$$

Obviously, for both variants $u^k \in U$ and $\xi^k \in U_A$. The iterative method is said to be convergent in U if the sequence of vectors u^k converges to some solution of system (1.1) for any initial $u^o \in U$.

Theorem 1. Under the assumptions made above the iterative methods (1.2), (1.3) and (1.2), (1.4) converge in space U and

$$\|u^k - u^*\|_D \le q^k \|u^o - u^*\|_D,$$

$$q = \frac{M - m}{M + m}, \qquad u^* = \sum_{i=1}^{s} \frac{(Hf, \Psi_i)_D}{\lambda_i} \Psi_i. \tag{1.5}$$

This estimate is unimprovable in the sense that there exists the vector $u^o \in U$ for which in (1.5) the equality holds for every $k \ge 0$.

Remark 1: Classic variants of the method (1.2), (1.4) were presented in [1] ÷ [3], generalizations of it are discussed in [4] ÷ [7].

Let $C_p(x)$ be Chebyshev's polynomial of order $p \ge 0$ and μ_k, $k = \overline{1, p}$ roots of the polynomial $C_p(\frac{M + m - 2\lambda}{M - m})$. Then Chebyshev's cyclic iterative method (first order Richardson's method) has the form

$$u^{k+1} = u^k - \alpha_k H(Au^k - f),$$

$$u^o \in U, \quad k = 0,1,\ldots, \tag{1.6}$$

where

$$\frac{1}{\alpha_k} = \mu_{k+1-p[\frac{k}{p}]}. \tag{1.7}$$

The following theorem is valid.

Theorem 2. The iterative method (1.6), (1.7) converges in the space U as geometrical progression, i.e.

$$\| u^k - u^* \|_D \leq K\, q^k \| u^o - u^* \|_D, \tag{1.8}$$

where

$$q = \frac{1}{[C_p(\varkappa)]^{1/p}}, \tag{1.9}$$

K is a constant, u^* is defined in (1.6) and $\varkappa = \frac{M + m}{M - m}$.

Remark 2: Many authors noted computing instability of the method (1.6), (1.7). This is connected with the fact that K in (1.8) depends not only on the values of m, M, p but also on the ordering of parameters α_k, $k = \overline{1, p}$. The problem of ordering of α_k in (1.6), (1.7) which guarantees its stability was solved in [8] + [10].

Remark 3. The estimate (1.8) with $K = 1$ is valid for Chebyshev's semi-iterative method (see [4]):

$$u^{k+1} = u^k - \alpha_k H(Au^k - f) - \beta_k(u^k - u^{k-1}), \tag{1.10}$$

$$u^o \in U, \qquad k = 0,1,\ldots ,$$

where

$$\alpha_k = \frac{4C_k(\varkappa)}{(M - m)C_{k+1}(\varkappa)}, \quad \beta_k = \begin{cases} 0, & k = 0, \\ -\dfrac{C_{k-1}(\varkappa)}{C_{k+1}(\varkappa)}, & k > 0, \end{cases} \tag{1.11}$$

To solve (1.1) let us apply the generalized conjugate gradient method

$$g_k = \begin{cases} H\,\xi^0, & \text{if} \quad k = 0, \\ H\,\xi^k - \beta_k g_{k-1}, & \text{if} \quad k > 0, \end{cases}$$

$$\alpha_k = \frac{(\xi^k, Ag_k)_{D_0}}{\|g_k\|_D^2}, \qquad \beta_k = \frac{(H\xi^k, g_{k-1})_D}{\|g_{k-1}\|_D^2}, \tag{1.12}$$

$$u^{k+1} = u^k - \alpha_k g_k, \quad k = 0,1,\ldots .$$

Theorem 3. The conjugate gradient method (1.12) converges in space U and estimate (1.8) with $K = 1$ is valid for any $k < s$.

Remark 4. Some variants of (1.12) were considered in [12]- [16]. More compact trinomial formulae of the method are presented in [17], [18] (see also the References there).

Remark 5 . The above estimates are unimprovable, i.e. for any $k < s$ there exist such eigenvalues $\lambda < \lambda < \ldots < \lambda_s$ and the initial approximation $u_0 \in U$ that for this k an exact equality is reached (see [15]).

In (1.9) the representation

$$C_p(\varkappa) = \frac{2\gamma^{2p}}{1 + \gamma^{2p}}, \quad \gamma = \frac{1 - (\frac{m}{M})^{1/2}}{1 + (\frac{m}{M})^{1/2}} . \tag{1.13}$$

is convenient. If matrices A and H are nonsingular, the estimate (1.10) may be replaced by

$$\|u^k - u^*\|_D \leq \frac{1}{(k+1)^2}\|H^{-1}(u^0 - u^*)\|_{D_0} . \tag{1.14}$$

If, additionally, the matrix D_0 is symmetric and positive definite in E_n , we have

$$\|u^k - u^*\|_D \le \frac{1}{2k+1} (H^{-1}A^{-1}\xi^o, \xi^o)^{1/2}_{D_o} . \qquad (1.15)$$

These and some other estimates for the conjugate gradient method are presented in [14], [16]and the papers cited there.

The only case considered below corresponds to $U = (I - HA)E_n$ and $U_A = (I - AH)E_n$ [15]. Obviously, this U any system $Av = g$ reduces to (1.1) with $v = u + Hg$ by means of substitution $f = (I - AH)g \in U_A$.

One of the effective approaches of the above methods is a special determination of H when matrix A can be represented as a sum of positive semi-definite A_i, $i = \overline{1,p}$. Beginning with the well-known investigations by Peaceman, Douglas and Rachford of 1955, iterative alternating direction methods, which use the inversion of matrices $I + \tau A_i$, $0 < \tau < \infty$ [19], have been further improved. Such methods are based on the following choice of matrix $H^{-1} \equiv H_\tau^{-1}$

$$H_\tau^{-1} = \frac{1}{\tau}(I + \tau A_1)(I + \tau A_2)\ldots(I + \tau A_p), \qquad (1.16)$$

where τ is an iterative parameter. Variations of such algorithms are developed by Yanenko, Birkhof , Varga, Habetler, Wachspress, Samarskii, Diakonov, Marchuk, etc. (see, e.g. [11], [15], [19], [20], [21]).

We will not discuss the theory and applications of the alternating directions method we will only establish an estimate for the number of iterations $n(\varepsilon)$ in (1.6), (1.7), (1.10), (1.11) and (1.12), necessary to meet the condition

$$\|u^n - u^*\|_D \le \varepsilon \|u^o - u^*\|_D \qquad (1.17)$$

if the matrix A is positive definite, $p = 2$ and $D = A^{-1}$ or the matrices A_1, A_2 are commutative.

Theorem 4. Let $A = A_1 + A_2$ be possitive definite matrices satisfying the condition $A_1 = A_2^\tau$

$$0 < \alpha \le (A_i u,u)/(u,u), \quad (A_i u, A_i u) \le \beta(A_i u,u), \quad i = 1,2. \; (1.18)$$

Then condition (1.17) is met in methods (1.6), (1.7) and (1.12), with H chosen according to (1.16) for $p = 2$ and $\tau = (\alpha\cdot\beta)^{-1/2}$ if the number of iterations is

$$n(\varepsilon) \leq \ln \frac{1-\sqrt{1-\varepsilon^2}}{\varepsilon} \Big/ \ln \frac{1-\sqrt{\gamma_0}}{1+\sqrt{\gamma_0}}, \tag{1.19}$$

$\gamma_0 = \dfrac{2\sqrt{\alpha/\beta}}{2+\alpha/\beta}$. From (1.19) we have for $\varepsilon \ll I$, $\alpha/\beta \ll I$

$$n(\varepsilon) \leq |\ln \varepsilon| / 2\sqrt{2\sqrt{\alpha/\beta}} \; .$$

3. Solution of variational difference equations

Let us consider the equation

$$Lu = -\sum_{i,j=1}^{2} \frac{\partial}{\partial x_i}\left(a_{ij}(x)\frac{\partial u}{\partial x_i}\right) + a(x)u = f(x), \tag{2.1}$$

$$x = (x_1, x_2) \in \Omega, \quad a(x) \geq 0$$

in the domain Ω with the boundary condition at Γ

$$u\Big|_\Gamma = 0. \tag{2.2}$$

Assume that operator L is strongly elliptic and Γ is a piecewise-smooth boundary. The domain Ω may be multiply connected and smoothness properties a_{ij}, a, f and Γ are assumed to satisfy the conditions of convergence of variational difference solutions to problem (2.1), (2.2) [24], [25].

Along with (2.1), (2.2) we consider two additional problems for the Poisson equation

$$\Delta u \equiv \frac{\partial^2 u}{\partial x_1^2} + \frac{\partial^2 u}{\partial x_2^2} = f(x) \tag{2.3}$$

under homogeneous boundary conditions of form (2.2) in the domain P, composed of rectangles P_k with sides parallel to the coordinate axes, and in the rectangle $R=\{0\leq x_1\leq c, 0\leq x_2\leq d\}$, circumscribed around P

Rectangles P_k are such that in R one can construct a sequence of rectangular nets $\omega_n=\{x_{1i}=ih, x_{2j}=jh, i=0,1,\dots,l+1, j=0,1,\dots,m+1, lh=c, mh=d\}$ so that the sides of P_k go along the lines of the net.

Introduce the domain $\hat{\Omega}$ circumscribed around Ω and represented by a curvilinear quadrangle partitioned into subdomains Q_k (also curvilinear quadrangles). Partioning of $\hat{\Omega}$ is assumed to be topologically equivalent to partioning of R into rectangular subdomains R_k (the latter coincides with P_k if $P_k \subset P$). This means that subdomains Q_k and P_k are in one-to-one correspondence and every Q_k is contiguous to subdomains Q_k which correspond to rectangles $P_{k'}$ contiguous to P_k . The boundary of Ω is assumed to run along the sides of the subdomains Q_k . We construct in each Q_k a sequence of nets, which for every h are topologically equivalent to nets in a respective rectangles P_k ; on subdomain boundaries the net points are taken common for contiguous nets. Construction of nets in domains with curvilinear boundaries, topologically equivalent to rectangular nets in rectangles, was considered in [27]-[29]. As a result for $\hat{\Omega}$ we obtain nets, topologically equivalent to rectangular in R and for Ω we have nets, topologically equivalent to those of the domain P . Carrying out triangulation of the domains (for simplicity we join the neighboring points in $\hat{\Omega}$ on curvilinear boundaries by straight lines we obtain net domains $\Omega_h, \hat{\Omega}_h, P_h, R_h$. We assume that there exist positive $\alpha_1, \alpha_2, \beta_1, \beta_2$ independent of $\hat{\Omega}_h$ such that lengths of the sides and the areas of triangles are, correspondingly, in the intervals $[\alpha_1 h \cdot \alpha_1 h]$, $[\beta_2 h^2, \beta_2 h^2]$. Triangulation of $\hat{\Omega}_h$ is supposed to be topologically equivalent to that of $\hat{P}_h$, the latter being a set of triangles resulting from intersection of lines and lines crossing the net points under the angles $\frac{\pi}{4}$ to the axes. Triangulation of $\hat{\Omega}_h$ is said to be uniform if it satisfies the above conditions.

We will consider methods of calculation of an approximate solution of (1.1)-(1.2) defined as a function $v_h \in H_{\Omega_h}$ which satisfies the integral relation

$$L(v, w) = \int_{\Omega_h} \left(\sum_{i,j=1}^{2} a_{ij} \frac{\partial v_h}{\partial x_i} \frac{\partial w_h}{\partial x_j} + a_o v_h w_h \right) d\Omega = \int_{\Omega_h} f w_h d\Omega \quad (2.4)$$

for any function $w_h \in \overset{o}{H}_{\Omega_h}$. Here $\overset{o}{H}_{\Omega_h}$ is a space of continuous -in Ω_h functions which are linear in every triangle, equal to zero on the boundary Ω_h and in $\Omega \backslash \Omega_h$. Problem (2.4) reduces to the system of linear algebraic equations

$$A_h v_h = f_h \tag{2.5}$$

with square matrix $A_h = \{L(v_i^h, v_j^h)\}$ and the known vector $f_h = \{(f, v_i)\}$ where v_i is a function equal to unity in some i-th net point and to zero in the others. Dimension of v_Ω^h, f_Ω is (N_Ω is the number of inner points in Ω_h).

Along with A_h , we consider matrices B_h and C_h arising from variational difference approximation of homogeneous boundary value problems of equation (2.3) in the domains P and R respectively. The orders of matrices B_h and C_h are N_P and N_R (N_P, N_R are equal to the number of inner net points in P_h and R_h respectively, $N_\Omega = N_P \leq N_R$ Obviously matrices A_h, B_h, C_h are symmetric and positive definite.

In accordance with [27], [28] we have

Theorem 5. If triangulation of $\hat{\Omega}_h$ is uniform, then matrices A_h and B_h are equivalent by spectrum, i.e. for any h and $u_h \neq 0$ the inequalities

$$\delta_0 (B_h u_h, u_h) \leq (A_h u_h, u_h) \leq \delta_1 (B_h u_h, u_h), \tag{2.6}$$

are valid. Here u_h, is any nontrivial N_Ω dimensional vector and δ_0, δ_1 are positive numbers independent of h .

Introduce into consideration the following matrices of order N_R:

$$\tilde{A} = \begin{bmatrix} A_h & \Theta_1 \\ \Theta_2 & \Theta_0 \end{bmatrix}, \qquad \tilde{B} = \begin{bmatrix} B_h & \Theta_1 \\ \Theta_2 & \Theta_0 \end{bmatrix}, \tag{2.7}$$

where Θ_0 is a square null matrix of the order $N_R - N_\Omega$ and Θ_1, Θ_2 are null rectangular matrices.

If $\tilde{u}_h$ is an N_R dimensional vector representable as $\tilde{u}_h^T = (u_h^T \cdot u_0^T)$, where u_h is of N_Ω dimension, then for any h and u_h from (2.6) we have the relations

$$\delta_0 (\tilde{B}_h \tilde{u}_h, \tilde{u}_h) \leq (\tilde{A}_h \tilde{u}_h, \tilde{u}_h) \leq \delta_1 (\tilde{B}_h \tilde{u}_h, \tilde{u}_h), \tag{2.8}$$

From [30], [31] we get similar inequalities for matrices $\tilde{B}_h$, C_h:

$$\gamma_o(C_h\tilde{u}_h, u_h) \le (\tilde{B}_h\tilde{u}_h, \tilde{u}_h) \le \gamma_1(C_h\tilde{u}_h, u_h). \qquad (2.9)$$

Here $\gamma_o = 1$, $\gamma_1 = 1 + \frac{s}{2\pi}$, where $2s$ is a maximal num - ber of intersections of the domain P boundary by any coordinate line of the net.

From (2.8), (2.9) for any h and u_h we have

$$\delta_o\gamma_o(C_h\tilde{u}_h, \tilde{u}_h) \le (\tilde{A}_h\tilde{u}_h, \tilde{u}_h) \le \delta_1\gamma_1(C_h\tilde{u}_h, \tilde{u}_h). \qquad (2.10)$$

Consider now an iterative process

$$\tilde{v}_h^{k+1} = \tilde{v}_h^k - \alpha_k C_h^{-1}(\tilde{A}_h\tilde{v}_h^k - \tilde{f}_h), \qquad (2.11)$$

where $f_h^T = (f_h^T, 0)$ and α_k are some iterative parameters.

The matrix $C_h^{-1}\tilde{A}_h$ is positive definite in space $\tilde{A}_h E_{N\Omega}$ in the sense of the scalar product $(\ , \ )_{\tilde{A}_h}$. Hence from Theorem 2 there follow convergence of the iterative process (1.6), (1.7) and correctness of estimate (1.8) at $D = \tilde{A}_h^+$, $m = \delta_o\gamma_o$, $M = \delta_1\gamma_1$. In this connection, in order to achieve accuracy of ε in successive approximations (i.e. to fulfil condition (1.17)) it is necessary to perform the following number of iterations:

$$n(\varepsilon) \le \ln \frac{1 - \sqrt{1 - \varepsilon^2}}{\varepsilon} \Big/ \ln \frac{1 - \sqrt{\frac{m}{M}}}{1 + \sqrt{\frac{m}{M}}}. \qquad (2.12)$$

Remark 6. As is shown in [30] definition of $\tilde{A}_h$ by (2.7) is an optimal implementation of the false component method in the sense that estimation of the number of iterations worsens if instead of Θ_o one takes any other positive definite matrix of order $N_R - N_\Omega$.

At each k^{-th} iteration in (2.11) it is required to solve a system of the simple five-point Poisson difference equation to which one can apply methods of cyclic reduction or fast Fourier transform (see [31] and the literature cited there). To solve (1.1), (1.2) to within $\varepsilon = Mh^2$ (M is a constant) one will require the number of arithmetic operations

$$k(\varepsilon) \leq \frac{k_o|\ln h|}{h^2} \ln \frac{1-\sqrt{1-\varepsilon^2}}{\varepsilon} / \ln \frac{1-\sqrt{\delta_o\gamma_o/\delta_1\gamma_1}}{1+\sqrt{\delta_o\gamma_o\ \delta_1\gamma_1}}, \qquad (2.13)$$

where k_o is independent of h.

In a similar way one can establish convergence of iterations applying formulae of conjugate gradient methods (1.12) with $A=\tilde{A}_h, D = A_h^+$ and $H = C_h^{-1}$. Estimation of the volume of calculation is of the form (2.13).

Remark 7. Similar considerations can be carried out for the condition of the third kind instead of (1.2). The case if C_h is defined from approximation of Neumann's problem in the rectangle R for the equation $\Delta u + u = f$ estimates of the number $n(\varepsilon)$ and of the volume of calculation are better because γ_1 becomes independent of h

For some simply connected domains Ω one can construct a uniform triangulation, topologically equivalent to a square net in the rectangle R . Efficiency of algorithms in this case increases because one needn't apply the false component method. Then in inequalities (2.10) $\gamma_o = \gamma_1 = 1$.

Next, we will consider iterative methods based on application of a sequence of nets. These methods were studied by R.P. Phedorenko [32] and N.S. Bakhvalov [33] for finite difference approximations of Dirichlet's problem in the rectangle and G.P. Astrakhantsev [34] for the third boundary value problem in the domain with a curvilinear boundary. The main result of these investigations is that the volume of calculation per one net point does not depend on when one uses the method of net sequence. We will discuss another approach to application of auxiliary nets for problem (1.1), (1.2), assuming that the domain is a polygon and Ω_{h_p} coincides with Ω . We construct triangulation Ω_{h_p} with half mesh sizes by partitioning every triangle of $\Omega_{h,p-1}$ into four ones, joining the middles of its sides. Similarly we construct nets $\Omega_{h_{p-l}}$, $h_{p-l} = h_p/2l$, $l = 1,2,\ldots,$.

In accordance with [35],[36] , we solve equation (2.5) for triangulation $\Omega_{h_o} = \Omega_h$, as follows. First, by means of some method we find variational difference solution for Ω_{h_1} setting its values in the points of Ω_{h_o} as the initial data for system (2.5). Then we make the necessary iterative correction, for instance, by means of (1.6). In turn to solve the equation for the net Ω_{h_1} one must employ the initial values from the solution of equations for Ω_{h_2} and so on.

Denoting by $\hat{v}_k$ piecewise linear fulfilment of net function v_{h_k} and by $(u)_h$ the values of solution of (1.1), (1.2) in net points and using the estimation of error of the variational difference solutions

$$L((\hat{u})_h, - \hat{v}_h, (\hat{u}_h) - v_h) = \|(\hat{u})_h - \hat{v}_h\|_1 \leq Gh^2, \qquad (2.14)$$

$$G = \text{const},$$

we have

$$\|(\hat{u})_o - \hat{v}_o\|_1 \leq \|(\hat{u})_h - \hat{v}_o\|_1 + \|\hat{v}_o - \hat{v}_o^{h_1}\|_A \leq$$

$$\leq Gh^2 + q\|\hat{v}_o - \hat{v}_o^o\| \leq (1 + q)Gh^2 + q\|(\hat{u})_1 - v_1^{h_2}\|_A. \qquad (2.15)$$

Here $v_o^{n_1}$ is the n_1 th step of method (1.10) with the initial value v_o^o and q the error decrease coefficient. $\hat{v}_o^o$ is assumed to be piecewise linear fulfilment, $\hat{v}_1^{n_2}$ the n_2 th step in method (1.6) for net Ω_{h_1}. If we choose the same q for all nets, we have, from (2.15),

$$\|(\hat{u})_o - \hat{v}_o^{n_1}\|_1 \leq Gh^2(1 + q)(\frac{1 - (4q)^p}{1 - 4q} + q^p\|(\hat{u})_p - (\hat{v}_p)\|_1) = \Psi_o. \qquad (2.16)$$

Putting $q < \frac{1}{4}$, $\|(\hat{u})_p - (\hat{v}_p)\|_1 = \Psi_1 = \text{const}$, from condition $\Psi_o \leq \varepsilon$ one can find values of h, q and p in order to ensure the necessary accuracy.

In this case the total number of arithmetic operations is

$$k = Lh^{-2}(n + \frac{n_2}{2^2} + \dots + \frac{n_{k-1}}{2^{(k-1)2}}) \leq \frac{4}{3} Lh^{-2}n_1, \qquad (2.17)$$

where Lh^{-2} is the number of operations needed to calculate one iteration for Ω_h. If one uses an alternating triangular method in (1.6) with optimal value of τ, then $n_1 = O(h^{-1/2})$ and $k = O(h^{-2,5})$, [24].

Remark 8. According to [33],[34] the number of arithmetic operations, necessary to obtain the variational difference solution, is $k = O(h^{-2})$ if Ω is a rectangle or if in Ω we have the third boundary value problem.

R e f e r e n c e s

I. Канторович Л.В. О методе наискорейшего спуска. - "Докл.АН СССР", т.56, №3, I947, с. 233-236.

2. Красносельский М.А., Крейн С.Г. Итерационный процесс с минимальными невязками. - "Мат. сб.", т. 3I (73), №2, I952, с. 3I5-334.

3. Фридман В.М. Метод минимальных итераций с минимальными ошибками для системы линейных алгебраических уравнений с симметричной матрицей. - "ЖВМ и МФ", т. 2, №2, I962, с. 34I-342.

4. Petryshyn W.V. Direct and iterative methods for the solution of linear operator equations in Hilbert space. - "Trans. Amer. Math. Soc.", v. 105, 1962, 136-175.

5. Кузнецов Ю.А. К теории итерационных процессов. - "Докл. АН СССР", т. I84, №2, I969, с. 274-277.

6. Самарский А.А. Итерационные двухслойные схемы для несамосопряженных уравнений. - "Докл. АН СССР", т. I86, №I, I969, с. 35-38.

7. Марчук Г.И., Кузнецов Ю.А. Итерационные методы решения систем линейных уравнений с особенными матрицами. - "Acta Univ. Garolinae. Math. et phys.", v. 15, 1974, 87-95.

8. Лебедев В.И., Финогенов С.А. О порядке выбора итерационных параметров в чебышевском циклическом итерационном методе. - "ЖВМ и МФ", т. II, №2, I97I, с. 425-438.

9. Николаев Е.С., Самарский А.А. Выбор итерационных параметров в методе Ричардсона. - "ЖВМ и МФ", т. I2, №4, I972, с. 960-973.

I0. Лебедев В.И. О бесконечно продолжаемых линейных оптимальных итерационных методах. Новосибирск, ВЦ СО АН СССР. Препринт № 20, I976.

II. Varga R.S. Matrix iterative analysis. Englewood Cliffs, N.y., Prentice-Hall, XIV, 322, 1962.

I2. Hestenes M.R., Stiefel E. Methods of conjugate gradients for solving linear system. -J. Res. Mat. Bur. Stand.", v.49, 1952 (1953), 409-436.

I3. Марчук Г.И., Кузнецов Ю.А. К вопросу об оптимальных итерационных процессах. -"Докл. АН СССР", т. I8I, №6, I968, с. I33I-I334.

I4. Годунов С.К., Прокопов Г.П. О решении разностного уравнения Лапласа. - "ЖВМ и МФ", т. 9, №2, I969, с. 462-468.

I5. Марчук Г.И., Кузнецов Ю.А. Итерационные методы и квадратичные функционалы. Новосибирск, "Наука", Сиб. отд., I972, с. 205. Methodes iteratives et fonctionnelles quadratiques. - In: Sur les methodes numeriques en sciences physiques et economiques. Paris, Dunod, 1974, 3-132.

I6. Ильин В.П. О некоторых оценках для методов сопряженных градиентов.

– "ЖВМ и МФ", т. I6, №4, I976, с. 847-855.

I7. Фаддеев Д.К., Фаддеева В.Н. Вычислительные методы линейной алгебры. Физматгиз, I963, с. 734.

I8. Reid J.K. On the method of conjugate gradients for the solution of large sfarse systems of linear equations. - In.: Large Sets Linear Equat., London-New York, Akad. Press, 1971, 231-252.

I9. Birkhoff G., Varga R., Young D. Alternating direction implicit methods. Advances in computers, v. 4, 1962, 140-274.

20. Самарский А.А. Введение в теорию разностных схем. М., "Наука", I97I, с. 552.

2I. Ильин В.П. Разностные методы решения эллиптических уравнений. Новосибирск, изд. НГУ, I970, с. 263.

22. Самарский А.А. Об одном экономичном алгоритме численного решения систем дифференциальных и алгебраических уравнений. – "ЖВМ и МФ", т. 4, №3, I964, с. 580-585.

23. Ильин В.П. О явных схемах переменных направлений. Известия СО АН, сер. технич., т. 3, №I3, I967, с.

24. Кучеров А.Б., Николаев Е.С. Переменно-треугольный итерационный метод решения сеточных эллиптических уравнений в произвольной области. – "ЖВМ и МФ", т. I7, №3, I977, с. 664-675.

25. Strang G., Fix G. An analysis of the finite element method. Prentice Hall, inc, Englewood Cliffs, N7, 1973.

26. Оганесян Л.А., Ривкинд В.Я., Руховец Л.А. Вариационно-разностные методы решения эллиптических уравнений. Часть П. Сб. Дифференциальные уравнения и их применение, Вильнюс, Ин-т Физики и математики АН ЛССР, вып. 8, I974, с. 9-3I7.

27. Мацокин А.М. Вариационно-разностный метод решения эллиптических уравнений в круге. Сб. Численные методы механики сплошной среды, Новосибирск, ВЦ СО АН СССР, т. 7, №7, I976.

28. Дьяконов Е.Г. О проекционно-разностных методах на сетках, топологически эквивалентных прямоугольным. Новосибирск, Препринт ВЦ СО АН СССР, №32, I976.

29. Дьяконов Е.Г. О некоторых топологических и геометрических задачах, возникающих при триангуляции области в проекционно-разностных методах. "Матем. заметки", т. 2I, вып. 3, I977, с. 427-442.

30. Мацокин А.М. К развитию метода фиктивных областей. "Вычислительные методы линейной алгебры", Новосибирск, ВЦ СО АН СССР, I973, с. 48-56.

3I. Ильин В.П., Короткевич В.А. О решении уравнения Пуассона в непрямоугольных областях. Сб. "Численные методы механики сплошной среды", Новосибирск, ВЦ СО АН СССР, т. 7, №7, I976, с. 30-44.

32. Кузнецов Ю.А., Мацокин А.М. Об оптимизации метода фиктивных компонент. - В кн: Вычисл. методы линейн. алгебры. Новосибирск, 1977, с. 79-86.
33. Dorr F.W. The direct solution of the discrete Poisson equation on a rectangle. SIAM review, v. 12, No 2, 1970, 248-263.
34. Федоренко Р.П. О скорости сходимости одного итерационного процесса. - "ЖВМ и МФ", т. 4, №3, 1964, с. 559-564.
35. Бахвалов Н.С. О сходимости одного релаксационного метода при естественных ограничениях на эллиптический оператор. - "ЖВМ и МФ", т. 6, №5, 1966, с. 861-883.
36. Астраханцев Г.П. Об одном итерационном методе решения сеточных эллиптических задач. - "ЖВМ и МФ", т. 11, №2, 1971, с. 439-448.
37. Ильин В.П., Свешников В.М. О разностных методах на последовательности сеток. Сб. "Численные методы механики сплошной среды", Новосибирск, ВЦ СО АН СССР, т. 2, №1, 1971, с. 43-54.
38. Дьяконов Е.Г. Об использовании последовательностей сеток при решении сильноэллиптических систем. Сб. "Вычислительные методы линейной алгебры", Новосибирск, ВЦ СО АН СССР, 1977, с. 146-160.

Вычислительный центр СО АН СССР, пр. Науки, 6, Новосибирск, 630090, СССР.

NUMERICAL METHODS FOR COMPLEMENTARITY PROBLEMS IN ENGINEERING AND APPLIED SCIENCE

Richard W. Cottle
Department of Operations Research
Stanford University
Stanford, California 94305/USA

1. Introduction. The complementarity problem comes up in the search for equilibrium points of one kind or another. The immediate source of the problem could be physical, economic, or purely technical. One of the most general formulations of the problem involves the following ingredients:

X	a real locally convex Hausdorff topological vector space
Y	a real vector space
F	a mapping from X to Y
$\langle \cdot,\cdot \rangle$	a bilinear form on $X \times Y$
K	a closed convex cone in X
K^*	the polar of K in Y

The complementarity problem (CP) is then to find a solution of the system

$$x \in K, \quad F(x) \in K^*, \quad \langle x,F(x) \rangle = 0 \tag{1}$$

or show that none exists. This problem will be denoted $C(F,K)$.

While this statement of the problem is useful in calling attention to the search for functionally related orthogonal vectors belonging to mutually polar cones, it is too general for numerical purposes. Accordingly, various specializations of the problem have been studied. In nearly all cases, X and Y are the n-dimensional real space R^n, and $\langle \cdot,\cdot \rangle$ is the ordinary inner product there. Four major problem types can be identified. If the closed convex cone K is something more general than the nonnegative orthant, R^n_+, then (1) is referred to as a generalized complementarity problem (GCP). If the mapping F is an affine transformation, then (1) is known as a linear complementarity problem (LCP). Otherwise, it is a nonlinear complementarity complementarity problem (NLCP).

To date, most of the numerical work (development and analysis of algorithms) on complementarity problems has been concerned with the ordinary LCP, although some can be found on the GNLCP. This paper briefly surveys some numerical methods for the CP and their applications to problems in engineering and applied science. The accent here is on solution methods for the ordinary (chiefly linear) complementarity problem and on applications of a physical type.

2. Equivalences. It may be helpful to recall that the generalized complementarity problem $C(F,K)$ is equivalent to a particular type of variational inequality problem. To be specific, let S be a convex set in X, and let $G:S \to Y$. Then $V(G,S)$ will

denote the variational inequality problem (VIP): Find a solution of

$$x \in S, \quad \langle z - x, G(x) \rangle \geqq 0 \quad \text{for all } z \in S \tag{2}$$

The conditions (2) are sometimes used as an optimality criterion in convex optimization problems. The equivalence we have in mind is for the case where $G = F$ and $S = K$, a closed convex cone. Then, according to KARAMARDIAN [43], $C(F,K)$ and $V(F,K)$ have the same solutions, if any.

Another equivalence concerns the VIP and the BROUWER fixed-point problem (BFP). Let S denote a closed convex set in a HILBERT space X with inner product $(\cdot,\cdot)$. If $H:S \to S$ is a given mapping, the corresponding BFP, denoted $B(H,S)$, calls for a solution of

$$x \in S, \quad H(x) = x \tag{3}$$

By setting $G(x) = x - H(x)$, it can be shown that $V(G,S)$ and $B(H,S)$ have the same solutions, if any.

Many complementarity problems come directly or indirectly from mathematical programming. In a program

$$\begin{aligned} \text{minimize} \quad & f(x) \\ \text{subject to} \quad & g(x) \leqq 0 \\ & x \geqq 0 \end{aligned} \tag{4}$$

for which the KUHN-TUCKER conditions [44] happen to be necessary and sufficient for optimality, the problem of finding a constrained minimum is equivalent to finding a solution of the complementarity problem $C(F,K)$ where, in terms of the LAGRANGIAN function $L(x,y) = f(x) + \langle y, g(x) \rangle$

$$F = L \quad \text{and} \quad K \doteq R_+^{n+m}$$

A case of considerable interest is the problem

$$\begin{aligned} \text{minimize } Q(x) = \tfrac{1}{2}\langle x, Dx \rangle + \langle c, x \rangle \\ \text{subject to} \qquad x \geqq 0 \end{aligned} \tag{5}$$

when D is symmetric and positive semi-definite this (convex) quadratic programming problem is equivalent to the linear complementarity problem

$$x \geqq 0, \quad c + Dx \geqq 0, \quad \langle x, c + Dx \rangle = 0 \tag{6}$$

It should be emphasized, however, that not every LCP corresponds to a (convex) quadratic program. Indeed not every affine transformation is the gradient of a convex quadratic function. Part of the strength of complementarity theory is its (occasional) ability to cope with problems without resorting to the use of objective function values.

3. Generalities about methods. The numerical methods for complementarity problems fall into two major categories: direct and iterative. Direct methods are those based on the process of pivoting, that is, exchanging the roles of dependent and independent variables in a system of equations. As applied to the LCP, direct methods generate a (finite) sequence of transformed representations of the underlying equation $y = c + Dx$. The object of the procedure is to bring about a particular configuration

relative to the system representation. Although the steps of such a procedure are called iterations, it would not be iterative as we use that term here. We think of iterative methods as those which produce a (possibly infinite) sequence of iterates (trial solutions) which converge to a solution. Moreover, the iterative methods typically modify the trial solutions rather than the problem data. Iterative algorithms are further described as point or block methods according to whether the successive iterates are updated one component at a time or by subvectors. In the case of block methods, the subvectors of which the iterates are formed may themselves be solutions of induced complementarity problems. If the latter are solved "directly", then the overall procedure can be viewed as a hybrid method.

The conditions under which numerical methods for the complementarity problem "work" are not all the same. To make effective use of these techniques, the analyst must know some properties of K and of the mapping F. For instance, in the ordinary LCP

$$z \geqq 0, \quad q + Mz \geqq 0, \quad \langle z, q + Mz \rangle = 0 \tag{7}$$

the cone $K = R^n_+$ (which is quite simple), and the mapping $F(z) = q + Mz$. For problems of this sort, much depends on the nature of M. Positive (semi-) definiteness is a desirable matrix-theoretic property and one which is commonly found in applications, but not all the methods are confined to this class of matrices. Other features are also of great importance. Some of these are the order of the matrix, its structure, its properties under pivotal transformation, and so forth.

4. Direct methods for the LCP. There are two main direct methods for solving the linear complementarity problem. One of these is the principal pivoting method [24], [15], [11]. The other is often called LEMKE's [45] almost-complementary pivoting method. We review these procedures here, but only quite briefly.

The principal pivoting method derives its name from the fact that it consists of a sequence of principal 1×1 or 2×2 pivots, i.e., exchanges of dependent and independent variables. Consider the LCP (7). We introduce the vector $w := q + Mz$ and reformulate the problem as

$$z \geqq 0, \quad w = q + Mz \geqq 0, \quad \langle z,w \rangle = 0 \tag{7'}$$

This problem can be denoted (q,M). For $i = 1,\ldots,n$ there is a pairing between the dependent variable w_i and the independent variable z_i. Together w_i and z_i form a complementary pair, and each is the complement of the other. For any solution of (7'), $z_i \geqq 0$, $w_i \geqq 0$, and $z_i w_i = 0$ $(i = 1,\ldots,n)$. Thus, at most one member of a complementary pair can be positive (although both can be zero).

The principal pivoting method applies to LCP's where $M \in P \cup D_0$. The class P consists of all real square matrices having positive principal minors, and the class D_0 consists of all positive semi-definite but not necessarily symmetric matrices. These two classes of matrices are invariant under principal rearrangement

$$M \to PMP' \quad P \text{ a permutation matrix}$$

and principal pivoting

$$M = \begin{pmatrix} M_{11} & M_{12} \\ M_{21} & M_{22} \end{pmatrix} \rightarrow \begin{pmatrix} M_{11}^{-1} & -M_{11}^{-1}M_{12} \\ M_{21}M_{11}^{-1} & M_{22} - M_{21}M_{11}^{-1}M_{12} \end{pmatrix} \quad M_{11} \text{ nonsingular} \qquad (8)$$

If the equation $w = q + Mz$ is partitioned as

$$w_1 = q_1 + M_{11}z_1 + M_{12}z_2$$

$$w_2 = q_2 + M_{21}z_1 + M_{22}z_2$$

it is possible to exchange the roles of w_1 and z_1 iff M_{11} is nonsingular, and when this is done, the result is

$$\begin{aligned} z_1 &= -M_{11}^{-1}q_1 && + M_{11}^{-1}w_1 && - M_{11}^{-1}M_{12}z_2 \\ w_2 &= q_2 - M_{21}M_{11}^{-1}q_1 && + M_{21}M_{11}^{-1}w_1 && + (M_{22} - M_{21}M_{11}^{-1}M_{12})z_2 \end{aligned} \qquad (9)$$

The coefficient matrix of the dependent variables in (9) is precisely the principal transform on the right-hand side of (8).

For any square matrix M, it is clear that if (7') has a solution and $q \not\geq 0$, then some principal transformation will result in an equivalent system in which the transform of q is nonnegative. Having this, one would set the independent variables to zero and read off the values of the corresponding dependent variables.

The principal pivoting method starts with $z = 0$, $w = q$. It executes a sequence of major cycles each of which is intended to make a particular "distinguished" negative dependent variable become nonnegative while maintaining the nonnegativity of the variables to the greatest extent possible. The invariance property mentioned above makes this both possible and desirable. For example, in the case where $M \in P$, the partial derivative of the i-th dependent variable (assumed negative) with respect to its complement is always positive. An independent variable selected for increase is called a driving variable. Its increase is said to be blocked when (and if) a dependent variable reaches a prescribed bound. In the present case, the driving variable is blocked when either the distinguished variable increases to zero or a nonnegative dependent variable decreases to zero. The first dependent variable to block the driving variable is called the blocking variable. Exchanging the blocking variable and its complement is a principal pivot. When $M \in P$, this pivotal transformation is possible and leads to an equivalent system also involving a P-matrix. (For $M \in D_0$, the procedure is similar but a little more complicated.) If the blocking variable was the distinguished variable, the major cycle is over. Otherwise it continues with further increase of the driving variable. Each completion of a major cycle decreases the total number of negative variables in the corresponding solution (i.e., that in which the dependent variables have value zero). Therefore only finitely many major cycles are required to obtain a solution. The finiteness of a major cycle is assured when the problem is nondegenerate, that is, when no solution of the equation $w = q + Mz$ has more than n (out of 2n) components equal to zero. When the problem is degenerate, the circling phenomenon (repetition of previously obtained representations of the equation) can occur. Such technical difficulties can be overcome by suitable perturbation of q. Recently, MURTY [53] and CHANG [5] have shown that least index rules can also be used to achieve finiteness in degenerate problems.

Except for special cases (such as M positive) LEMKE's method uses an auxiliary problem in solving the given one. Consider the system

$$z_0 \geqq 0, \quad z \geqq 0, \quad w = q + z_0 d + Mz \geqq 0, \quad \langle z,w \rangle = 0 \tag{10}$$

in which the vector d is positive. These conditions are easily satisfied. The object is to find a triple $(\bar{z}_0,\bar{z},\bar{w})$ satisfying (10) and having $\bar{z}_0 = 0$, for then $(\bar{z},\bar{w})$ satisfies (7').

LEMKE's algorithm is simply stated. Assuming $q \ngeqq 0$, one starts by letting $z = 0$ and increasing z_0 to the value $\max_i -q_i/d_i$. In the nondegenerate case, this maximum is attained for a unique index r. The first pivot is an exchange of w_r and z_0, the former being viewed as the first blocking variable. The first driving variable is its complement, z_r. All dependent variables, including z_0 are eligible blocking variables, their lower bounds being zero. If no dependent variable blocks the driving variable, the procedure terminates "on a ray." If the driving variable is blocked, an exchange (pivot) is made between the blocking variable and the driving variable. If z_0 is the blocking variable, the procedure terminates with a solution. Otherwise, the complement of the blocking variable becomes the new driving variable, and the procedure continues.

Besides the simplicity of its statement, this algorithm has the virtue of being able to process the same problems as the principal pivoting method and several others as well. See [27], [33], [47] for examples. Relations between the two methods described in this section have been studied by VAN DE PANNE [60].

5. <u>Special direct methods</u>. Often the LCP to be solved exhibits special structure, particularly in the matrix M. To achieve computational efficiency (in storage and execution time) it is considered important to exploit this special structure in the construction of corresponding numerical methods. In this connection, we point out that pivoting methods can turn sparse matrices into dense ones and thereby create serious storage problems for matrices of large order.

5.1 The use of finite difference approximation schemes may lead to LCP's in which M has nonpositive off-diagonal elements. (Following FIEDLER and PTAK [30], we denote the class of such matrices by Z.) A simple, but important observation pertaining to this class of problems is

<u>Theorem</u> 1 (CHANDRASEKARAN [4]). If (q,M) has a solution $(\bar{z},\bar{w})$ and $M \in Z$, then

$$q_i < 0 \quad \text{implies} \quad z_i > 0 \quad \text{and} \quad w_i = 0 \tag{11}$$

There is clearly no other way to satisfy the inequalities of the problem. The fact that certain inequalities must hold as equations suggests using fast linear equation solvers (perhaps repeatedly) to determine trial solutions. CHANDRASEKARAN has laid down an algorithm along the following lines. Given (q,M), let $I = \{i \mid q_i \leqq 0\}$, and let J be the complement of I in $\{1,\ldots,n\}$. One can begin by attempting to solve the subproblem (q_I, M_{II}) or, more precisely, the equation $M_{II}z_I = -q_I$. If its solution exists and is $\bar{z}_I$, set $\bar{z}_J = 0$ and calculate $\bar{w}_J = q_J + M_{JI}\bar{z}_I$. If $\bar{w}_J \geqq 0$, a solution has been found. Otherwise, I is enlarged to include the indices i for which $\bar{w}_i < 0$, and the process is repeated. At most n linear systems need be solved, and because they are nested, special schemes can be used to reduce the computational effort. See DIAMOND [26].

5.2 An even more special structure--again with Z matrices--is the tridiagonal case. In this instance, it is possible to use principal pivoting with column suppression. Once a column has been used for pivoting, it can be disregarded, for the effect of principal pivoting on the sign configuration is such that the pivot column will never again be needed. This is true for Z-matrices in general, but it is particularly interesting for tridiagonal ones of large order since the scheme just indicated preserves their sparsity. See [20].

5.3 There is also a parametric version of the complementarity problem and in particular the LCP. Given vectors q and p (nonzero), the square matrix M and an interval $A \subset R$, one seeks solutions for the system

$$z \geqq 0, \qquad w = q + \alpha p + Mz \geqq 0, \qquad \langle z,w \rangle = 0, \qquad \alpha \in A \tag{12}$$

Problems of this sort arise in structural mechanics where the parametrization relates to proportionally increasing loads; they also come up as a technical device in some algorithms. As a matter of fact, the LCP itself can be expressed in parametric form. For example, given (q,M), let d be a (positive) vector such that $q + d > 0$. Then (q,M) can be attacked by way of the parametric LCP $((q + d) - \alpha d, M)$ with $\alpha \in A = [0,1]$.

Typically, one needs a value of α in A for which the ordinary LCP $(q + \alpha p, M)$ has a solution. The central problem can be reformulated so that A is of the form $[0, \bar{\alpha}]$ with $\bar{\alpha} < \infty$ or else $[0, \infty)$. By reversing the sign of the "direction" vector p, one can handle nonpositive values of α with the same method.

A parametric principal pivoting algorithm is given in [12] for the case where $M \in P \cup D_0$. One starts with the assumption $q \geqq 0$ and uses $z = 0$ as the solution to $(q + 0 \cdot p, M)$. The next step is to determine a critical value of α, namely the largest number $\alpha = \alpha_1$ in A for which $q + \alpha p \geqq 0$. For $\alpha > \alpha_1$, some component of w would become negative. This suggests a principal pivot on the corresponding diagonal entry of M. This is always possible when $M \in P$. When $M \in D_0$, the matter is more delicate and may require a principal pivot of order 2. After the pivot, a new (larger) critical value is to be determined and another principal pivot is to be performed. The various possibilities for stopping are:

1° α reaches $\bar{\alpha}$;

2° α can be made arbitrarily large without forcing a dependent variable to become negative;

3° α reaches a value $\tilde{\alpha} < \bar{\alpha}$ such that for $\alpha > \tilde{\alpha}$ there is no solution to $(q + \alpha p, M)$.

The third possibility cannot occur when $M \in P$ by virtue of a theorem of SAMELSON, THRALL, and WESLER [56] to the effect that (r,M) has a unique solution for all $r \in R^n$ iff $M \in P$.

5.4 A specialized method of parametric principal pivoting has been developed by PANG [55]. It applies to problems (q,M) in which $M = D + GG^T$ where D is diagonal and positive definite, and G is $n \times m$ where $m < n$. Thus, M is symmetric and positive definite. In many applications of the LCP, the matrix M is given in just such a form. Some examples are mentioned by PANG, KANEKO, and HALLMAN [56]. The complete details of this algorithm are too involved for presentation here. In general, these parametric methods find critical values of the parameter and corresponding solutions of the nonparametirc problem. PANG's approach makes use of the matrix structure to obtain this information efficiently. It is particularly effective when the ratio m/n is small. In some problems, n is large but m is just 1.

5.5 It is easily demonstrated that if (7) has a solution then it has a solution which is an extreme point of the polyhedral convex set given by the inequalities $z \geqq 0$, $q + Mz \geqq 0$. Thus, if a suitable linear form (objective function) $\langle p,z \rangle$ could be found, the problem could be <u>solved as the linear program</u>

$$\text{minimize } \langle p,z \rangle \quad \text{subject to } z \geqq 0, \quad q + Mz \geqq 0 \tag{13}$$

In general, the aim of these studies has been to relate p to the problem data. It is easy to see from the duality theory of linear programming that when (13) has an optimal solution, the vector p is expressible in the form $p = r + M^T s$ for some $r,s \geqq 0$. Attention is therefore restricted to vectors of this form. MANGASARIAN [50] has produced the following remarkable characterization:

<u>Theorem</u> 2. The linear complementarity problem (7) has a solution if and only if the linear program (13) with $p = r + M^T s$ is solvable for some $r,s \in R^n$ which must satisfy the following conditions

$$MZ_1 = Z_2 + qc^T$$
$$M^T(Y - sc^T) \leqq rc^T$$
$$r^T Z_1 + s^T Z_2 + q^T(sc^T - Y) \geq 0$$
$$c,r,s \geqq 0, \quad Y \geqq 0, \quad Z_1, Z_2 \in Z$$

for some vector $c \in R^n$ and matrices Y, Z_1, $Z_2 \in R^{n \times n}$. Furthermore, each solution of the linear program solves the LCP.

Although this result looks hopelessly complicated, it can be brought to bear on some special cases. MANGASARIAN [50] lists matrices of various kinds for which it is not too difficult to find p. The simplest case of all is that in which $M \in P \cap Z$ (i.e., M is a MINKOWSKI matrix) where it suffices to let p be any positive vector. This fact, known for some time, can be seen from another point of view. When $M \in P \cap Z$, every LCP (q,M) has a unique solution which is the least element in the vector sense of the polyhedral convex set of all solutions of the inequalities $z \geqq 0$, $q + Mz \geqq 0$. This least element minimizes every positive linear form $\langle p,z \rangle$ over that set. See COTTLE and VEINOTT [21]. For a reinterpretation of MANGASARIAN's results in terms of polyhedral sets with least elements, see COTTLE and PANG [18], [19].

Once an LCP is formulated as a linear program, its solution by DANTZIG's simplex method immediately comes to mind. But, at least in principle, this is not the only alternative. One can express the necessary and sufficient conditions of optimality of <u>any</u> linear program as a linear inequality system and then attempt to solve it by iterative methods for (convex) inequality systems. See [50], [19]. Often the LCP's solvable as linear programs (without an inordinate amount of work to define p) can be handled by other, more efficient specialized methods. See [19]. So, at this stage, the computational value of the linear programming approach to the LCP remains unclear.

6. <u>Iterative methods</u>. One of the oldest iterative methods related to the LCP is due to HILDRETH [38], [39]. The procedure was in fact designed to solve a strictly convex quadratic program. HILDRETH stated its KUHN-TUCKER conditions and used the nonsingularity of the Hessian matrix of the objective function to eliminate the primal variables. What remains after this operation is a linear complementarity problem (q,M) in which the variables are LAGRANGE multipliers and the matrix M is symmetric and positive semi-definite. Solving such an LCP is equivalent to minimizing a convex quadratic function $\psi(z) = \langle q,z\rangle + \frac{1}{2}\langle z,Mz\rangle$ on the nonnegative orthant R^n_+. It should be noted that if M is symmetric and positive semi-definite, then $m_{ii} = 0$ implies $m_{ij} = m_{ji} = 0$ for all j. In the context of the quadratic program above, this means the objective function ψ has a finite minimum on R^n_+ only if $q_i \geqq 0$ for all i such that $m_{ii} = 0$. In this case, one can take the corresponding $\bar{z}_i = 0$ in a solution (if one exists). Accordingly, it is not restrictive to assume $m_{ii} > 0$ for all i. In addition, we shall assume for the moment that (q,M) does in fact have a solution. (It is known [9] that when $M \in D_0$, the problem (q,M) has a solution if and only if the inequalities $z \geqq 0$, $q + Mz \geqq 0$ are consistent; when this is the case and M is also symmetric, any two solutions of (q,M) differ by an element of the null space of M. See [40], [10].

HILDRETH's iterative method uses the point GAUSS-SEIDEL iterative scheme with projection. Starting from an initial estimate $z^0 \geqq 0$, and $\nu = 0$, one first calculates

$$\tilde{z}_i^{\nu+1} = \frac{-1}{m_{ii}}\left(\sum_{j=1}^{i-1} m_{ij}z_j^{\nu+1} + \sum_{j=i+1}^{n} m_{ij}z_j^{\nu} + q_i\right) \tag{14}$$

and then

$$z_i^{\nu+1} = \max\,\{0,\tilde{z}_i^{\nu+1}\}\ . \tag{15}$$

The iteration implied by the right-hand side of (14) is just that of the GAUSS-SEIDEL method. It can be interpreted as the unconstrained minimization of the univariate function $\psi(z_1^{\nu+1},\ldots,z_{i-1}^{\nu+1},\cdot,z_{i+1}^{\nu},\ldots,z_n^{\nu})$. But, because the problem is actually constrained, the projection (15) is used to assure the nonnegativity of the iterates.

Essentially the same algorithm (for the symmetric positive definite case) is stated by FRIDMAN and CHERNINA [32] as a solution technique for the finite-dimensional contact problem. (See Section 7.)

A more general iterative method attributed to CHRISTOPHERSON [6] has been analyzed and clarified by CRYER [22], [23]. In the literature, it is often cited as CRYER's method. It is a successive over-relaxation (SOR) method proposed for the solution of the free-boundary problem for journal bearings by means of finite differences. As stated it deals with the symmetric positive definite case. For such problems, the solution exists and is unique. In the application alluded to above, additional structure is also present. For the infinite (finite) journal bearing problem, M is also a (block) tridiagonal Z-matrix. Nevertheless, we state the method in its general form.

Let $\omega \in (0,2)$, $\varepsilon > 0$ (small), and $z^0 \in R^n_+$ be given. Set $\nu = 0$. For $i = 1,\dots,n$ define $z_i^{\nu+1}$ by (14), and then set

$$z_i^{\nu+1} = \max\{0,\ z_i + \omega(\tilde{z}_i^{\nu+1} - z_i^{\nu})\} \tag{16}$$

When the relaxation parameter ω equals 1, this process is just HILDRETH's. For $\omega < 1$, it is called under-relaxation. When $\omega > 1$, it becomes an over-relaxation method. The step indicated in (16) first involves a relaxation and then a projection. The process terminates if

$$\max_{i\in\mathcal{J}} |(q + Mz^{\nu+1})_i| < \varepsilon \tag{17}$$

where

$$\mathcal{J} = \{i \mid z_i^{\nu+1} > 0\} \cup \{i \mid z_i^{\nu+1} = 0,\ (q + Mz^{\nu+1})_i < 0\} \tag{18}$$

Otherwise, it continues with ν increased by 1.

CRYER's method has been extended by ECKHARDT [28] to the positive semi-definite case. We have already noted that for this class of problems, the existence of a solution is not assured, nor is uniqueness. Making the nonrestrictive assumption that M has a positive diagonal, ECKHARDT shows that it is unnecessary to assume solvability since the iterates generated by the method can be used to reveal the nonexistence of a solution.

MANGASARIAN [51] has introduced a robust iterative method for solving symmetric LCP's. The algorithm has many options. Some of its realizations include a JACOBI-like algorithm, a generalization of the SOR method, and a generalization of the symmetric SOR (i.e., SSOR) method.

Closely related to the convex quadratic minimization problem mentioned above is that in which the constraints are of the form $0 \leqq z \leqq b$ where $b > 0$ is given. This sort of problem comes up in several contexts, one of which (elasto-plastic torsion of a cylindrical bar) is described and solved by CEA and GLOWINSKI [2]. See also Chapter 3 of GLOWINSKI, LIONS, and TREMOLIERES [34]. The CEA-GLOWINSKI point method starts with a trial solution z^0 such that $0 \leqq z^0 \leqq b$, a relaxation parameter $\omega \in (0,2)$, a tolerance $\varepsilon > 0$, and $\nu = 0$. For $i = 1,\dots,n$ one defines $z_i^{\nu+1}$ as in (14). This is followed by relaxation

$$\hat{z}_i^{\nu+1} = z_i^{\nu} + \omega(z_i^{\nu+1} - z_i^{\nu}) \tag{19}$$

and then the projection:

$$z_i^{\nu+1} = P_{b_i}(\hat{z}_i^{\nu+1}) \tag{20}$$

is the point of the interval $[0,b_i]$ closest to $\hat{z}_i^{\nu+1}$. A convergence criterion for this method is

$$\|z^{\nu+1} - z^{\nu}\|_1 < \varepsilon.$$

A few words now about block methods. Suppose the matrix M is partitioned into N^2 blocks M_{ij} of order $n_i \times n_j$. Let the vectors q and z be partitioned conformably. This partitioning may be a "natural" consequence of the problem formulation, but it may also be induced. The idea is to work cyclically on the blocks which correspond to subproblems of the same general type--that is, LCP's. This approach has been used by COTTLE, GOLUB, and SACHER [17], COTTLE and GOHEEN [16], and by CEA and GLOWINSKI [2]. The algorithms introduced in [17] and [16] use direct LCP methods on the subproblems followed by "constrained relaxation", i.e., after the constrained minimum is found in the (LCP) subproblem, the solution is relaxed but kept within the admissible range. These methods are therefore hybrids. The CEA-GLOWINSKI block method uses unconstrained minimization

$$\tilde{z}_i^{\nu+1} = - M_{ii}^{-1}(q_i + \sum_{j=1}^{i-1} M_{ij}z_j^{\nu+1} + \sum_{j=i+1}^{N} M_{ij}z_j^{\nu})$$

followed by relaxation

$$\hat{z}_i^{\nu+1} = z_i^{\nu+1} + \omega(z_i^{\nu+1} - z_i^{\nu})$$

and projection into the interval $[0,b_i]$ with respect to the norm induced by the i-th diagonal block M_{ii}. The projection then becomes a problem of the same sort; a purely iterative method would solve it by an iterative scheme.

Block partitioned problems sometimes have special structure that contributes to their efficient solution. A case in point is the block tridiagonal MINKOWSKI structure that shows up in the finite difference approximation to elliptic partial differential inequalities over a rectangular region. In this case one gets diagonal blocks M_{ii} which are themselves tridiagonal; the only nonzero off-diagonal blocks are $M_{i,i+1}$ and $M_{i+1,i}$ $(i = 1,\ldots,N-1)$, and these are diagonal. Thus, even though the order of M may be very large (several thousand), its rows have at most 5 nonzero entries. Direct methods like principal pivoting could easily lead to a prohibitive amount of "fill-in" and thereby become completely impractical.

In addition to the advantages of compact storage, iterative methods make it possible to exploit a good user-supplied starting solution. On the negative side, it happens that relaxation methods are significantly affected by the choice of the relaxation parameter ω. In the case of linear complementarity problems, there is at present little theory to guide one's selection of this important parameter.

Much less has been done on methods for solving the nonlinear complementarity problem. Algorithms for the computation of fixed points have been used in several instances. Another approach is the iterative scheme of HABETLER and PRICE [36], and [37] developed for the generalized nonlinear complementarity problem.

7. <u>On applications</u>. It is quite impossible to give a comprehensive summary of the widespread applications of the complementarity problem in just a few pages even by narrowing the discussion to the LCP in engineering and applied science. Accordingly the author disavows any claim of completeness and apologizes for the omissions necessitated by brevity.

7.1 The <u>contact problem</u> comes up repeatedly in the literature and is often stated as an LCP. The specific application, the attendant formulation, and the methodological discussion may vary from one article to the next, but significant common features remain. The following statement of the problem follows FRIDMAN and CHERNINA [32]. Consider two elastic bodies in contact at a finite number of points. Let z_i denote the contact stress at the i-th point. The matrix components m_{ij} of M give the effect of the j-th stress on the relative deflection at the i-th point of the surface of the bodies in contact, and the quantities $-q_i$ represent the distances that would exist between corresponding i-th points of the surfaces if free penetration were permitted. The clearance condition is expressed by the linear unilateral constraints. The contact stresses z_i are nonnegative and the important contact condition says that the contact stress z_i can be positive only if the bodies are actually in contact at the i-th point, i.e., $(q + Mz)_i = 0$. Thus, the LCP.

At this point, the discussion typically turns to the positive semi-definiteness of M and from there to the equivalence of the problem at hand to a convex quadratic programming problem. This in turn brings up extremal principles as well as algorithms for quadratic programming, usually WOLFE's [61] rather than the special purpose methods for large-scale problems. See FICHERA [29], CONRY and SEIRIG [8], FISCHER [31], CHAND, HAUG and RIM [3], PANAGIOTOPOULOS [54], and KALKER [41].

7.2 <u>Engineering plasticity</u> is another area where the linear complementarity problem is of considerable interest. The developments in this application are largely due to MAIER [49]. They deal with the elasto-plastic analysis of discrete structures (e.g., trusses) subjected to external loads. Many technical considerations are involved including the nature of the applied loads and the behavioral characteristics of the material. The latter refers to holonomic v. nonholonomic constitutive laws and to the various types of work hardening possible in the plastic range.

Yield functions are used to describe the set of loads which leave a structural element in the elastic range. This set is taken to be convex and polyhedral. The yield conditions are given by linear inequalities. To be determined are the plastic multipliers (and possibly displacements) through which other quantities of interest are obtained. Roughly speaking, the complementarity enters through the condition that a plastic multiplier cannot be positive in the elastic range.

STRANG [59] has given an extremely readable account of the subject as developed by MAIER [49]. In the deformation theory (the static case) one has

- compatibility of the strains ε and displacements u: $\varepsilon = Bu$
- equilibrium of stresses σ with external forces f: $B^T\sigma = f$
- a splitting of strains into elastic and plastic parts: $\varepsilon = S^{-1}\sigma + p$
- plastic strains related to normals to piecewise linear yield surfaces: $p = N\lambda$, $\lambda \geqq 0$
- the yield condition with hardening matrix H: $\phi = N^T\sigma - H\lambda - k \leqq 0$
- the complementarity condition: $\phi^T\lambda = 0$.

What emerges is a "partial LCP"

$$-\phi = (H + N^TSN)\lambda - N^TSBu + k \geqq 0$$

$$-B^TSN\lambda + B^TSBu + f = 0$$

$$\lambda \geqq 0, \text{ u free}$$

$$\phi^T\lambda = 0$$

The stiffness matrix B^TSB is symmetric and positive semi-definite. In the positive definite case, u can be eliminated and a standard LCP is obtained. Different devices can be used in the semi-definite case. It is possible for a problem such as this, that a solution fails to exist. This would correspond to collapse due to the inability of the structure to "support" the load f.

When the load f is replaced by αf for given f and $\alpha \in A$, we obtain a parametric LCP. However, the problem of nonholonomic (or path-dependent) behavior now enters the picture. The solution could in fact depend on the loading history. A special (and desirable) case is regularly progressive yielding in which a yield surface once reached is not left. Even more special is the absence of local unloading: isotonic behavior of λ with respect to the parameter α. See [12]. When this is not present, an incremental theory is used. See [49] and [59]. KANEKO [42] has recently considered this matter in the following form: Find a solution to the system

$$w = q + tp + Mz$$

$$w \geqq 0, \quad \dot{z} \geqq 0, \quad t \geqq 0$$

$$w^T\dot{z} = 0$$

where M is a P-matrix and $\dot{z}$ denotes the derivative of z with respect to t.

7.3 Some free-boundary problems have been solved using the methods described here. The free-boundary problem for journal bearings has already been mentioned. See CRYER [22], [23], COTTLE and SACHER [20], and COTTLE, GOLUB and SACHER [17]. To this can be added the much-discussed (model) problem of fluid flow through a porous medium. See BAIOCCHI, COMINCIOLI, GUERRI and VOLPI [1]. For reports on some computational experience with these and other large structured problems, see [13] and [14]. The problem of elasto-plastic torsion in a cylindrical bar and computational work thereon is reported by CEA and GLOWINSKI [2] and COTTLE and GOHEEN [16].

7.4 So far, the problems mentioned have been equivalent to convex quadratic programs albeit sometimes of very large scale. The list of applications in this category is very long indeed and cannot be included. We close this section and the paper by mentioning other problems which are not equivalent to convex quadratic programs.

One of the great accomplishments in the "early" history of the LCP was the invention by LEMKE and HOWSON [48] of a constructive algorithm for calculating a NASH equilibrium point for a bimatrix game. This work led to many significant developments including LEMKE's method itself. It is the foundation for algorithms that seek economic equilbria. See SCARF [58], MATHIESEN [52], and DANTZIG, EAVES, and GALE [25], for instance. For problems in this class, the robustness of LEMKE's method (vis-a-vis matrix classes) is especially useful. One may very well wish to avoid the assumptions of symmetry and positive (semi-)definiteness as made by KENNEDY [44].

Still another application is found in the theory of optimal stopping. ÇINLAR [7, p.208] following DYNKIN and YUSHKEVICH, considers a "game" in which one has a MARKOV chain X with state space E, a transition matrix P, and a bounded function f on E. One observes the process X as long as desired; if one then decides to stop it and the process is in state j, a payoff f(j) results and the game ends. If the process is never stopped, there is no payoff. The problem is to optimize the payoff.

For $i \in E$, let $v(i) := \sup_T E_i[f(X_T)]$ where E_i denotes the expected value given that the process starts in state i. The function v is called the value of the game. The problem is to compute v and find T_0 such that $v(i) = E_i[f(X_{T_0})]$, $i \in E$. When the state space is finite, a theorem shows that the value function is given by the solution to the linear program

$$\begin{aligned} &\text{minimize} \quad \sum_{i \in E} v(i) \\ &\text{subject to} \quad v(i) \geqq \sum_{j \in E} P(i,j)v(j) \\ &\qquad\qquad\quad v(i) \geqq g(i) := \max\{0, f(i)\} \end{aligned}$$

Letting $z = v - g$, $M = I - P$, and $q = -Mg$, we have the equivalent linear program

$$\begin{aligned} &\text{minimize} \quad e^T z \\ &\text{subject to} \quad q + Mz \geqq 0 \\ &\qquad\qquad\quad z \geqq 0 \end{aligned} \tag{21}$$

with $M \in Z$. The coefficients of the objective function $e^T z$ being ones, it follows that if this linear program has a solution, it is also obtainable via methods for the LCP (q,M). Another theorem shows how to compute the optimal stopping time.

REFERENCES

[1] C. BAIOCCHI, V. COMINCIOLI, L. GUERRI, and G. VOLPI, Free boundary problems in the theory of fluid flow through porous media: a numerical approach, Calcolo 10 (1973), 1-86.

[2] J. CEA and R. GLOWINSKI, Sur des méthodes d'optimisation par relaxation, RAIRO R-3 (1973), 5-32.

[3] R. CHAND, E.J. HAUG, and K. RIM, Analysis of unbonded contact problems by means of quadratic programming, J. Optimization Theory and Appl. 20 (1976), 171-189.

[4] R. CHANDRASEKARAN, A special case of the complementary pivot problem, Opsearch 7 (1970), 263-268.

[5] Y.Y. CHANG, On degeneracy in the linear complementarity problem, Technical Report Department of Operations Research, Stanford University, forthcoming.

[6] D.G. CHRISTOPHERSON, A new mathematical method for the solution of film lubrication problems, Inst. Mech. Engrs. J. Proc. 146 (1941), 126-135.

[7] E. ÇINLAR, Introduction to stochastic processes, Prentice-Hall, Inc., Englewood Cliffs, New Jersey, 1975.

[8] T.F. CONRY and A. SEIRIG, A mathematical programming method for design of elastic bodies in contact, J. Applied Mechanics 38 (1971), 387-392.

[9] R.W. COTTLE, Note on a fundamental theorem in quadratic programming, J. Soc. Indust. Appl. Math. 12 (1964), 663-665.

[10] __________, On a problem in linear inequalities, J. London Math. Soc. 43 (1968), 378-384.

[11] __________, The principal pivoting method of quadratic programming, in Mathematics of the decision sciences, Part 1 (G.B. DANTZIG and A.F. VEINOTT, Jr., eds.) American Mathematical Society, Providence, Rhode Island, 1968.

[12] __________, Monotone solutions of the parametric linear complementarity problem, Math. Programming 3 (1972), 210-224.

[13] __________, Complementarity and variational problems, Symposia Mathematica 19 (1976), 177-208.

[14] __________, Computational experience with large-scale linear complementarity problems, in Fixed points: algorithms and applications (S. KARAMARDIAN, ed.) Academic Press, New York, 1977.

[15] __________ and G.B. DANTZIG, Complementary pivot theory of mathematical programming, Linear Algebra Appl. 1 (1968), 103-125.

[16] __________ and M.S. GOHEEN, A special class of large quadratic programs, Technical Report SOL 76-7, Department of Operations Research, Stanford University.

[17] __________, G.H. GOLUB, and R.S. SACHER, On the solution of large, structured linear complementarity problems, Technical Report 73-5, Department of Operations Research, Stanford University.

[18] __________ and J.S. PANG, On solving linear complementarity problems as linear programs, to appear in Mathematical Programming Studies.

[19] ____________________, A least element theory of solving linear complementarity problems as linear programs, Technical Report 76-30, Department of Operations Research, Stanford University.

[20] R.W. COTTLE and R.S. SACHER, On the solution of large, structured linear complementarity problems: the tridiagonal case, to appear in J. Appl. Math. and Optimization.

[21] __________ and A.F. VEINOTT, Jr., Polyhedral sets having a least element, Math. Programming 3 (1972), 238-249.

[22] C.W. CRYER, The method of Christopherson for solving free boundary problems for infinite journal bearings by means of finite differences, Math. of Computation 25 (1971), 435-443.

[23] __________, The solution of a quadratic programming problem using systematic overrelaxation, SIAM J. Control 9 (1971), 385-392.

[24] G.B. DANTZIG and R.W. COTTLE, Positive (semi-)definite programming, in Nonlinear Programming (J. ABADIE, ed.), North-Holland Publishing Co., Amsterdam, 1967.

[25] G.B. DANTZIG, B.C. EAVES, and D. GALE, An algorithm for a piecewise linear model of trade and production with negative prices and bankruptcy, Technical Report SOL 76-19, Department of Operations Research, Stanford University.

[26] M.A. DIAMOND, The solution of a quadratic programming problem using fast methods to solve systems of linear equations, Int. J. Systems Sci. 5 (1974), 131-136.

[27] B.C. EAVES, The linear complementarity problem, Management Sci. 17 (1971), 68-75.

[28] U. ECKHARDT, Quadratic programming by successive overrelaxation, Technical Report Jül-1064-MA, Kernforschungsanlage Jülich, April 1974.

[29] G. FICHERA, Problemi elastostatici con vincoli unilaterali; il problema di Signorini con ambigue condizioni al contorno, Mem. Accad. Naz. dei Lincei 8 (1964), 91-140.

[30] M. FIEDLER and V. PTAK, On matrices with nonpositive off-diagonal elements and positive principal minors, Czech. Math. J. 12 (1962), 382-400.
[31] F.D. FISCHER, Zur Lösung des Kontaktproblems elastischer Körper mit ausgedehnter Kontaktfläche durch quadratische Programmierung, Computing 13 (1974), 353-384.
[32] V.M. FRIDMAN and V.S. CHERNINA, An iteration process for the solution of the finite-dimensional contact problem, Zh. Vychisl. Mat. Mat. Fiz. 7 (1967), 160-163.
[33] C.B. GARCIA, Some classes of matrices in linear complementarity theory, Math. Programming 5 (1973), 299-310.
[34] R. GLOWINSKI, J.L. LIONS, and R. TREMOLIERES, Analyse numérique des inéquations variationnelles, vol. 1, Dunod, Paris, 1976.
[35] R.L. GRAVES, A principal pivoting simplex method for linear and quadratic programming, Operations Res. 15 (1967), 482-494.
[36] G.J. HABETLER and A.L. PRICE, Existence theory for generalized nonlinear complementarity problems, J. Optimization Theory and Appl. 7 (1971), 223-239.
[37] ______________________, An iterative method for generalized nonlinear complementarity problems, J. Optimization Theory and Appl. 11 (1973), 36-48.
[38] C. HILDRETH, Point estimates of ordinates of concave functions, J. Amer. Statist. Assoc. 49 (1954), 598-619.
[39] _________, A quadratic programming procedure, Naval Res. Logist. Quart. 4 (1957), 79-85.
[40] A. INGLETON, A problem in linear inequalities, Proc. London Math. Soc. 16 (1966), 519-536.
[41] J.J. KALKER, A survey of the mechanics of contact between solid bodies, Z. angew. Math. Mech. 57 (1977), T3-T17.
[42] I. KANEKO, A parametric linear complementarity problem involving derivatives, Technical Report WP 77-23, Department of Industrial Engineering, University of Wisconsin-Madison, July 1977.
[43] S. KARAMARDIAN, Generalized complementarity problem, J. Optimization Theory Appl. 8 (1971), 161-168.
[44] M. KENNEDY, An economic model of the world oil market, Bell J. Econ. Management Sci. 5 (1974), 540-577.
[45] H.W. KUHN and A.W. TUCKER, Nonlinear programming, in Second Berkeley symposium on mathematical statistics and probability (J. NEYMAN, ed.), University of California Press, Berkeley and Los Angeles, 1951.
[46] C.E. LEMKE, Bimatrix equilibrium points and mathematical programming, Management Sci. 11 (1965), 681-689.
[47] _________, Recent results on complementarity problems, in Nonlinear programming (J.B. ROSEN, O.L. MANGASARIAN, and K. RITTER, eds.), Academic Press, New York, 1970.
[48] _________ and J.T. HOWSON, Jr., Equilibrium points of bimatrix games, J. Soc. Indust. Appl. Math. 12 (1964), 413-423.
[49] G. MAIER, A matrix structural theory of piecewise linear elastoplasticity with interacting yield planes, Meccanica 5 (1970), 54-66.
[50] O.L. MANGASARIAN, Characterization of linear complementarity problems as linear programs, to appear in Mathematical Programming Studies.
[51] ______________, Solution of symmetric linear complementarity problems by iterative methods, Technical Report #275, Department of Computer Sciences, University of Wisconsin-Madison, July 1976.
[52] L. MATHIESEN, Efficiency pricing in a linear programming model: a case of constraints on the dual variables, Technical Report SOL 74-18, Department of Operations Research, Stanford University.
[53] K.G. MURTY, Note on a Bard-type scheme for solving the complementarity problem, Opsearch 11 (1974), 123-130.
[54] P.D. PANAGIOTOPOULOS, Stress-unilateral analysis of discretized cable and membrane structure in the presence of large displacements, Ing. Archiv 44 (1975) 291-300.
[55] J.S. PANG, Some new and efficient algorithms for portfolio analysis, Technical Report # 1738, Mathematics Research Center, University of Wisconsin-Madison, August 1977.
[56] ________, I. KANEKO, and W.P. HALLMAN, On the solution of some (parametric) linear complementarity problems with applications to portfolio analysis, structural engineering and graduation, Technical Report WP 77-27, Department of Industrial Engineering, University of Wisconsin-Madison, August 1977.

[57] H. SAMELSON, R.M. THRALL, and O. WESLER, A partition theorem for Euclidean n-space, Proc. Amer. Math. Soc. 9 (1958), 805-807.
[58] H. SCARF, The computation of economic equilibria, Yale University Press, New Haven, Connecticut, 1973.
[59] G. STRANG, Discrete plasticity and the complementarity problem, to appear in Proceedings of the U.S.-Germany symposium on finite elements, MIT Press, Cambridge, Massachusetts, 1977.
[60] C. VAN DE PANNE, A complementary variant of Lemke's method for the linear complementarity problem, Math. Programming 7 (1974), 283-310.
[61] P. WOLFE, The simplex method for quadratic programming, Econometrica 27 (1959), 382-398.

OPTIMISATION NON DIFFERENTIABLE: METHODES DE FAISCEAUX

C. Lemarechal
International Institute for Applied Systems Analysis
2361 Laxenburg (Autriche)

Abstract. We define a class of descent methods to minimize a nondifferentiable function. These methods are based on a representation of the objective which combines a quadratic approximation and the usual approximation by a piecewise linear function. Hence, they realize a synthesis between quasi-Newton methods and cutting plane methods. In addition, they have the particularity of requiring no sophisticated line search. They are also presented in ref. [7] (in English).

1. MOTIVATION.

Soit f une fonction convexe et lipschitzienne, définie sur l'espace Euclidien réel habituel. Ses sous-gradients sont donc bornés. Le problème est de minimiser f.

Supposons qu'un algorithme d'optimisation est arrivé en x_n, et il s'agit de déterminer le nouvel itéré $x_{n+1} = x_n + d_n$.

A tout x et tout $g \in \partial f(x)$, on associe le nombre positif

$$\alpha(x,g,x_n) = f(x_n) - [f(x) + (g, x_n - x)] \tag{1}$$

qui n'est autre que l'erreur faite en remplaçant $f(x_n)$ par la valeur en x_n de la fonction linéaire supportant f au point x.

Il est reconnu que les algorithmes d'optimisation non différentiable doivent tenir compte de l'information accumulée au cours des précédentes itérations. Nous supposons donc qu'on dispose d'un "faisceau" d'information, constitué de

$$g_1, \ldots, g_n \quad \text{où} \quad g_i \in \partial f(x_i)$$

et

$$\alpha_1, \ldots, \alpha_n \quad \text{où} \quad \alpha_i = \alpha(x_i, g_i, x_n) \quad ,$$

c'est-à-dire

$$\alpha_i = f(x_n) - f(x_i) - (g_i, x_n - x_i) \quad . \tag{2}$$

On notera que, lorsque x_n est changé en x_{n+1}, chaque α_i est changé en $\alpha_i + f(x_{n+1}) - f(x_n) - (g_i, x_{n+1} - x_n)$, si bien que les x_i et $f(x_i)$ n'ont pas à être stockés.

Par convexité, on sait que

$$f(x_{n+1}) \geq f(x_i) + (g_i, x_{n+1} - x_i)$$

ce qui, utilisant (2), peut s'écrire:

$$f(x_{n+1}) \geq f(x_n) - \alpha_i + (g_i, d_n) \qquad i = 1,\ldots,n \quad .$$

Par conséquent, $f(x_{n+1})$ ne peut être inférieur à $f(x_n)$ que si

$$-\alpha_i + (g_i, d_n) < 0 \qquad i = 1,\ldots,n \quad .$$

On appellera "méthode de faisceaux" toute méthode dans laquelle d_n est choisi de façon que le nombre

$$\text{(3)} \qquad v(d) = \max \{-\alpha_i + (g_i, d) \mid i = 1,\ldots,n\}$$

soit aussi négatif que possible, dans un sens à préciser. Des exemples de telles méthodes sont connus:

(i) Méthodes de plans sécants [1], [4] dans lesquelles d_n minimise $v(d)$. Cependant, un tel d_n n'existe pas à distance finie, à moins que 0 n'appartienne à conv $\{g_1,\ldots,g_n\}$; de plus, lorsque la fonction coût est régulière, si x_n est proche de l'optimum, alors x_{n+1} en est loin; ces méthodes sont instables.

(ii) La méthode "Boxstep" [8] où on choisit un paramètre $t_n > 0$, et d_n minimise $v(d)$ sur la "boîte" de taille t_n: $\{d \mid |d|_\infty \leq t_n\}$.

(iii) La méthode définie dans [6], qui est une méthode boxstep dans laquelle on considère une boîte euclidienne $\{d \mid |d|_2 \leq t_n\}$.

(iv) Les méthodes dites de sous-gradient conjugué [5], [9], [11], qui sont essentiellement (iii) avec $t_n = +\infty$. On peut considérer aussi que les méthodes de type Demjanov [2] sont (iii) avec t_n très petit.

Ainsi, les méthodes (ii) et (iii) sont des tentatives de stabilisation de (i). Les normes ℓ_∞ et ℓ_2 sont choisies "off line", et on peut aussi bien formaliser ces méthodes dans un cadre plus général: à l'itération n, on choisit une norme p_n, et on détermine d_n qui minimise $v(d)$ sur la boule $\{d \mid p_n(d) \leq t_n\}$. Ceci peut s'écrire

$$\begin{cases} \min v \\ p_n(d) \leq t_n \\ -\alpha_i + (g_i, d) \leq v \end{cases}$$

Soit $\mu_n \geq 0$ le multiplicateur de Lagrange (inconnu) associé à la 1ère contrainte. Alors, d_n est aussi solution de

$$\begin{cases} \min \mu_n p_n(d) + v \\ -\alpha_i + (g_i, d) \leq v \end{cases} .$$

Pour d'évidentes raisons de commodité, il y a intérêt à choisir p_n "au pire" quadratique. D'autre part, il convient de normaliser p_n de façon à guider automatiquement le choix de t_n ou de μ_n. (Remarquons ici que si tous les α_i sont nuls, alors d_n dépend homothétiquement de t_n; c'est ce qui explique que les méthodes (iv) ont retenu un temps l'attention.)

C'est pourquoi nous supposons qu'une matrice symétrique définie positive A_n est également donnée, de telle façon que la fonction quadratique

$$(4) \qquad \tfrac{1}{2}(d, A_n d) + (g_n, d) + f(x_n)$$

est une approximation de $f(x_n + d)$. Alors, nous choisirons p_n de façon que

$$\mu_n p_n(d) = \tfrac{1}{2}(d, A_n d) \quad .$$

Ce choix peut être justifiée en remarquant que, si le faisceau est réduit à un élément g_n avec $\alpha_n = 0$, alors la méthode de (quasi-) Newton consiste à prendre $d_n = A_n^{-1}\ d_n$, qui est justement la solution de

$$\begin{cases} \min \tfrac{1}{2}(d, A_n d) + v \\ (g_n, d) \leq v \end{cases}$$

2. ALGORITHME

Pour des raisons informatiques, le faisceau peut être trop encombrant pour être stocké en entier, et on devra en extraire une partie. Il est possible de ne garder que les g_i tels que $\alpha_i \leq \varepsilon$ pour un $\varepsilon > 0$ choisi à notre convenance. Nous continuerons à noter g_i, α_i $i = 1,\dots,n$ le sous-faisceau obtenu, et d_n est ainsi solution de

$$(5) \qquad \begin{cases} \min \frac{1}{2}(d, A_n d) + v \\ -\alpha_i + (g_i, d) \leq v \end{cases}$$

Plutôt que A_n, il est intéressant de considérer $H_n = A_n^{-1}$; on résout alors le programme dual de (5)

$$(6) \qquad \begin{cases} \min \quad (\Sigma\lambda_i g_i, \Sigma\lambda_i H_n g_i) + \Sigma\lambda_i\alpha_i \\ \lambda_i \geq 0 \quad , \quad \Sigma\lambda_i = 1 \end{cases}$$

et on pose $s_n = \Sigma\lambda_i g_i$, $\varepsilon_n = \Sigma\lambda_i\alpha_i$. Ainsi $s_n \in \partial_{\varepsilon n} f(x_n)$, et la solution de (5) est

$$(7) \qquad d_n = -H_n s_n = -\Sigma\lambda_i H_n g_i \quad , \quad v_n = -(s_n, H_n s_n) - \varepsilon_n \quad .$$

Il est clair que v_n est négatif; s'il était nul, on aurait $s_n = 0$, $\varepsilon_n = 0$, $0 \in \partial f(x_n)$, et x_n serait optimal. L'algorithme d'optimisation doit donc s'efforcer de faire tendre v_n vers 0 (par valeurs inférieures).

Théorème 1. Si $v_n \geq -\eta$ et si H_n est définie positive $(H_n s,s) \geq c|s|^2$, alors x_n vérifie la condition d'optimalité approchée $f(x_n) \leq f(\bar{x}) + \eta + \sqrt{\eta/c}|x_n - \bar{x}|$ où $\bar{x}$ minimise f.

Démonstration. Si $v_n \geq -\eta$, alors $c|s_n|^2 \leq (H_n s_n, s_n) \leq \eta$ et $\varepsilon_n \leq \eta$.

Donc $s_n \in \partial_\eta f(x_n)$ et $|s_n| \leq \sqrt{\eta/c}$ entraînent

$$\forall y \quad , \quad f(y) \geq f(x_n) + (s_n, y - x_n) - \eta \geq f(x_n) - \sqrt{\eta/c}|y - x_n| - \eta \ //$$

Comme dans [11], v_n sera forcé vers 0 en n'effectuant la descente de x_n à x_{n+1} que si

$$(8) \qquad f(x_{n+1}) \leq f(x_n) + m\, v_n$$

où m est un coefficient positif fixé. Toute suite infinie ainsi construite sera une suite minimisante de f:

Théorème 2. Si une suite infinie est construite satisfaisant (0), alors $f(x_n) \to -\infty$ ou $v_n \to 0$. Si, de plus, H_n reste uniformément définie positive, alors x_n est une suite minimisante.

Démonstration. (8) montre directement que

$$f(x_{n+p}) \leq f(x_n) + m \sum_{i=1}^{p-1} v_{n+i} \quad .$$

Si f est bornée inférieurement, la série Σv_n converge. Le résultat suit d'après le théorème 1. //

En toute généralité, si $x_{n+1} = x_n + t_n d_n$, on posera $m = m_2 t_n$ où $m_2 \in]0,1[$. Si une recherche linéaire est effectuée (ce qui, en toute rigeur, n'est pas utile), il faudra donc restreindre t_n à être borné inférieurement par un nombre positif t_0.

Dans le cas où la descente ne peut pas s'effectuer, c'est-à-dire lorsqu'il est impossible de trouver $t \geq t_0$ tel que

$$f(x_n + t\, d_n) \leq f(x_n) + m_2 t\, v_n$$

alors on considéra que l'algorithme est provisoirement bloqué. Il y aura lieu de modifier la direction d_n en enrichissant le faisceau d'informations. Le fait remarquable est que l'algorithme se débloquera si on ajoute dans le faisceau n'importe quel couple g, α calculé en $x_n + t\, d_n$ pourvu que t soit inférieur à 1. La clef de la démonstration tient dans le résultat suivant:

Théorème 3. Soit

$$t \in]0,1], \ g(t) \in \partial f(x_n + td_n), \ \alpha(t) = f(x_n) - f(x_n + td_n) + t(g(t), d_n)$$

et supposons que

$$f(x_n + td_n) > f(x_n) + m_2 t\, v_n \quad .$$

Alors

$$(9) \qquad -\alpha(t) + (g(t), d_n) > m_2 v_n \quad .$$

Démonstration. $-\alpha(t) + (g(t),d_n) = f(x_n + td_n) - f(x_n) + (1-t)(g(t),d_n)$

$$> m_2 t v_n + (1-t)(g(t),d_n) \quad .$$

D'autre part, la convexité de f entraîne que $\alpha(t) \geq 0$, soit

$$(g(t),d_n) \geq \frac{f(x_n + td_n) - f(x_n)}{t} > m_2 v_n \quad .$$

On obtient donc

$$-\alpha(t) + (g(t),d_n) \geq m_2 t\, v_n + (1-t)\, m_2 v_n = m_2 v_n \quad . \quad //$$

Ainsi, en opérant successivement des recherches linéaires le long de directions d_n <u>issues du même point x</u>, et en ajoutant dans le faisceau successivement g_{n+1}, α_{n+1} calculés à chaque recherche linéaire et satisfaisant (9), on peut montrer que l'algorithme finit par être débloqué pourvu que H_n soit borné.

<u>Théorème 4</u>. Considérons la suite des problèmes (5) où le $(n+1)^e$ problème se déduit du n^e par modification de H_n et addition d'une contrainte

$$-\alpha_{n+1} + (g_{n+1},d) \leq v \quad \text{telle que} \quad -\alpha_{n+1} + (g_{n+1},d_n) \geq m_2 v_n \quad .$$

Supposons H_n bornée: $|H_n s| \leq C|s|$ $\forall s, \forall n$. Alors, $v_n \to 0$.

Si, de plus, H_n reste uniformément définie positive, alors l'algorithme se débloque à moins que x ne soit optimal.

Démonstration. Les vecteurs s_n sont bornés en tant que combinaisons convexes de sous-gradients bornés. Donc d_n est borné. De même ε_n est borné supérieurement d'après (6). Donc $v_n = (s_n,d_n) - \varepsilon_n$ est borné inférieurement (et supérieurement par 0).

Soit une sous-suite d_n et v_n tendant vers d et $v \leq 0$, et soit p le successeur de n dans cette sous-suite. On a donc

$$-\alpha_{n+1} + (g_{n+1},d_p) \leq v_p \quad .$$

D'autre part, par hypothèse:

$$\alpha_{n+1} - (g_{n+1},d_n) \leq -m_2 v_n \quad .$$

On en déduit par addition:

$$(g_{n+1},d_p - d_n) \leq v_p - m_2 v_n \quad .$$

Passons à la limite: $0 \leq (1-m_2)v$. Donc $v \geq 0$. Le résultat suit d'après le théorème 1. //

Nous pouvons donner maintenant l'algorithme dans sa forme la plus simple, c'est-à-dire sans recherche linéaire:

x_1, $g_1 \in \partial f(x_1)$ sont donnés. La tolérance $\eta > 0$ est donnée, $m_2 = 0.1$.
Poser $n_0 = 1$, $\alpha_1 = 0$, $J = \{1\}$, H = Identité.

Etape 1. Résoudre (6) où i parcourt J. Obtenir ainsi d,s,ε,v.
Si $-(d,s) + \varepsilon \leq \eta$. STOP.

Etape 2. Poser $x = x_{n_0} + d$. Calculer $f(x)$, $g_{n+1} \in \partial f(x)$.
Si $f(x) \leq f(x_{n_0}) + m_2 v$ aller en 4.
Sinon, aller en 3.

Etape 3. (déblocage). Calculer $\alpha_{n+1} = f(x_{n_0}) - f(x) + (g_{n+1},d)$
$\gamma = g_{n+1} - g_{n_0}$ et aller en 5.

Etape 4. (descente).
Pour chaque $i \in J$, changer α_i en $\alpha_i + f(x) - f(x_{n_0}) - (g_i,d)$.
Poser $x_{n_0} = x$, $\alpha_{n+1} = 0$, $\gamma = g_{n+1} - g_{n_0}$, $n_0 = n + 1$.

Etape 5. (update). Faire $n = n+1$. Ajouter n à J.
Modifier H en utilisant d et γ.
Réduire éventuellement J en effaçant au choix des indices
$i < n_0$. Aller en 1. //

Dans cet algorithme, n_0 représente la suite des itérations de descente et $\alpha_{n_0} = 0$. C'est la suite x_{n_0} qui est censée minimiser f. La réduction de J pourra s'effectuer chaque fois que le nombre de sous-gradients est trop grand pour la mémoire allouée. Ceci pourra s'effectuer en remplissant le nouveau J d'abord avec les $i \geq n_0$, puis avec les autres en commençant par les plus petits α_i, jusqu'à ce que la taille maximum soit atteinte.

3. REMARQUES

Le fait que H se comporte "comme il faut" dépendra de la formule d'update choisie. On peut prévoir que l'étude dans ce domaine sera difficile puisqu'il a fallu attendre 15 ans le premier théorème de convergence globale dans les méthodes de quasi-Newton [10] (pour des fonctions-coût differentiables, bien sûr). On ignore si la propriété que H reste doublement bornée est réalisable ou non.

Toute fois, on peut accepter que H est non borné en inhibant l'update dans le cas d'un déblocage. Alors, le théorème 4 s'appliquera (puisque la suite des iterations de déblocage sera faite avec H fixée).

Dans ces conditions, l'algorithme terminera après un nombre fini d'itérations, mais on ignorera la valeur de la constante C du théorème 1. (Elle peut dépendre de η). On pourra éventuellement affiner la convergence en continant les itérations avec H fixé, et en prenant comme nouveau test d'arrêt $|s|^2 + \varepsilon \leq \eta$.

D'autre part on peut effectuer une recherche linéaire embryonnaire, ce qui aura éventuellement deux effets bénéfiques:

(i) Diminuer t à l'étape 3, de façon à accepter moins d'itérations de déblocage. Ceci est loisible à condition de toujours choisir $t \geq t_0 > 0$.

(ii) Augmenter t à l'étape 4, ce qui pourra améliorer la descente. De plus cela pourra améliorer aussi le comportement de H. On sait par exemple que la formule dite de BFS [3]

$$H \to H - (d\tilde{\gamma}H + H\gamma\tilde{d})/(d,\gamma) + [t + (\gamma,H\gamma)/(d,\gamma)]d\tilde{d}/(d,\gamma)$$

(d est la direction, t le pas, γ la variation du gradient) conserve la coercivité de H pourvu que $t(d,\gamma)$ soit positif. Ceci peut être réalisé en incorporant un test sur (g_{n+1},d) dans l'étape 4:

<u>Théorème 5</u>. Soit $m_1 \in]0,1[$ (par exemple $m_1 = 0.7$), et supposons que l'étape 4 produise $t \geq 1$ et $g_{n+1} \in \partial f(x_{n_0} + td)$ tels que

$$(g_{n+1},d) \geq m_1 v.$$

Alors $t(\gamma,d) > 0$.

<u>Démonstration</u>. Comme $\alpha_{n_0} = 0$, on a $(g_{n_0},d) \leq v$.

On en déduit par addition:

$$(\gamma,d) = (g_{n+1},d) - (g_{n_0},d) \geq (m_1 - 1)v > 0 \quad . //$$

BIBLIOGRAPHIE

[1] E.W. Cheney, A.A. Goldstein, Newton's methods for convex programming and Tchebycheff approximation, *Numerische Mathematik*, 1(1959), 253-268.

[2] V.F. Demjanov, Algorithms for some minimax problems, *Journal of Computer and Systems Sciences*, 2(1968), 342-380.

[3] R. Fletcher, A new approach to variable metric algorithms, *The Computer Journal*, 13, 3(1970), 317-322.

[4] J.E. Kelley, The cutting plane method for solving convex programs, *Journal of the SIAM*, 8(1960), 703-712.

[5] C. Lemarechal, An extension of Davidon methods to nondifferentiable problems, *Mathematical Programming Study*, 3(1975), 95-109.

[6] ————————, Combining Kelley's and conjugate gradient methods. Abstracts, IX. International Symposium on Mathematical Programming (Budapest,1976), 158-159.

[7] ————————, Nondifferentiable Optimization and descent methods, International Institute for Applied Systems Analysis, Laxenburg, Austria.

[8] R.E. Marsten, W.W. Hogan, J.W. Blankenship, The Boxstep method for large-scale optimization, *Operations Research*, 23, 3(1975), 389-405.

[9] R. Mifflin, An algorithm for constrained optimization with semi-smooth functions, *Mathematics of Operations Research* (1977) à paraître.

[10] M.J.D. Powell, Some global convergence properties of a variable metric algorithm for minimization without exact line searches, AERE, Harwell, Working Paper CSS 15(1975).

[11] P. Wolfe, A method of conjugate subgradients for minimizing nondifferentiable functions, *Mathematical Programming Study*, 3(1975), 145-173.

Variable metric methods for constrained optimization

by

M.J.D. Powell

(D.A.M.T.P., University of Cambridge, England)

1. Variable metric methods

This paper considers variable metric methods for calculating the vector of variables $\underline{x}$ in R^n that minimizes a real valued objective function $F(\underline{x})$, subject to the constraints

$$\left.\begin{aligned} c_i(\underline{x}) &= 0, \quad i = 1,2,\ldots,m' \\ c_i(\underline{x}) &\geqslant 0, \quad i = m'+1,\ldots,m \end{aligned}\right\} \tag{1}$$

on the values of the variables. It is assumed that first and second derivatives of the objective and constraint functions exist, and that the functions and their first derivative vectors $\underline{\nabla} F(\underline{x})$ and $\underline{\nabla} c_i(\underline{x})$ $(i = 1,2,\ldots,m)$ can be computed for any $\underline{x}$.

When there are no constraints on the variables the most successful algorithms that use function and gradient information are of three types, namely variable metric, quasi-Newton and conjugate gradient methods. Detailed descriptions of these methods are given in a book edited by Murray (1972). They are all iterative, each iteration begins with a starting point $\underline{x}_k$ in the space of the variables, a search direction $\underline{d}_k$ is calculated and the iteration replaces $\underline{x}_k$ by the vector

$$\underline{x}_{k+1} = \underline{x}_k + \alpha_k \underline{d}_k, \tag{2}$$

where α_k is a step-length whose value is chosen so that a reduction in $F(\underline{x})$ is obtained.

The main feature of a <u>variable metric</u> method for unconstrained optimization is that the calculation of $\underline{d}_k$ depends on a matrix B_k that is forced to be positive definite. When $\underline{x}_k$ is near the required solution, it is usual to regard B_k as an approximation to the second derivative matrix of $F(\underline{x})$. Away from the solution, however, it is better to view B_k as a matrix that is chosen automatically and that usually gives good search directions. Specifically $\underline{d}_k$ is the vector

$$\underline{d}_k = -B_k^{-1} \underline{\nabla} F(\underline{x}_k), \tag{3}$$

which is just the value of $\underline{d}$ that minimizes the quadratic function

$$Q(\underline{d}) = F(\underline{x}_k) + \underline{d}^T \underline{\nabla} F(\underline{x}_k) + \tfrac{1}{2}\underline{d}^T B_k \underline{d}. \tag{4}$$

In the extension to the constrained case, a variable metric method retains the important property that the definition of the search direction $\underline{d}_k$ depends only on a positive definite matrix B_k of dimension n and on function and first derivative values that are calculated at the starting point of the iteration $\underline{x}_k$. The technique is to let $\underline{d}_k$ be the vector $\underline{d}$ that minimizes the quadratic function (4) subject to the linear constraints

$$\left.\begin{aligned} c_i(\underline{x}_k) + \underline{d}^T \underline{\nabla} c_i(\underline{x}_k) &= 0, \quad i = 1,2,\ldots,m' \\ c_i(\underline{x}_k) + \underline{d}^T \underline{\nabla} c_i(\underline{x}_k) &\geqslant 0, \quad i = m'+1,\ldots,m \end{aligned}\right\} \tag{5}$$

(Garcia-Palomares and Mangasarian, 1976), which are approximations to the conditions (1) at the point $\underline{x} = \underline{x}_k + \underline{d}$. Therefore the calculation of $\underline{d}_k$ is a convex quadratic programming problem. Now it is advisable, even when $\underline{x}_k$ is close to the required solution, to regard B_k not as a second derivative matrix but as a convenient positive definite matrix. It is of course useful to ask what properties B_k must have in order that the final rate of convergence is superlinear. They depend on second derivatives of the Lagrangian function and they are discussed in Section 2.

In order to complete the definition of a variable metric algorithm for constrained optimization, it remains to specify a method that generates the sequence of matrices B_k $(k = 1,2,3,\ldots)$, and to choose a procedure that fixes the step-length α_k that occurs in equation (2). Most of the available methods for generating the positive definite matrices are based on formulae that are used in the unconstrained case. Suitable techniques for both operations are given by Han (1975, 1976) and by Powell (1977a). The numerical results that are reported later were obtained by Powell's algorithm.

Some of these results are given in Section 3, where the algorithm is demonstrated by applying it to an interesting new problem of seven variables and seven constraints. In Section 4 a discussion compares variable metric methods with some other algorithms that are currently in use for constrained optimization calculations. The discussion is supported by numerical results taken from Powell (1977a). We find that it is usual for variable metric methods to require far fewer function and gradient evaluations than other algorithms.

2. The convergence of variable metric methods

The main convergence question is whether it is sensible to force every matrix B_k $(k = 1,2,3,\ldots)$ to be positive definite. It is a sensible strategy in the unconstrained case because it ensures that each search direction (3) is downhill and, in the usual case when the true second derivative matrix of $F(\underline{x})$ is positive definite at the required solution, it allows a superlinear rate of convergence. However the situation is less clear when constraints are present.

On the question of downhill search directions, we first have to choose an objective function that we wish to decrease when $\underline{x}_k$ is replaced by $\underline{x}_{k+1}$. It must take account not only of $F(\underline{x})$ but also of the constraint functions. Han (1975) proposes a useful objective function for which $\underline{d}_k$ is necessarily downhill. He suggests a choice of α_k and proves that, if the matrices B_k $(k = 1,2,3,\ldots)$ and B_k^{-1} $(k = 1,2,3,\ldots)$ are uniformly bounded, then the sequence $\underline{x}_k$ $(k = 1,2,3,..)$ converges to a Kuhn-Tucker point of the optimization calculation. Therefore the use of positive definite matrices is suitable for achieving convergence when constraints are present.

The question of the rate of convergence is considered by Garcia-Palomares and Mangasarian (1976). They prove that, if the sequence $\underline{x}_k$ $(k = 1,2,3,\ldots)$ converges to the required solution, $\underline{x}^*$ say, and if the matrices B_k $(k = 1,2,3,\ldots)$ satisfy the condition

$$\lim_{k\to\infty} \frac{\| (G^* - B_k)\, \underline{d}_k \|}{\| \underline{d}_k \|} = 0 , \qquad (6)$$

then the rate of convergence is superlinear. The matrix G^* is the second derivative matrix at $\underline{x}^*$ of the Lagrangian function

$$L(\underline{x}, \underline{\lambda}^*) = F(\underline{x}) - \sum_{i=1}^{m} \lambda_i^* \, c_i(\underline{x}) , \qquad (7)$$

where the Lagrangian parameters λ_i^* $(i = 1,2,\ldots m)$ are defined by complementarity conditions and by the equation

$$\underline{\nabla} F(\underline{x}^*) - \sum_{i=1}^{m} \lambda_i^* \, \underline{\nabla} c_i(\underline{x}^*) = 0 . \qquad (8)$$

Thus they extend the Dennis and Moré (1974) theorem to constrained optimization calculations.

However, in contrast to the case when there are no constraints, the matrix G^* may have some negative eigenvalues. In this case it may happen, even when k is large, that the search direction $\underline{d}_k$ is a direction of negative curvature of the Lagrangian function. Thus condition (6) may conflict with the requirement that every matrix B_k be positive definite, which is the main reason for asking whether it is sensible to use a variable metric method.

A numerical experiment reported by Powell (1976) suggests that, even when G^* has some large negative eigenvalues, superlinear convergence can still be achieved by variable metric algorithms. What happens usually, but not always, is that, due to the linear approximations to the constraints, the constraint violations approach zero more quickly than the rate at which $\underline{x}_k$ $(k = 1,2,3,\ldots)$ tends to $\underline{x}^*$. Thus the search directions tend to be orthogonal to the gradient vectors of the active constraints. Along these directions the curvature of the Lagrangian function is usually positive, due to the 'second-order necessary conditions' (see Fiacco and McCormick, 1968, for instance). Hence condition (6) can usually be satisfied, even though every matrix B_k $(k = 1,2,3,\ldots)$ is positive definite.

Because of the less usual cases, however, condition (6) is sometimes not obtained in practice. Therefore Powell (1977b) gives further consideration to the rate of convergence. He finds a relation between G^* and B_k that can be satisfied by variable metric algorithms and that gives the two-step superlinear convergence result

$$\lim_{k \to \infty} \frac{\| \underline{x}_{k+2} - \underline{x}^* \|}{\| \underline{x}_k - \underline{x}^* \|} = 0. \tag{9}$$

This relation assumes that the strict complementarity condition is satisfied at $\underline{x}^*$ and that the gradients $\{\underline{\nabla} c_i(\underline{x}^*);\ i \in I\}$ are linearly independent, where I is the set of indices of the active constraints. It depends on the symmetric matrix, P say, that projects orthogonally into the subspace of vectors $\underline{y}$ that satisfy the condition

$$N^T \underline{y} = 0, \tag{10}$$

where N is the matrix whose columns are the vectors $\{\underline{\nabla} c_i(\underline{x}^*);\ i \in I\}$. Therefore P is the matrix

$$P = I - N(N^T N)^{-1} N^T. \tag{11}$$

It states that, if the matrices B_k $(k = 1,2,3,\ldots)$ satisfy the condition

$$\lim_{k \to \infty} \frac{\| (PG^*P - PB_k P) \underline{d}_k \|}{\| \underline{d}_k \|} = 0, \tag{12}$$

then equation (9) holds.

The main advantage of condition (12) over condition (6) is that, although the matrix G^* may be indefinite, the second-order necessary condition at $\underline{x}^*$ states that PG^*P is positive semi-definite. Therefore condition (12) can be satisfied by a sequence of positive definite matrices B_k $(k = 1,2,3,\ldots)$. Thus we have found a favourable answer to the question whether variable metric algorithms for constrained optimization can provide a superlinear rate of convergence when G^* has some negative eigenvalues.

Another important consequence of equation (12) is that the rank of the matrix that contains the second derivative information that is needed for fast convergence, namely PG^*P, is at most the number of variables minus the number of active constraints. Thus, even when the constraints are nonlinear, they decrease the size of the problem of estimating second derivative information. Experience with variable metric algorithms shows that usually the presence of constraints does reduce the number of iterations that are needed to calculate the required vector of variables.

In addition to showing the importance of condition (12), Powell (1977b) studies the convergence of the algorithm that gives the numerical results that are reported in Sections 3 and 4. He proves that its rate of convergence is R-superlinear, which means that it converges faster than any sequence that has a linear rate of convergence. This theorem is less strong than equation (9), but numerical results suggest that condition (9) is obtained.

3. A numerical example

This section describes the progress of a variable metric algorithm for constrained optimization when it is applied to the problem of calculating the triangle of smallest area that contains two circular discs of unit radius. The problem can be expressed in terms of seven variables and nine constraints. Specifically we let the vertices of the triangle be $(0,0)$, $(x_1, 0)$ and (x_2, x_3), and we let the centres of the discs be (x_4, x_5) and (x_6, x_7), so we have to minimize the function

$$F(\underline{x}) = x_1 x_3. \tag{13}$$

We keep $F(\underline{x})$ positive by requiring both x_1 and x_3 to be non-negative. We keep the discs apart by imposing the condition

$$(x_4 - x_6)^2 + (x_5 - x_7)^2 - 4 \geq 0. \tag{14}$$

The disc whose centre is at (x_4, x_5) is inside the triangle if the inequalities

$$x_4 - 1 \geq 0, \tag{15}$$

$$\frac{x_3x_4 - x_2x_5}{\left[x_2^2 + x_3^2\right]^{\frac{1}{2}}} - 1 \geq 0, \tag{16}$$

and

$$\frac{-x_3x_4 + (x_2 - x_1)\,x_5 + x_1x_3}{\left[x_3^2 + (x_2 - x_1)^2\right]^{\frac{1}{2}}} - 1 \geq 0 \tag{17}$$

are obtained. Similar inequalities must be satisfied by the centre at (x_6, x_7). Note that it is inadvisable to multiply the left-hand side of expression (16) by the factor $\left[x_2^2 + x_3^2\right]^{\frac{1}{2}}$ and to multiply the left-hand side of expression (17) by $\left[x_3^2 + (x_2 - x_1)^2\right]^{\frac{1}{2}}$, because then it is possible to reduce $F(\underline{x})$ to zero and to satisfy the constraints by setting x_1, x_2 and x_3 to zero.

Initially we let the triangle have the vertices (0,0), (3,0) and (0,2) and we let the centres of the discs be at the points (-1.5, 1.5) and (5.0,0). These starting conditions are shown in Figure 1. Moreover we let B_1 be the unit matrix. Even though the starting values are far from the solution, Powell's (1977a) algorithm sets all the step-lengths α_k $(k = 1,2,3,\ldots)$ to one. As is usual in variable metric methods, on each iteration it calculates B_{k+1} by using B_k and the differences

$$\underline{\delta}_k = \underline{x}_{k+1} - \underline{x}_k \tag{18}$$

and

$$\underline{\gamma}_k = \underline{\nabla} L(\underline{x}_{k+1}, \underline{\lambda}_k) - \underline{\nabla} L(\underline{x}_k, \underline{\lambda}_k), \tag{19}$$

where $L(\underline{x}, \underline{\lambda})$ is the Lagrangian function (7), and where $\underline{\lambda}_k$ is the vector of Lagrange multipliers at the solution of the quadratic programming problem that defines the search direction $\underline{d}_k$. The matrix B_{k+1} is defined by the B-F-G-S formula

$$B_{k+1} = B_k - \frac{B_k \underline{\delta}_k \underline{\delta}_k^T B_k}{\underline{\delta}_k^T B_k \underline{\delta}_k} + \frac{\underline{\eta}_k \underline{\eta}_k^T}{\underline{\eta}_k^T \underline{\delta}_k}, \tag{20}$$

where $\underline{\eta}_k$ is the vector

$$\underline{\eta}_k = \theta_k \underline{\gamma}_k + (1-\theta_k) B_k \underline{\delta}_k . \tag{21}$$

The scalar θ_k is set automatically to a value from the interval $[0,1]$ that maintains positive definiteness. The value $\theta_k = 1$ is chosen except on the second, third and sixth iterations, when the values of θ_k are 0.8155, 0.5099 and 0.8555 respectively.

This example was chosen because, although there are seven variables, the progress of the calculation can be described by diagrams. Figures 2-6 show the triangle vertices and disc positions at the end of the first, second, third, fifth and eighth iterations. We note that the optimal solution is approached quite rapidly. It occurs when the triangle is right-angled and isoceles and when the active constraints are the ones that are clear in Figure 6. Twelve iterations are required to reduce the residuals of all the active constraints to less than 10^{-6}.

It is hoped that this example may be helpful to other work because, as well as the property that the results can be displayed easily, it has fewer active constraints than variables, and the final second derivative matrix of the Lagrangian function, which we call G^*, has some negative eigenvalues.

4. Comparison with other algorithms

Variable metric methods for constrained optimization have six advantages that are not all obtained by any other algorithms that treat nonlinear constraints. They are as follows.

(i) Except sometimes at the beginning of the calculation, linear constraints are satisfied exactly, without the need to distinguish between linear and non-linear constraints. Moreover a variable metric method treats efficiently any constraints that are almost linear.

(ii) Any active constraints automatically reduce the number of degrees of freedom that depend on estimates of second derivatives, in contrast with the augmented Lagrangian method that requires unconstrained minimizations in the full space of the variables.

(iii) No penalty terms are needed, because it is unnecessary to augment the Lagrangian function in order to make its second derivative matrix positive definite.

(iv) Curved constraint boundaries need not be followed closely, because the objective function that is used in the line search to determine α_k can provide automatically a suitable balance between reductions in the objective function and constraint violations.

(v) The quadratic programming calculation of each iteration predicts automatically which constraints are active at the solution. Therefore no active set strategies are required by a variable metric algorithm. Each quadratic programming problem is strictly convex.

(vi) The calculation is independent of changes of scale of the constraint functions. Moreover, as in variable metric methods for unconstrained optimization, we can obtain invariance under changes in scale of the variables, except for the initial choice of B_k.

There are also, however, some disadvantages and some unanswered questions. The most serious drawback is the work of solving a quadratic programming problem on each iteration. Little research has been done so far on reducing this overhead, but there is hope of substantial savings because usually successive quadratic programming calculations have much in common. For example, the set of active constraints may remain the same, the directions of constraint normals often change only slowly, and the differences between successive variable metric matrices are of low rank.

When solving problems in very large numbers of variables, it is important to reduce the number of matrix elements that occur in the calculation. In the case when all constraints are linear, some powerful techniques exist already. For example Buckley (1975) shows that, by using an active set strategy, the dimension of the variable metric matrix that needs to be stored can be reduced to the number of variables minus the number of active constraints. Because the stored information corresponds to the matrix PB_kP in equation (12), it may be possible to extend Buckley's technique to nonlinear constraints. We would have to give up advantage (v), but we would be able to make good use of any sparsity in the matrix N of constraint gradients.

One unanswered question is the behaviour of variable metric algorithms in degenerate situations. For example, the strict complementarity condition may not hold at $\underline{x}^*$ and the gradients of the active constraints at $\underline{x}^*$ may be linearly dependent. It is important to investigate these situations in order to ensure the robustness of the numerical method.

In normal situations, however, variable metric methods are usually very efficient. This fact is shown well by Colville's (1968) first three problems. His third problem is the easiest. There are five variables and sixteen inequality constraints. Three linear and two nonlinear constraints are active at the solution, so they fix the required vector $\underline{x}^*$, and the search direction $\underline{d}_k$ becomes independent of B_k. Thus a variable metric algorithm reduces to Newton's method for solving the equations that come from the active constraints. When Colville's value of $\underline{x}_1$ is used, the three active linear constraints are satisfied as equalities at the end of the first iteration. Only one more iteration is needed to reduce the residuals of the two active nonlinear constraints to 10^{-6}. In Colville's first problem there are five variables and fifteen linear constraints, four of which are active at $\underline{x}^*$. They are satisfied at the end of the second iteration. Therefore at this early stage the variable metric algorithm has reduced the calculation to the minimization of a function of only one variable. In Colville's second problem there are fifteen variables and twenty constraints. Therefore four degrees of freedom have to be taken up by the part of the calculation that depends most strongly on the matrix B_k.

These examples show that the presence of constraints can be very helpful in practice, but in most other algorithms constraints increase the amount of work that has to be done. Table 1 gives the number of function and gradient evaluations when Colville's first three problems are solved by a variable metric algorithm, by the augmented Lagrangian method (Fletcher, 1975), by REQP (Biggs, 1972) and by the best of the methods mentioned in Colville's report that state the number of function and gradient evaluations that are required. The table is taken from Powell (1977a) and the algorithms that are compared are among the best that are in current use. The table shows that, when several constraints are active at the required solution and when most of the total computer time is taken by the evaluation of functions and gradients, then variable metric methods are much more efficient than other algorithms.

Problem	Algorithm			
	Colville	REQP	Aug. Lagr.	V. Metric
Colville 1	13	8	39	6
Colville 2	112	47	149	17
Colville 3	23	10	64	3

Table 1

A comparison of algorithms

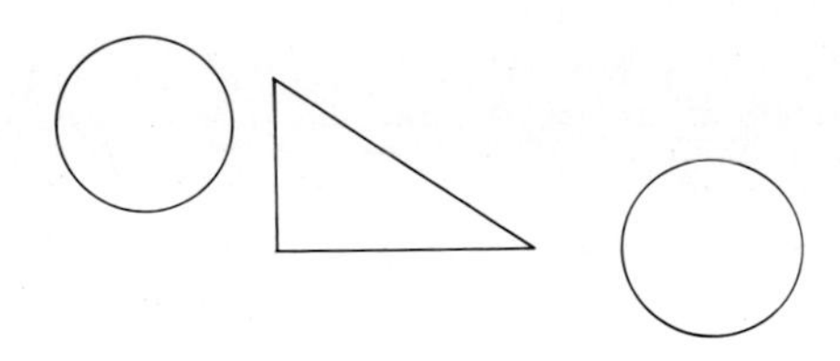

Figure 1: Starting Conditions.

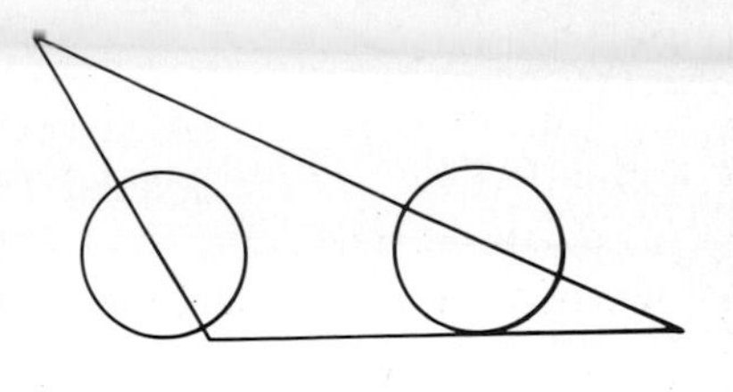

Figure 2: Iteration 1.

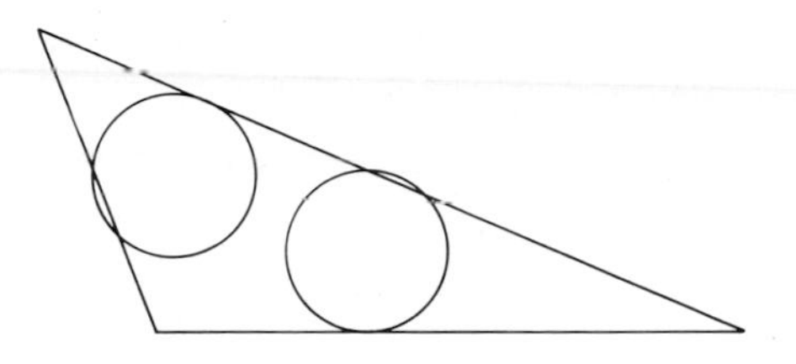

Figure 3: Iteration 2.

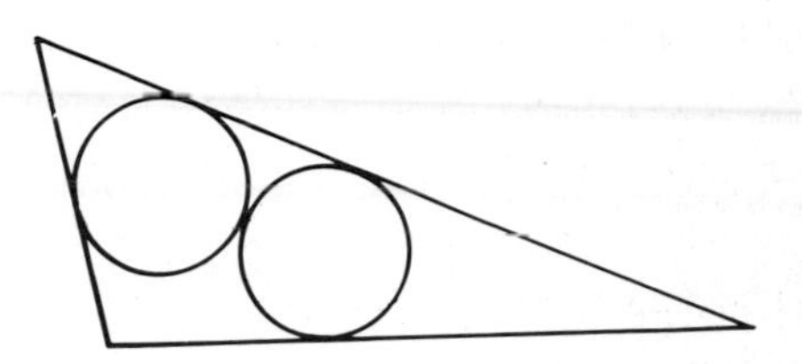

Figure 4: Iteration 3.

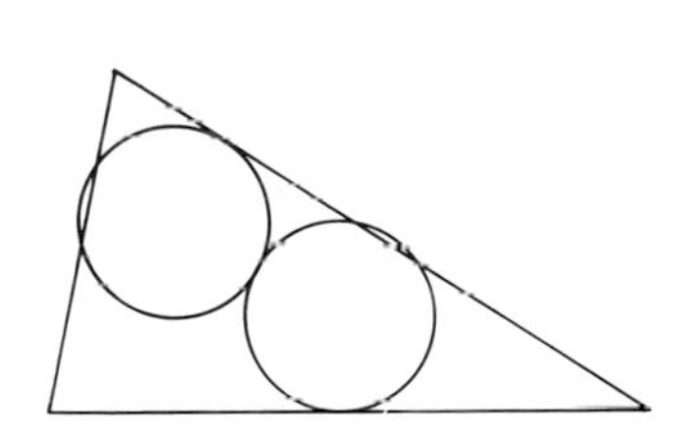

Figure 5: Iteration 5.

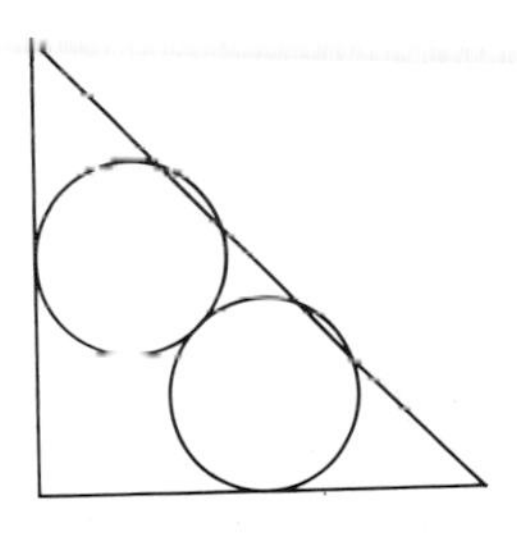

Figure 6: Iteration 8.

References

Biggs, M.C. (1972) "Constrained minimization using recursive equality quadratic programming" in Numerical methods for nonlinear optimization, ed. F.A. Lootsma, Academic Press (London).

Buckley, A.G. (1975) "An alternate implementation of Goldfarb's minimization algorithm", Math. Prog., Vol. 8, pp. 207-231.

Colville, A.R. (1968) "A comparative study on nonlinear programming codes", Report No. 320-2949 (IBM New York Scientific Center).

Dennis, J.E. and Moré, J.J. (1974) "A characterization of superlinear convergence and its application to quasi-Newton methods", Maths. of Comp., Vol. 28, pp. 549-560.

Fiacco, A.V. and McCormick, G.P. (1968) Nonlinear programming: sequential unconstrained minimization techniques, Wiley (New York).

Fletcher, R. (1975) "An ideal penalty function for constrained optimization", J. Inst. Maths. Applics., Vol. 15, pp. 319-342.

Garcia Palomares, U.M. and Mangasarian, O.L. (1976) "Superlinearly convergent quasi-Newton algorithms for nonlinearly constrained optimization problems", Math. Prog., Vol. 11, pp. 1-13.

Han, S-P. (1975) "A globally convergent method for nonlinear programming", Report No. 75-257 (Dept. of Computer Science, Cornell University).

Han, S-P. (1976) "Superlinearly convergent variable metric algorithms for general nonlinear programming problems", Math. Prog., Vol. 11, pp. 263-282.

Murray, W. (1972) Numerical methods for unconstrained optimization, Academic Press (London)

Powell, M.J.D. (1976) "Algorithms for nonlinear constraints that use Lagrangian functions", presented at the Ninth International Symposium on Mathematical Programming, Budapest.

Powell, M.J.D. (1977a) "A fast algorithm for nonlinearly constrained optimization calculations", presented at the 1977 Dundee conference on Numerical Analysis.

Powell, M.J.D. (1977b) "The convergence of variable metric methods for nonlinearly constrained optimization calculations", presented at Nonlinear Programming Symposium 3, Madison, Wisconsin.

FINITE ELEMENTS

ELEMENTS FINIS

CONSTANT STRAIN FINITE ELEMENTS
FOR ISOCHORIC STRAIN FIELDS

John H. Argyris and Padraic C. Dunne
Institut für Statik und Dynamik
(University of Stuttgart)

1. Introduction

As the finite element method is increasingly finding application in non-linear problems it is clear that the simpler finite elements are coming back into favour. This is because of the enormously increased complexity of programming the higher order elements and the realization that the problem definition is often too approximate to justify very refined modelling. This trend has been particularly noticeable in relation to shell elements and also in the field of large strains and plasticity.

However, in the case of incompressible or nearly incompressible material problems it is well known that the constant strain simplex elements are not suitable except with very special arrangements of the elements and then only for small strains [1, 2, 3]. Even with the higher order elements the satisfaction of the full point-wise incompressibility condition may lead to an over constrained assembly of finite elements. In these elements various methods have been used to relax the point-wise incompressibility condition. Although differing in details all these methods effectively reduce the number of parameters representing the hydrostatic stress or strain, either by weighted residual methods [2, 4] or by reduced integration techniques [5, 6]. Clearly they are not applicable to the constant strain simplex elements which can have only one hydrostatic stress parameter per element.

There have also been suggested methods which use groups of simplex elements and apply constraint only on the total area or volume of the group [1] and a method [7] which proposes the use of an area or volume constraint on the total of all elements connected to a given node. This latter method would apparently have the ideal ratio 1/2 of (constraints / freedoms) for TRIM3 and 1/3 for TET4. However it may be shown to be equivalent to the use of a conforming hydrostatic field with linear variation over the elements and in the latter form has been discussed previously by Oden [8] . Another procedure for improving the performance of finite elements in an isochoric field is to choose "a small but non-zero value of the compressibility for which the 'compressibility error' is on the same level as the other sources of error" [5] . However this method is only effective in higher order elements where the reduced integration or other techniques previously mentioned can be used to evaluate the dilatational strain energy. It also has the disadvantage of resulting in badly conditioned stiffness matrices and in very high frequency vibration modes which reduce the efficiency of both modal and step-by-step dynamic calculations.

The only other constant strain simplex element known to the authors is the non-conforming triangle in which the nodal points are at the mid-sides [7] . This element has been proved to be convergent [9] in the incompressible fluid applications for which it was designed but for elasticity problems it would appear to be unsuitable because of the possibility of mechanisms forming with commonly encountered element arrangements and boundary conditions.

From the above discussion it is clear that if the constant strain simplex elements are to remain available for incompressible applications some new procedure must be developed. It is the purpose of this article to describe such a development which may even lead to a better behaviour of the TRIM3, TRIAX3 and TET4 elements in the compressible regime.

2. Mechanical representation of TRIM-3

Although not essential to the development of the modified simplex elements it is useful to consider the mechanical or physically lumped representation of the TRIM3 element. In fact this representation was suggestive of the measures adopted to improve the behaviour of this element. It may be shown [10] that the stiffness matrix of the TRIM3 element whether isotropic or anisotropic may be represented by a triangle of three bars linked by rotational springs at the vertices. The axial stiffness of the bars and the rotational stiffness of the springs are easily determined and it is instructive to examine their behaviour in terms of the elastic constants of the membrane. For this purpose we consider the isotropic membrane with Young's modulus E and Poisson's ratio ν .

Note that if the membrane is really a sheet of isotropic material the value of ν corresponding to in-plane incompressibility depends on whether we are regarding the material to be under plane stress or plane strain.

Thus for an equilateral triangle of side a, height h, thickness t and area Ω we find,

Axial stiffness of sides,

$$EB = \frac{Eht}{3(1-\nu)}$$

Corner spring rate,

$$k = \frac{2E\Omega t(1-3\nu)}{9(1-\nu^2)} \tag{1}$$

We note that for isochoric behaviour $\nu \to 1$ and both EB and k become infinite with k negative. This suggests that the model and hence the TRIM3 element will behave badly in this case.

The next step in our argument is to consider the models arising from the separation of the deviatoric and dilational strain energies.

The strain invariants for membrane strain are

$$F_1 = \varepsilon_I + \varepsilon_{II} \quad ; \quad F_2 = \varepsilon_I \, \varepsilon_{II} \tag{2}$$

where ε_I , ε_{II} are the principal strains. Then for an isotropic membrane the superficial strain energy density may be written for unit thickness,

$$\bar{\Phi} = \frac{E}{2(1-\nu^2)}\left[F_1^2 - 2(1-\nu)F_2\right] = \frac{1}{2}E_h F_1^2 + \frac{1}{2}G\left(F_1^2 - 4F_2\right). \qquad (3)$$

where $E_h = E/2(1-\nu)$ is the dilatation modulus and $G = E/2(1+\nu)$ is the shear modulus.

The expression $\frac{1}{2}G\left(F_1^2 - 4F_2\right)$ is the deviatoric energy density and by inspection of (3) this may be represented by a membrane for which

$$\frac{E}{1-\nu^2} = G \quad \text{and} \quad \frac{2E}{1+\nu} = 4G \qquad (4)$$

Thus a material having only the deviatoric strain energy has $E = 0$, $\nu = -1$ and in the model of eq. (1).

$$EB = 0 \quad \text{and} \quad k = 8\Omega Gt/9 \qquad (5)$$

We see that the deviatoric part gives the well behaved model of fig. 1 in which there are only corner springs of finite stiffness.

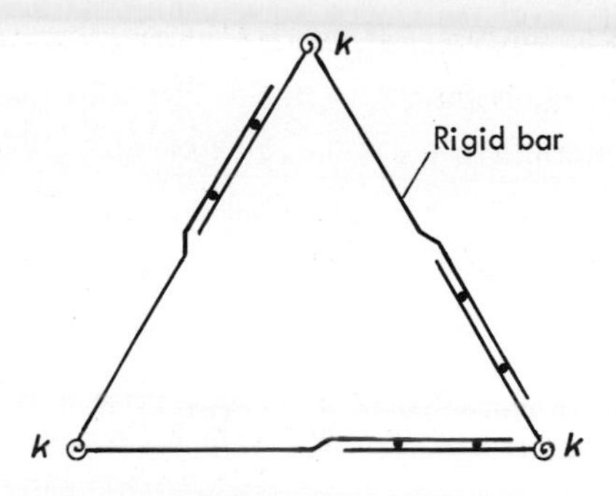

Fig. 1 DEVIATORIC MODEL

Instead of the axially stiff rods and corner-springs the dilational energy will be represented by another model consisting of three telescopic rods enclosing a fluid with compressibility modulus E_h which becomes infinite for incompressible material; see fig. 2.

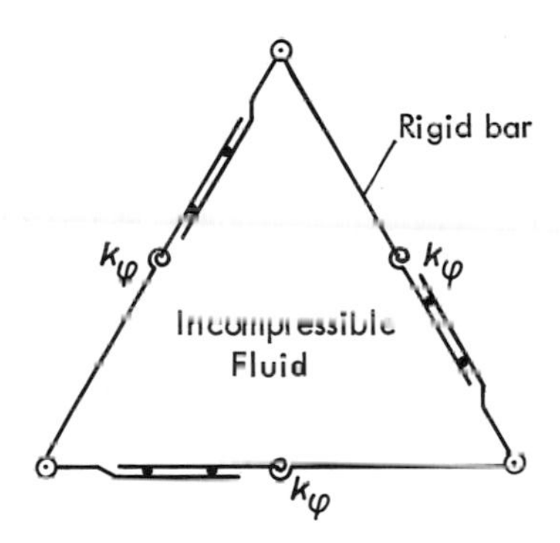

Fig. 2 DILATATION MODEL

However this model will have no advantage if the rods are rigid. By supposing the rods to be hinged at their centres with springs of rate k_φ we shall have a model which relaxes the excessive global constraints of the ordinary TRIM3 element without having to suppose that the enclosed fluid is compressible. It is clear that the magnitude of the stiffness k_φ will govern the convergence with mesh refinement and also that the required k_φ on a free boundary will have to be of greater magnitude then internal values.

Before considering the problem of convergence it is necessary to decide whether the present procedure is a special case of existing methods based on variational principles. That this is not so with respect to those variational principles, such as the Hellinger-Reissner, which are expressed at continuum level, may be seen from the fact that they can contain no terms corresponding to the stiffness k_φ. At the element level it is, of course, possible to add terms to the functional which will lead to the same equations as those obtained from the equilibrium of each side directly. This however is hardly a good reason to couch the problem in variational terms unless it can lead to a useful generalization of the method.

2.1 Calculation of k_φ

The convergence will depend on k_φ and we require a reasonable criterion for its determination. A method which is applicable also to other elements in which a simple mechanical model is not possible is to suppose that the strain energy due to the "bubble" or bulge mode of a side is of the form,

$$\frac{1}{2}\lambda G t(\delta\Omega/\Omega)^2\Omega \tag{6}$$

where λ is a factor independent of the side and $\delta\Omega$ is the area of the "bubble". In terms of the angle φ and length l of a side,

$$\delta\Omega = \varphi l^2/8 \tag{7}$$

Hence equating strain energies,

$$\frac{1}{2}k_\varphi\varphi^2 = \frac{1}{2}\lambda G t\varphi^2 l^4/64\Omega$$

or

$$k_\varphi = \lambda G t l^4/64\Omega \tag{8}$$

Thus for a triangle of given shape the variation of k_φ is like that of the deviatoric corner springs which are proportional to $G\Omega t$. It remains to determine a suitable value for λ and for this purpose we use an heuristic argument based on the strain energy density of the element in an isochoric field with linearly varying stress and strain.

2.2 Convergence

It was shown in reference [1] that the dilational energy density of a TRIM3 element with the displacements of an arbitrary quadratic isochoric field is proportional to,

$$E_h \Omega \tag{9}$$

This was established for an isosceles right-angled element but may be shown to be so for an arbitrary shape. Although this result demonstrates that the TRIM3 element is theoretically convergent for any finite E_h it is clear that for a nearly incompressible material the convergence will be too slow for practical applications and that at complete incompressibility there will be no convergence to the exact solution. Although, as mentioned in the introduction, there are special groupings of TRIM3 elements which behave very much better than arbitrary arrangements, this solution is not applicable for large strains and has not, so far as is known to the authors, a three-dimensional counterpart [1, 2, 3] .

The procedure for estimating λ may be described as follows. From the nodal displacements at the vertices as given by the quadratic displacement field calculate $\delta\Omega$. From $\delta\Omega$ for each "bubble" calculate the strain energy stored in the three sides. Convert this to an equivalent superficial strain energy density. Next estimate the difference in the hydrostatic stress σ_h between two elements. Use this to calculate the strain energy in the walls and the corresponding strain energy density. The two energy densities will be convergent to zero as the mesh is refined. However in the first case the energy increases with λ and in the other case decreases. Thus there is a value of λ which makes the two energies equal and this will be considered as the optimum value. Since in general the approximate quadratic field and the element mesh is dependent on the problem the optimum λ will vary over the field and will generally require to be much larger on free boundaries than internally. In practice it will be inconvenient to make λ dependent on the solution and in general we shall take it as infinite on the boundary and constant internally. Numerical experiments show that in some problems λ may range over an order of magnitude without greatly influencing the results. This desirable behaviour is in marked contrast to that of the method using a large E_h.

The general isochoric quadratic displacement field may be expressed in cartesion co-ordinates as,

$$\left.\begin{aligned} u &= axy + \tfrac{1}{2}bx^2 + cy^2 \\ v &= -bxy - \tfrac{1}{2}ay^2 + dx^2 \end{aligned}\right\} \qquad (10)$$

Then for an arbitrarily placed TRIM3 element with vertex co-ordinates x , y etc. we may show that, neglecting higher order terms the bulging of side 12 is,

$$\delta\Omega_{12} = -\tfrac{1}{4}bx_{12}^2y_{12} - \tfrac{1}{4}ax_{12}y_{12}^2 - \tfrac{1}{6}cy_{12}^3 + \tfrac{1}{6}dx_{12}^3 \qquad (11)$$

where $x_{12} = x_2 - x_1$ etc., and similar expressions hold for sides 23 and 31. Then from eq. (6) the equivalent strain energy density over the triangle is,

$$\tfrac{1}{2}\lambda Gt\left[\delta\Omega_{12}^2 + \delta\Omega_{23}^2 + \delta\Omega_{31}^2\right]/\Omega^2 \qquad (12)$$

This energy density converges to zero as the square of the mesh size. Also the deviatoric energy density as calculated from the TRIM3 elements will converge to the exact value and so the total strain energy will converge correctly.

In order to estimate the least value we can assign to λ it is necessary to have an approximation to the difference in σ_h between adjacent elements.

From the displacement field of eq. (10) the stresses arising from the deviatoric strain energy are,

$$\left.\begin{aligned} \sigma_{Dxx} &= -\sigma_{Dyy} = 2G(ay + bx) \\ \sigma_{Dxy} &= G\left[(a + 2d)x + (2c - b)y\right] \end{aligned}\right\} \qquad (13)$$

If we assume that there are no body forces equilibrium requires that,

$$\left.\begin{aligned}\frac{\partial}{\partial x}\left(\sigma_h + \sigma_{Dxx}\right) + \frac{\partial}{\partial y}\sigma_{Dxy} &= 0\\ \frac{\partial}{\partial y}\left(\sigma_h + \sigma_{Dyy}\right) + \frac{\partial}{\partial x}\sigma_{Dxy} &= 0\end{aligned}\right\} \qquad (14)$$

Hence

$$\left.\begin{aligned}\frac{\partial}{\partial x}\sigma_h &= -G(2c + b)\\ \frac{\partial}{\partial y}\sigma_h &= G(a - 2d)\end{aligned}\right\} \qquad (15)$$

If the co-ordinates of the centroids of adjacent triangles I and II are x_I , y_I etc. the pressure difference across the common side is

$$\sigma_{hI\,II} = -G\left[(2c + b)x_{I\,II} - (a - 2d)y_{I\,II}\right] \qquad (16)$$

Considering that half the pressure difference is resisted by the side belonging to each triangle (this is strictly true only when the areas of I and II are equal) we see from (6) that the strain energy density arising from k_φ for the triangle I may be written,

$$t\left[\sigma^2_{hI\,II} + \sigma^2_{hI\,III} + \sigma^2_{hI\,IV}\right]/8\lambda G \qquad (17)$$

where III and IV signity the other triangles adjacent to triangle I. Thus we again have an energy convergence proportional to the square of the mesh length. This shows that the form (6) chosen for the "bulge" energy is consistent. The optimum λ may be estimated by equating (12) and (17). Thus,

$$\lambda_{OPT} = \frac{\Omega \left[\begin{array}{l} (2c+b)^2 (x_{I\,II}^2 + x_{I\,III}^2 + x_{I\,IV}^2) + (a-2d)^2 (y_{I\,II}^2 + y_{I\,III}^2 + y_{I\,IV}^2) \\ -2(2c+b)(a-2d)(x_{I\,II} y_{I\,II} + x_{I\,III} y_{I\,III} + x_{I\,IV} y_{I\,IV}) \end{array} \right]^{1/2}}{2 \left[\delta\Omega_{12}^2 + \delta\Omega_{23}^2 + \delta\Omega_{31}^2 \right]^{1/2}} \tag{18}$$

Note that λ_{OPT} is independent of mesh size when the mesh geometry is similar. To see what are typical values for λ_{OPT} we now examine the special case of an equilateral triangle mesh. Since a , b , c , d are arbitrary we lose nothing by taking the case of fig. 3 with side = 2 and height = $\sqrt{3}$.

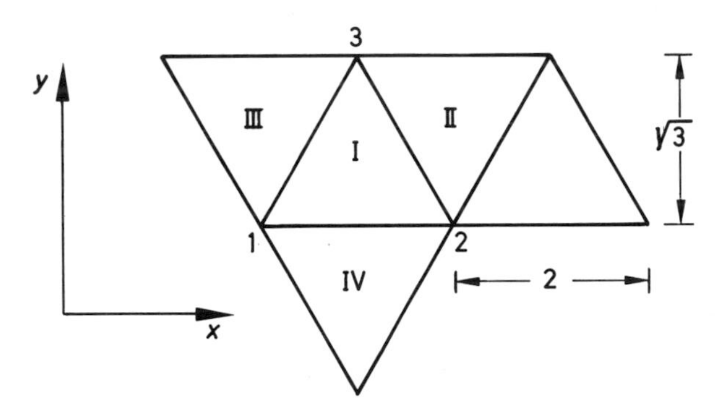

Fig. 3 EQUILATERAL TRIANGLE MESH

For triangle I (1 2 3) we have,

$$x_{12} = 2 \; ; \quad x_{23} = x_{31} = -1 \; ; \quad \Omega = \sqrt{3}$$

$$y_{12} = 0 \; ; \quad y_{23} = \sqrt{3} \; ; \quad y_{31} = \sqrt{3}$$

$$\delta\Omega_{12} = 4d/3 \; ; \quad \delta\Omega_{23} = -\sqrt{3}\,b/4 + 3a/4 - \sqrt{3}\,c/2 - d/6$$

$$\delta\Omega_{31} = \sqrt{3}\,b/4 + 3a/4 + \sqrt{3}\,c/2 - d/6$$

$$x_{I\,II} = 1 \; ; \quad x_{I\,III} = -1 \; ; \quad x_{I\,IV} = 0$$

$$y_{I\,II} = \sqrt{3}/3 \; ; \quad y_{I\,III} = \sqrt{3}/3 \; ; \quad y_{I\,IV} = -2\sqrt{3}/3$$

Substituting in eq. (18),

$$\lambda_{OPT} = 2\left[\frac{(2c+b)^2 + (a-2d)^2}{(2c+b)^2 + (a-2d)^2 + 2(a+2d/3)^2}\right] \tag{19}$$

Thus unless $2c + b = 0$

$$0 \leq \lambda_{OPT} \leq 2 \tag{20}$$

Otherwise λ_{OPT} may become indeterminate when both $2c + b = 0$; $a = d = 0$. This is because under these conditions the hydrostatic stress is constant and the displacement field is isochoric for the standard TRIM3 element so that no bulging is required. If

$2c + b = a - 2d = 0$ then $\lambda_{OPT} = 0$

because the condition of zero hydrostatic stress is combined with no bulging.

Another common mesh is the isosceles right-angled triangle of fig. 4

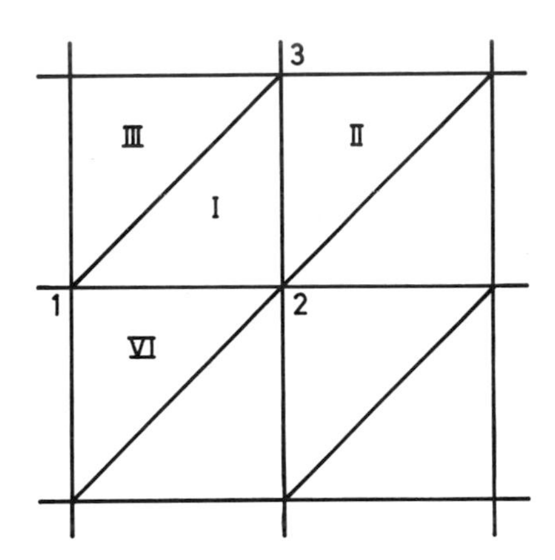

Fig. 4 RIGHT-ANGLED TRIANGLE MESH

Then for element I with side unity,

$$x_{12} = 1 \quad ; \quad x_{31} = 0 \quad ; \quad x_{31} = -1$$

$$y_{12} = 0 \quad ; \quad y_{23} = 1 \quad ; \quad y_{31} = -1$$

$$\delta\Omega_{12} = d/6 \quad ; \quad \delta\Omega_{31} = -c/6 \quad ; \quad \Omega = 1/2$$

$$\delta\Omega_{31} = b/4 + a/4 + c/6 - d/6$$

$$x_{12} = 2/3 \quad ; \quad x_{23} = -1/3 \quad ; \quad x_{31} = -1/3$$

$$y_{12} = 1/3 \quad ; \quad y_{23} = 1/3 \quad ; \quad y_{31} = -2/3$$

$$\lambda_{OPT} = 2\sqrt{3}\left[\frac{(2c+b)^2 + (a-2d)^2 + (2c+b-a+2d)^2}{(2c-2d+3a+3b)^2 + 4d^2 + 4c^2}\right]^{1/2} \tag{21}$$

In this case λ_{OPT} may range from zero to infinity, the latter because when the displacement field is isochoric for TRIM3 the hydrostatic stress is not constant. Typical values of λ_{OPT} are shown in the following table. From this table it appears that a suitable λ is in the range 0.5 to 3.

Coefficients of displacement field of equation (10)				λ_{OPT}	
a	b	c	d	△ (equilateral)	◺ (right-angled)
1	0	0	0	1.15	1.6
0	1	0	0	2	1.6
0	0	1	0	2	3.5
0	0	0	1	1.8	3.5
1	1	1	1	1.6	2.7
1	-1	-1	-1	2.0	3.5
1	-1	-1	1	1.6	2.6
1	-1	1	1	1.0	3.0
1	-1	1	-1	2.0	2.2
1	1	1	-1	2.0	1.4
1	1	-1/2	-1/2	0	0
0	1	-1/2	0	Indet.	0
1	-1	0	0	1.4	∞

Table 1 VARIATION OF λ_{OPT} WITH STRAIN FIELD

When λ has its optimum value the energy density may be written,
For equilateral triangle,

$$\bar{\Phi}_{\varphi\, OPT} = \frac{Gt\Omega}{8\sqrt{3}}\left[\left[(2c+b)^2+(a-2d)^2\right]\left[(2c+b)^2+(a-2d)^2+2(a+2d/3)^2\right]\right]^{1/2} \quad (22)$$

For right-angled triangle,

$$\bar{\Phi}_{\varphi\, OPT} = \frac{Gt\Omega}{12\sqrt{3}}\left[\left[(2c+b)^2+(a-2d)^2+(2c+b-a+2d)^2\right]\left[4d^2+4c^2+(3b+3a+2c-2d)^2\right]\right]^{1/2} \quad (23)$$

Since the coefficients a, b etc. are of the order of the strain gradients $\bar{\Phi}_{\varphi}$ will generally be small compared with the deviatoric energy density $\bar{\Phi}_D$. In fact one has approximately,

$$\bar{\Phi}_{\varphi} \Big/ \bar{\Phi}_D = 0\left(\tfrac{1}{20}(\Delta\varepsilon/\varepsilon)^2\right) \quad (24)$$

where $\Delta\varepsilon$ represents the increment of strain across an element and ε is the strain in the element.

2.2 Non-isotropic case

The deviatoric strain energy will be of the form,

$$\bar{\Phi}_D = A(\varepsilon_{xx} - \varepsilon_{yy})^2 + 2B(\varepsilon_{xx} - \varepsilon_{yy})\varepsilon_{xy} + C\varepsilon_{xy}^2 \tag{25}$$

The TRIM3 mechanical model will again require only corner springs, since the terms coupling the areal strain with the shear strain do not exist because of the incompressibility constraint. However the values of k in fig. 1 will now depend on the orientation of the element. On the other hand the k_φ springs may be taken as in the isotropic case but with

$$G = [A\,C]^{1/2} \tag{26}$$

in equations (6), (8), (16), (17), (22) and (23).

3. The modified TET4.

In this case expression (6) becomes,

$$\frac{1}{2}\lambda G(\delta V/V)^2 V \tag{27}$$

where V is the volume of the element and δV the volume of the "bubble" formed by the bulging face of the TET4.

An estimate of the convergence of the "bulge" energy density may be made as before and is again proportional to the square of the mesh size. Explicit expressions for λ_{OPT} will involve very tedious calculations but by analogy with TRIM3 will be in the same range.

4. The modified TRIAX3 element

The TRIAX3 ring element is applicable to bodies of revolution under axi-symmetrical or periodic loading. We consider at present only the modification for use with axi-symmetrical loading. Then eq. (27) may be used if it is understood that V and δV correspond to the volume of one radian of the ring formed by the element of area Ω and one radian of the bulge $\delta\Omega_{12}$. See fig. 5.

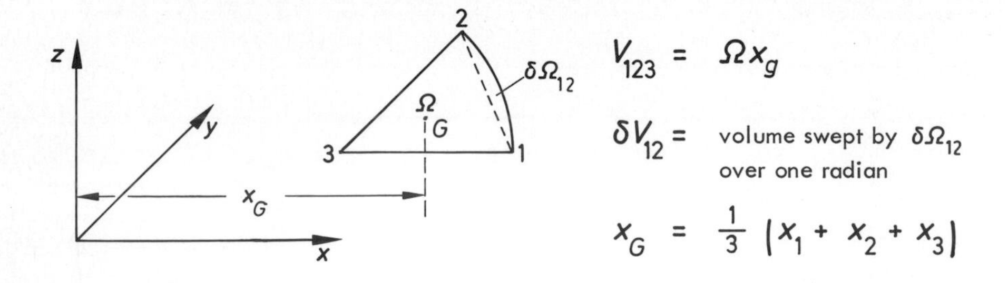

Fig. 5 TRIAX3 ELEMENT WITH BULGE

Note that there is no need to define the form of $\delta\Omega_{12}$ since it enters only through δV_{12} which is common to the two elements separated by 12.

5. Value of λ on boundary.

At an unsupported boundary the value of λ need not necessarily be infinite. It is possible to use a finite value provided the bulge energy density is maintained at the same order for boundary elements as for internal elements. This will be so if,

$$\lambda_B = \left(\sigma_{hB} \Big/ \frac{\partial\sigma_h}{\partial n}\right)^2 \frac{1}{h^2}\lambda \tag{28}$$

where σ_{hB} = value of σ_h on boundary, $\frac{\partial\sigma_h}{\partial n}$ = gradient of σ_h normal to boundary h = mesh size. Since we do not know σ_{hB} and $\frac{\partial\sigma_h}{\partial n}$ initially, the value of λ_B may be taken as,

$$\lambda_B = (H/h)^2\lambda \tag{29}$$

where H is the mean dimension of the domain. In any case λ_B becomes very large with mesh refinement and then may as well be infinite.

6. Formation of equilibrium and constraint equations

The nodal displacement vector is denoted by $\boldsymbol{u}$ and the bulge vector by $\boldsymbol{\varphi}$. The hydrostatic stress is σ_h. Then regarding σ_h as given the total potential will be,

$$\frac{1}{2}\boldsymbol{u}^t\boldsymbol{K}_u\boldsymbol{u} + \frac{1}{2}\boldsymbol{\varphi}^t\boldsymbol{K}_\varphi\boldsymbol{\varphi} + \boldsymbol{\sigma}_h^t\left[\boldsymbol{H}_u^t\boldsymbol{u} + \boldsymbol{H}_\varphi^t\boldsymbol{\varphi}\right] - \boldsymbol{f}^t\boldsymbol{u} \tag{30}$$

where $\boldsymbol{K}_u$ is the stiffness matrix arising from the material strain energy which is entirely deviatoric, $\boldsymbol{K}_\varphi$ is the diagonal stiffness matrix from the bulges, $\boldsymbol{H}_u^t\boldsymbol{u}$ is the volume increment vector for all elements due to $\boldsymbol{u}$, $\boldsymbol{H}_\varphi^t\boldsymbol{\varphi}$ is the corresponding increment due to $\boldsymbol{\varphi}$ and $\boldsymbol{f}$ is the nodal loading vector. If the coordinate $\boldsymbol{\varphi}$ is taken as the actual area or volume increment, $\boldsymbol{H}_\varphi$ will consist only of elements ± 1, or $\pm t$ for TRIM-3, and the element of $\boldsymbol{K}_\varphi$ corresponding to the boundary between two elements I and II will have the form,

$$\lambda G t\left(\frac{1}{\Omega_{\mathrm{I}}} + \frac{1}{\Omega_{\mathrm{II}}}\right) \quad \text{for TRIM3}$$

$$\lambda G\left(\frac{1}{V_{\mathrm{I}}} + \frac{1}{V_{\mathrm{II}}}\right) \quad \text{for TET4 and TRIAX3} \tag{31}$$

The elements of $\boldsymbol{H}_u^t$ corresponding to the vector $\{u_1\ u_2\ u_3\ v_1\ v_2\ v_3\}$

of a TRIM3 element 1 2 3 will be,

$$\frac{1}{2}t\left[\ -y_{23}\quad -y_{31}\quad -y_{12}\quad x_{23}\quad x_{31}\quad x_{12}\ \right] \tag{32}$$

For the TET4 with nodes 1 2 3 4 the terms of $\boldsymbol{H}_u^t$ corresponding to u_1, v_1, w_1 will be

$$\frac{1}{3}\left[\ \Omega_{234}^{YZ}\quad \Omega_{234}^{ZX}\quad \Omega_{234}^{XY}\ \right] \tag{33}$$

where Ω_{234}^{YZ} etc. are the projections of the area Ω_{234} on plane YZ etc. with due attention to sign.

In the case of TRIAX3, with the notation of fig. 5, instead of expression (32) one has

$$\frac{1}{6}\left[(2\Omega-3x_G y_{23})(2\Omega-3x_G y_{31})(2\Omega-3x_G y_{12})\ 3x_G x_{23}\ 3x_G x_{31}\ 3x_G x_{12}\right] \quad (34)$$

The equations of equilibrium are, from (30),

$$\left.\begin{aligned} \boldsymbol{K}_u\,\boldsymbol{u} + \boldsymbol{H}_u\,\boldsymbol{\sigma}_h &= \boldsymbol{f} \\ \boldsymbol{K}_\varphi\,\boldsymbol{\varphi} + \boldsymbol{H}_\varphi\,\boldsymbol{\sigma}_h &= \boldsymbol{o} \end{aligned}\right\} \quad (35)$$

Since we suppose the area or volume of each element, including the bulges, to remain constant, $\boldsymbol{\sigma}_h$ is an unknown vector to be determined by (35) and the constraint condition

$$\boldsymbol{H}_u^t\,\boldsymbol{u} + \boldsymbol{H}_\varphi^t\,\boldsymbol{\varphi} = \boldsymbol{o} \quad (36)$$

If, on the other hand, we suppose the material to be compressible with bulk modulus E_h then,

$$\boldsymbol{\sigma}_h = E_h\,\boldsymbol{V}^{-1}\left[\boldsymbol{H}_u^t\,\boldsymbol{u} + \boldsymbol{H}_\varphi^t\,\boldsymbol{\varphi}\right] \quad (37)$$

where $\boldsymbol{V}$ is the diagonal matrix of element volumes. The equations of equilibrium are then, from (35) and (37),

$$\begin{aligned} \left[\boldsymbol{K}_u + E_h\boldsymbol{H}_u\,\boldsymbol{V}^{-1}\boldsymbol{H}_u^t\right]\boldsymbol{u} + E_h\,\boldsymbol{H}_u\,\boldsymbol{V}^{-1}\boldsymbol{H}_\varphi^t\,\boldsymbol{\varphi} &= \boldsymbol{f} \\ E_h\,\boldsymbol{H}_\varphi\,\boldsymbol{V}^{-1}\boldsymbol{H}_u^t\,\boldsymbol{u} + \left[\boldsymbol{K}_\varphi + E_h\,\boldsymbol{H}_\varphi\,\boldsymbol{V}^{-1}\boldsymbol{H}_\varphi^t\right]\boldsymbol{\varphi} &= \boldsymbol{o} \end{aligned} \quad (38)$$

When $\lambda \to \infty$, $\boldsymbol{\varphi} \to 0$ and we have the standard TRIM3, TET4 or TRIAX3 formulation.

7. Solution of the incompressible system.

Since $\boldsymbol{K}_\varphi$ is diagonal it is easy to obtain, from the second of (35),

$$\boldsymbol{\varphi} = \boldsymbol{K}_\varphi^{-1} \boldsymbol{\sigma}_h \tag{39}$$

and then from the first of (35) and (36),

$$\begin{bmatrix} \boldsymbol{K}_u & \boldsymbol{H}_u \\ \boldsymbol{H}_u^t & -\boldsymbol{F} \end{bmatrix} \begin{bmatrix} \boldsymbol{u} \\ \boldsymbol{\sigma}_h \end{bmatrix} = \begin{bmatrix} \boldsymbol{f} \\ \boldsymbol{o} \end{bmatrix} \tag{40}$$

with

$$\boldsymbol{F} = \boldsymbol{H}_\varphi^t \boldsymbol{K}_\varphi^{-1} \boldsymbol{H}_\varphi \tag{41}$$

There are various ways of solving this system depending on whether $\boldsymbol{K}_u$ and/or $\boldsymbol{F}$ are singular. If the supports constrain both rigid body and pure dilation movements $\boldsymbol{K}_u$ will be invertible, but $\boldsymbol{F}$ will be invertible only when there are free φ nodes on the boundary. If as we shall generally suppose the boundary λ's are infinite $\boldsymbol{F}$ is not invertible. Two cases have to be considered - boundary held everywhere and boundary free or partly free.

7.1 Boundary fixed

In this case equation (40) has a simple singularity due to the indeterminacy of $\boldsymbol{\sigma}_h$ in a closed region. We may then reduce equation (40) by omitting $\boldsymbol{\sigma}_{h1}$ and the corresponding constraint eq. for element 1. Thus if

$$\bar{\boldsymbol{\sigma}}_h = \left\{ \sigma_{h2} \ \sigma_{h3} \cdots\cdots\cdots \sigma_{hn} \right\} \tag{42}$$

where n is the number of elements, eq. (40) becomes

$$\begin{bmatrix} \boldsymbol{K}_u & \bar{\boldsymbol{H}}_u \\ \bar{\boldsymbol{H}}_u^t & -\boldsymbol{F} \end{bmatrix} \begin{bmatrix} \boldsymbol{u} \\ \boldsymbol{\sigma}_h \end{bmatrix} = \begin{bmatrix} \boldsymbol{f} \\ \boldsymbol{0} \end{bmatrix} \tag{43}$$

where

$$\bar{\boldsymbol{F}} = \bar{\boldsymbol{H}}_\varphi^t \boldsymbol{K}_\varphi^{-1} \boldsymbol{H}_\varphi \tag{44}$$

$\bar{\boldsymbol{F}}$ will be positive definite and sparse and we may obtain easily,

$$\left[\boldsymbol{K}_u + \bar{\boldsymbol{H}}_u \bar{\boldsymbol{F}}^{-1} \bar{\boldsymbol{H}}_u^t \right] \boldsymbol{u} = \boldsymbol{f} \tag{45}$$

Then,

$$\bar{\boldsymbol{\sigma}}_{h1} = \bar{\boldsymbol{F}}^{-1} \bar{\boldsymbol{H}}_u^t \boldsymbol{u} \tag{46}$$

and then minimizing the hydrostatic strain energy with a supposed finite E_h

$$\sigma_{h1} = -\sum \bar{\sigma}_{hr} V_r \Big/ \sum_{r=1}^{n} V_r \tag{47}$$

and

$$\sigma_{hr} = \sigma_{h1} + \bar{\sigma}_{hr} \tag{48}$$

Equation (45) is suitable also for use in dynamic calculations with a lumped mass matrix.

7.2 Boundary free or partly free.

The singularity of the system (40) is now due to the constraint of the boundary displacements which have to satisfy the condition that the total volume is constant. This condition may be written

$$\mathbf{L}^t \boldsymbol{u} = 0 \tag{49}$$

where

$$\mathbf{L}^t \quad = \quad \text{sum of rows of } \quad \boldsymbol{H}_u^t \; .$$

If we again write $\bar{\boldsymbol{\sigma}}_h$ as in eq. (42) the equilibrium and constraint equations become,

$$\begin{aligned} \boldsymbol{K}_u \boldsymbol{u} + \bar{\boldsymbol{H}}_u \bar{\boldsymbol{\sigma}}_h + \mathbf{L}\sigma_{h1} &= \boldsymbol{f}_1 \\ \boldsymbol{K}_\varphi \boldsymbol{\varphi} + \bar{\boldsymbol{H}}_\varphi \bar{\sigma}_h &= \mathbf{0} \\ \boldsymbol{H}_u^t \boldsymbol{u} + \boldsymbol{H}_\varphi^t \boldsymbol{\varphi} &= \mathbf{0} \\ \mathbf{L}^t \boldsymbol{u} &= \mathbf{0} \end{aligned} \tag{50}$$

Then,

$$\begin{bmatrix} \bar{\boldsymbol{K}}_u & \mathbf{L} \\ \mathbf{L}^t & 0 \end{bmatrix} \begin{bmatrix} \boldsymbol{u} \\ \sigma_{h1} \end{bmatrix} = \begin{bmatrix} \boldsymbol{f} \\ \mathbf{0} \end{bmatrix} \tag{51}$$

where,

$$\bar{\boldsymbol{K}}_u = \left[\boldsymbol{K}_u + \bar{\boldsymbol{H}}_u \bar{\boldsymbol{F}} \bar{\boldsymbol{H}}_u^t \right] \tag{52}$$

We must now eliminate one of the boundary displacements which is preferably one which enters strongly into the constraint condition (49). Supposing this to be u_1, we take,

$$\bar{\boldsymbol{u}} = \left\{ u_2 \quad u_3 \cdots\cdots\cdots u_N \right\} \tag{53}$$

where N is the number of freedoms.

Then if

$$\mathbf{L} = \{ l_1 \quad l_2 \cdots\cdots\cdots l_N \} \tag{54}$$

we have,

$$\boldsymbol{u} = \begin{bmatrix} \bar{\mathbf{L}}^t \\ \boldsymbol{I}_{N-1} \end{bmatrix} \boldsymbol{u} = \boldsymbol{T}\bar{\boldsymbol{u}} \tag{55}$$

where,

$$\bar{\mathbf{L}} = -\left\{ \frac{l_2}{l_1} \quad \frac{l_3}{l_1} \cdots\cdots\cdots \frac{l_N}{l_1} \right\} \tag{56}$$

Then from (51),

$$\widetilde{\boldsymbol{K}}\bar{\boldsymbol{u}} = \widetilde{\boldsymbol{f}} \tag{57}$$

where

$$\widetilde{\boldsymbol{K}} = \boldsymbol{T}^t\bar{\boldsymbol{K}}\boldsymbol{T} \tag{58}$$

$$\widetilde{\boldsymbol{f}} = \boldsymbol{T}^t\boldsymbol{f} \tag{59}$$

If we use a lumped mass matrix,

$$\boldsymbol{M} = \lceil m_1 \quad m_2 \cdots\cdots\cdots m_N \rfloor \tag{60}$$

the equation of motion will be,

$$\widetilde{\boldsymbol{M}}\ddot{\boldsymbol{u}} + \widetilde{\boldsymbol{K}}\bar{\boldsymbol{u}} = \widetilde{\boldsymbol{f}} \tag{61}$$

where,

$$\tilde{\boldsymbol{M}} = \boldsymbol{T}^t\boldsymbol{M}\boldsymbol{T} \tag{62}$$

Note that if we choose u_1 at a point which has little movement we may take $f_1 = m_1 = 0$ and then,

$$\bar{\boldsymbol{M}}\ddot{\bar{\boldsymbol{u}}} + \tilde{\boldsymbol{K}}\bar{\boldsymbol{u}} = \bar{\boldsymbol{f}} \tag{63}$$

where,

$$\bar{\boldsymbol{M}} = \lceil m_2 \quad m_3 \qquad m_N \rfloor \tag{64}$$

and

$$\bar{\boldsymbol{f}} = \{ f_2 \quad f_3 \cdots\cdots\cdots f_N \} \tag{65}$$

8. Freedom/constraint ratio

A criterion that has been suggested to judge the efficiency of various finite elements for use with incompressible materials is the ratio of constraints to freedoms [1,2] . For a large area of two-dimensional elements it was suggested in ref. [2] that this should be about 1/2 or for a large volume of three-dimensional elements about 1/3. These ratios suppose the constraints to be linearly independent and it is not always easy to detect these dependencies in advance. However, since the dependencies always represent a relaxation of constraints, the ratio is still useful as an upper limit. It is therefore interesting to calculate the ratio for the present modifications of the constant strain elements. The results are given in TABLE 2 with those for the point-wise incompressible TRIM-6 and TET-10 for comparison. In the latter linear dependencies are possible.

	Rectangle of 2 pq elements		Block of 6 pqr elements	
	Modified TRIM 3	TRIM 6	Modified TET 4	TET 10
Nodal freedoms	2 pq	8 pq	3 pqr	24 pqr
φ freedoms	3 pq	-	12 pqr	-
Total freedoms n	5 pq	8 pq	15 pqr	24 pqr
Elements	2 pq	2 pq	6 pqr	6 pqr
Constraints c	2 pq	6 pq	6 pqr	24 pqr
c/n	2/5	3/4	2/5	1

Table 2 CONSTRAINT/FREEDOM RATIOS FOR LARGE ARRAY OF ELEMENTS

9. Solution of the compressible system

Equation (38) is positive definite and sparse. Thus unless E_h is very large it is probable that a standard Cholesky reduction is most efficient. If E_h is large, the system approaches the incompressible case and it may be better to rewrite the equations as follows.

$$\begin{bmatrix} \boldsymbol{K}_u & \boldsymbol{O} & \boldsymbol{H}_u \\ \boldsymbol{O} & \boldsymbol{K}_\varphi & \boldsymbol{H}_\varphi \\ \boldsymbol{H}_u^t & \boldsymbol{H}_\varphi^t & -E_h^{-1}\boldsymbol{V} \end{bmatrix} \begin{bmatrix} \boldsymbol{u} \\ \boldsymbol{\varphi} \\ \boldsymbol{\sigma}_h \end{bmatrix} = \begin{bmatrix} \boldsymbol{f} \\ \boldsymbol{o} \\ \boldsymbol{o} \end{bmatrix} \qquad (66)$$

This system may be reduced to

$$\left[\boldsymbol{K}_u + \boldsymbol{H}_u \left[\boldsymbol{F} + E_h^{-1}\boldsymbol{V}\right]^{-1} \boldsymbol{H}_u^t \right] \boldsymbol{u} = \boldsymbol{f} \qquad (67)$$

where $\boldsymbol{F}$ is as in (41). Although $\boldsymbol{F}$ is singular when the boundary λ's are infinite, the inversion of the matrix $\left[\boldsymbol{F} + E_h^{-1}\boldsymbol{V} \right]$ will present no special numerical difficulties. Equation (67) is the form necessary for the treatment of dynamic problems with lumped masses at the vertices and the elimination of φ is accomplished with less work than directly from (38).

10. Extension of method to large strains

In large strain finite elements the incompressibility constraint is even more restrictive. Whereas for small strains the higher order TRIM elements have a (constraint/freedom) ratio tending to the ideal value of 1/2 as we increase the order of the element, the corresponding ratio for the large strain elements is 2/1. See TABLE 3.

Element Polynomial degree	TRIM 3 1	TRIM 6 2	TRIM 10 3	TRIM 15 4	TRIM ∞ ∞
Natural Modes s	3	9	17	27	
Constraints Small Strains c_{ss}	1	3	6	10	
Constraints Large Strains c_{ls}	1	6	15	28	
Element $(c/s)_{ss}$	1/3	1/3	6/17	10/27	1/2
Element $(c/s)_{ls}$	1/3	2/3	15/17	28/27	2/1
Large array $(c/s)_{ss}$	1	3/4	2/3	5/8	1/2
Large array $(c/s)_{ls}$	1	3/2	5/3	7/4	2/1

Table 3 CONSTRAINTS/FREEDOMS FOR SMALL AND LARGE STRAINS

This table assumes that all constraints are independent which is known not to be so in certain cases (See [1]). However, it is clear enough that pointwise incompressibility is impossible to satisfy in a practical large strain TRIM element. The same will be true for the large strain TET elements.

The modified TRIM3 element will give the same (constraint/freedom) ratio for large as for small strains - see TABLE 2. However, the complete separation of deviatoric and dilatational strain energy and stresses is no longer possible. In the small strain case it is not essential to make this separation - we may, for example, use in equation (30) a $\boldsymbol{K}_u$ derived from a material with a finite E_h but the same G as the actual material. The standard simplex elements will then give the same result, apart from numerical error, for any E_h. (See [3]). The modified simplex elements will give a different result for each E_h and from the numerical examples it appears that $E_h = 0$ is indeed the best value to use. Assuming that the same is desirable in the large strain case, we conclude that it is advantageous to express the strain energy density function in such a way that it gives zero hydrostatic stress for zero dilatation. We now show how this may be done for those isotropic materials whose strain energy density is given as a polynomial in terms of the principal Green strains ϵ_I, ϵ_{II}, ϵ_{III} or equivalently in terms of the principal extensions λ_I, λ_{II}, λ_{III} noting that,

$$\epsilon_I = \frac{1}{2}\left(\lambda_I^2 - 1\right) \qquad \text{etc.} \tag{68}$$

Suppose that the energy density is written as a polynomial in λ_I^2 etc. of degree n which is equivalent to a polynomial with terms up to degree n in ϵ_I etc. Let P_m denote all the terms of degree m in λ^2. Then the expression,

$$\bar{\Phi} = P_m\left(\lambda_I^2, \lambda_{II}^2, \lambda_{III}^2\right) \tag{69}$$

may be replaced by,

$$\bar{\Phi} = P_m\left(\frac{m+3}{3} - \frac{m}{3} I_3\right) \tag{70}$$

where I_3 is the strain invariant $\lambda_I^2 \lambda_{II}^2 \lambda_{III}^2$. When the incompressibility condition is satisfied, the principal stresses arising from the strain energy are,

$$\sigma_I = \lambda_I \frac{\partial \bar{\Phi}}{\partial \lambda_I} \qquad \text{etc.} \tag{71}$$

Note that σ_I is the true stress - not that based on the original area.
It follows that when $\bar{\Phi}$ is a polynomial of degree m in λ_I^2 etc. the hydrostatic stress σ_h' due only to $\bar{\Phi}$ is,

$$\sigma_h' = \frac{2m}{3}\bar{\Phi} \tag{72}$$

Now using expression (70) for $\bar{\Phi}$ we get, with $I_3 = 1$,

$$\sigma_h' = \frac{2P_m}{9}\left[m(m+3)-(m+3)m\right] = 0 \tag{73}$$

Also (70) may be written,

$$\bar{\Phi} = P_m + \frac{m}{3}P_m\left(1 - I_3\right) \tag{74}$$

The second term on the right of (74) gives,

$$\sigma_{\mathrm{I}} = \frac{m}{3}\lambda_{\mathrm{I}}\frac{\partial P_m}{\partial\lambda_{\mathrm{I}}}\left(1 - I_3\right) - \frac{m}{3}P_m 2 I_3 \tag{75}$$

This is purely hydrostatic when $I_3 = 1$ which justifies the use of (70) in place of (69).

10.1 The special case of Mooney material

As an example consider the Mooney material

$$\bar{\Phi} = C_1\left(I_1 - 3\right) + C_2\left(I_2 - 3\right) \tag{76}$$

Then

$$\left.\begin{aligned}
P_0 &= -3\left(C_1 + C_2\right)\\
P_1 &= C_1\left(\lambda_{\mathrm{I}}^2 + \lambda_{\mathrm{II}}^2 + \lambda_{\mathrm{III}}^2\right)\\
P_2 &= C_2\left(\lambda_{\mathrm{I}}^2\lambda_{\mathrm{II}}^2 + \lambda_{\mathrm{II}}^2\lambda_{\mathrm{III}}^2 + \lambda_{\mathrm{III}}^2\lambda_{\mathrm{I}}^2\right)
\end{aligned}\right\} \tag{77}$$

Then from (70), the new form of $\bar{\Phi}$ is,

$$\bar{\Phi} = C_1 I_1\left(\frac{4}{3} - \frac{1}{3}I_3\right) + C_2 I_2\left(\frac{5}{3} - \frac{2}{3}I_3\right) - 3\left(C_1 + C_2\right) \tag{78}$$

If we prefer to work in terms of the Green strain invariants,

$$\left.\begin{aligned}
J_1 &= \epsilon_{\mathrm{I}} + \epsilon_{\mathrm{II}} + \epsilon_{\mathrm{III}}\\
J_2 &= \epsilon_{\mathrm{I}}\epsilon_{\mathrm{II}} + \epsilon_{\mathrm{II}}\epsilon_{\mathrm{III}} + \epsilon_{\mathrm{III}}\epsilon_{\mathrm{I}}\\
J_3 &= \epsilon_{\mathrm{I}}\,\epsilon_{\mathrm{II}}\,\epsilon_{\mathrm{III}}
\end{aligned}\right\} \tag{79}$$

the expression (78) becomes,

$$\bar{\Phi} = -\frac{4}{3}C_1\left(J_1^2 + 3J_2 + 2J_1J_2 + 6J_3 + 4J_1J_3\right)$$

$$-\frac{4}{3}C_2\left(3J_2^2 + 4J_1^2 + 12J_1J_2 + 12J_3 + 8J_2^2 + 16J_1J_3 + 16J_2J_3\right) \tag{80}$$

10.3 Nearly incompressible large strain case

A nearly incompressible material will be for example one containing a term of the form

$$\frac{1}{8}E_h\left(I_3 - 1\right)^2 \tag{81}$$

in the energy density expression $\bar{\Phi}$, where E_h has a large value. Equation (80) reduces to

$$\frac{1}{2}E_h F_1^2$$

in the small strain case where

$$F_1 = \varepsilon_I + \varepsilon_{II} + \varepsilon_{III} \tag{82}$$

The stresses due to expression (80) contain no deviatoric components and at element level will give a dilatational energy of,

$$\frac{1}{2}E_h V\left[\frac{\delta V}{V} + \left(\frac{\delta V}{2V}\right)^2\right]^2 \tag{83}$$

Since the material is supposed nearly incompressible the term $(\delta V / 2V)^2$ in the bracket may be neglected. If we now include the contribution of the bulge co-ordinates φ to δV we will obtain an expression analogous to (37) for the hydrostatic stress σ_h'' . The expression for $\boldsymbol{H}_u^t \boldsymbol{u}$ will be replaced by a non-linear function of the vector $\boldsymbol{u}$ but $\boldsymbol{H}_\varphi$ may be retained in its original form. Note, however, that the remaining parts of $\bar{\Phi}$ will also contribute to a hydrostatic stress σ_h' and the total will then be,

$$\sigma_h = \sigma_h' + \sigma_h'' \tag{84}$$

10.4 Incompressible membranes with large strain

We understand an isotropic membrane to be a two-dimensional elastic body in which the strain energy density may be expressed only in terms of the in-plane strain invariants

$$I_1 = \lambda_I^2 + \lambda_{II}^2 \quad ; \quad I_2 = \lambda_I^2 \lambda_{II}^2 \tag{85}$$

Examples of problems which may be treated as membranes are plane strain and plane stress. In the former there will be a principal stress and in the latter a principal strain through the thickness obtainable from the solution for λ_I and λ_{II} . For plane strain it is easy to obtain $\bar{\Phi}$ from the three dimensional energy density function by placing $\lambda_{III} = 1$. In this case an incompressible solid material becomes an isochoric membrane. The equations corresponding to (69) and (70) are now,

$$\bar{\Phi} = P_m \left(\lambda_I^2 , \lambda_{II}^2 \right) \tag{86}$$

and

$$\bar{\Phi} = \frac{1}{2} P_m \left(m + 2 - m I_2 \right) \tag{87}$$

The Mooney material defined in (76) when under plane strain becomes, with $I_2 = 1$

$$\bar{\Phi} = C_1 \left(I_1 - 2 \right) \tag{88}$$

and when idealised by the modified TRIM3 we use,

$$\bar{\Phi} = \frac{1}{2} C_1 I_1 \left(3 - I_2 \right) - 2C_1 = -2C_1 \left(J_1^2 + 2J_2 + 2J_1 J_2 \right) \tag{89}$$

where,

$$J_1 = \epsilon_1 + \epsilon_2 \quad ; \quad J_2 = \epsilon_1 \epsilon_2 \tag{90}$$

In plane stress if we ignore the stress variation through the thickness the energy density based on an original unit thickness may also be expressed in the form (86). Thus, for example the Mooney material of (76) becomes the membrane with energy density,

$$\bar{\Phi} = C_1 \left(I_1 + I_2^{-1} - 3 \right) + C_2 \left(I_2 + I_1 I_2^{-1} - 3 \right) \tag{91}$$

Since this membrane is neither isochoric nor nearly so, the normal TRIM3 element may be used. However, $\bar{\Phi}$ is no longer a simple polynomial in λ or ϵ . Alternatively we may use the original three dimensional $\bar{\Phi}$ in conjunction with the modified TRIM3 but the thickness is then an additional element variable which greatly increases the total of degrees of freedom.

11. Numerical examples

The numerical examples are at present only for linear problems. In comparing the results with those for higher order elements it should be remembered that the objective has been to retain the use of the constant strain elements over the whole range from low bulk modulus to incompressible. Simplex constant strain elements are capable only of engineering accuracy in the compressible regime - we should be content therefore if the modified elements can give similar accuracy for the isochoric case. All calculations are made with single precision on the UNIVAC 1108 computer (about 8 decimals) except for the comparative calculations made with higher order elements at near incompressibility which require double precision.

11.1 Symmetrically loaded equilateral triangle membrane

This example (figs. 6 and 7) was used previously [1,3] as a test case because the central stress is independent of Poisson's ratio. We note that the convergence is good over a wide range of λ , is comparable to TRIM6 at near incompressibility and is much better than the normal TRIM3 with $\nu = 1/3$. The value $\lambda = 1$ appears as a suitable value to assume.

11.2 Deep cantilever membrane

The application of the loads and the support conditions are approximately those appropriate to the engineers' theory of bending except that warping is prevented at the support. TABLE 4 demonstrates the variation of certain stresses with λ and it again appears that the value unity is suitable. The engineers' theory would give a deflection of 41.6 of which 9.6 is shear deflection. Fig. 8 shows that the stress distribution using the average of the stresses each side of a diagonal is very close to the engineers' theory. With normal TRIM3 elements it is necessary to use crossed-diagonal elements to obtain similar accuracy [3] .

11.3 Thick cylinder under internal pressure

The example shown in fig. 9 was calculated with and without the crossed-diagonal arrangement of elements. In this example the normal TRIAX3 element with volume constraint only (corresponding to $\lambda = \infty$) gives reasonable results with the uncrossed arrangement but the crossed arrangement is very badly conditioned because of the near linear dependence of the constraints. With the modified element the crossed arrangement is a little more accurate and gives no numerical troubles.

ACKNOWLEDGEMENT

The special programs for the solution of the numerical examples were devised by M. Müller. The manuscript was prepared by G. Grimm, K. Mai and G. Singh.

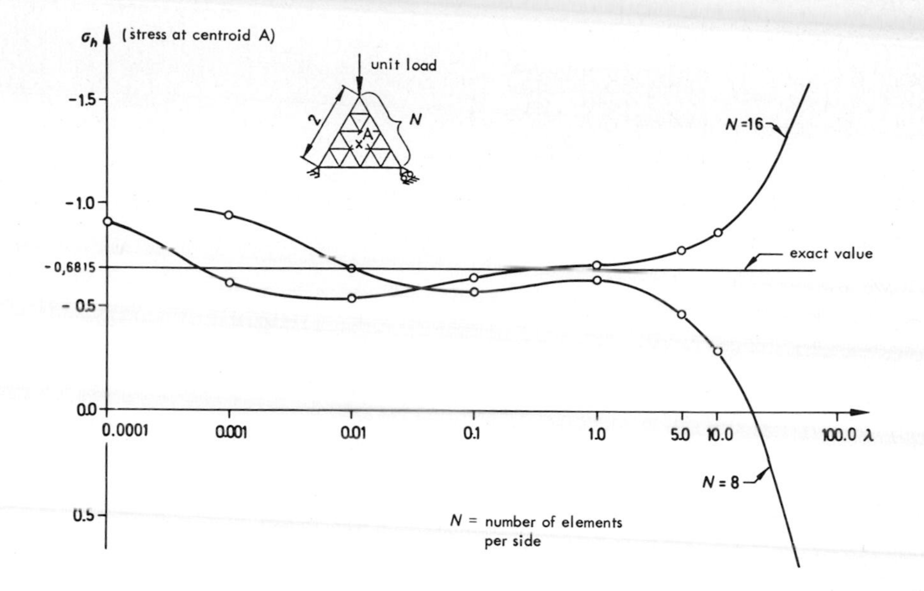

Fig. 6 VARIATION OF CENTRE STRESS WITH λ AND N

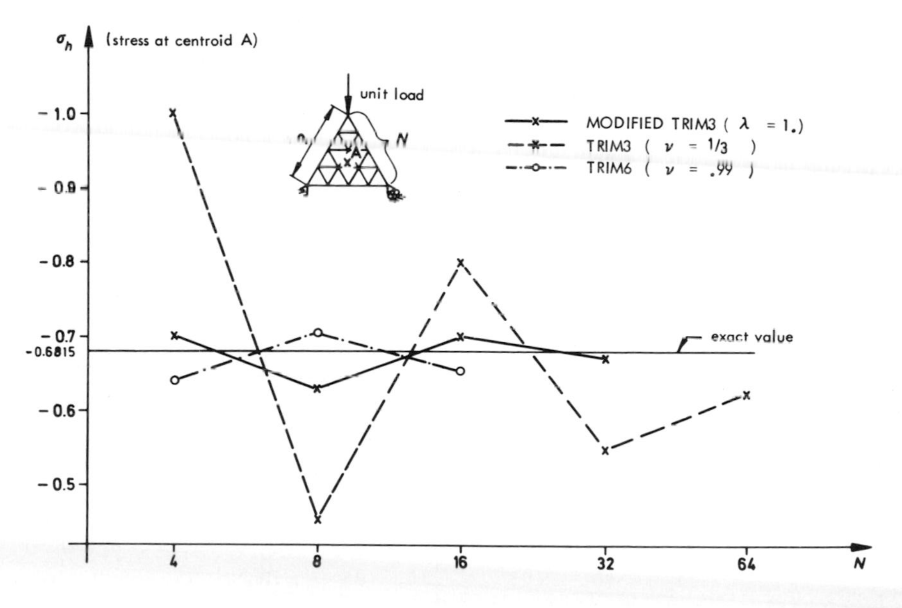

Fig. 7 VARIATION OF CENTRE STRESS WITH N FOR TRIM3, TRIM6 and MODIFIED TRIM3 (λ = 1.)

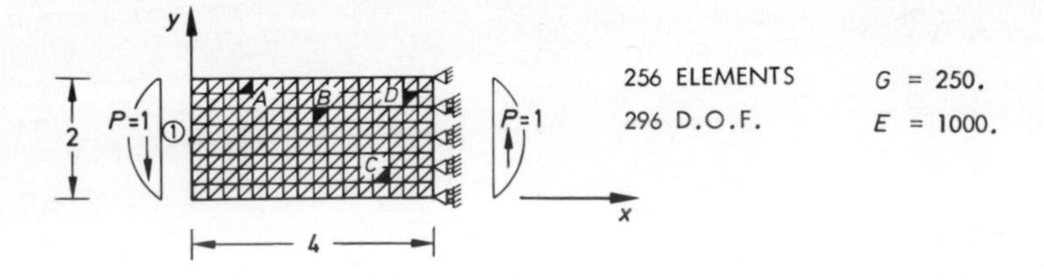

256 ELEMENTS $G = 250.$
296 D.O.F. $E = 1000.$

	ELEMENT A		ELEMENT B		ELEMENT C		ELEMENT D		Vertical Displacement at Node ①
λ	σ_x	τ_{xy}	σ_x	τ_{xy}	σ_x	τ_{xy}	σ_x	τ_{xy}	$(\times 10^3)$.
.01	0.6158	0.2529	1.7812	0.3303	-3.6450	0.7820	4.3344	0.1117	-122.093
.1	0.9847	0.2411	1.5236	0.4998	-3.3815	0.5696	3.8855	0.3514	- 47.999
1.	1.0935	0.2534	1.3827	0.5228	-3.2537	0.5953	3.6874	0.3414	- 39.479
5.	0.9709	0.2655	1.4724	0.5219	-3.3066	0.5927	3.8325	0.3587	- 38.222
10.	0.7912	0.2750	1.6162	0.5187	-3.4158	0.5822	4.1197	0.3863	- 37.573
100.	-1.2533	0.3921	3.4656	0.4915	-4.7210	0.6097	7.6137	0.5980	- 30.880

Table 4 VARIATION OF STRESSES AND DISPLACEMENTS WITH λ. (MODIFIED TRIM 3)

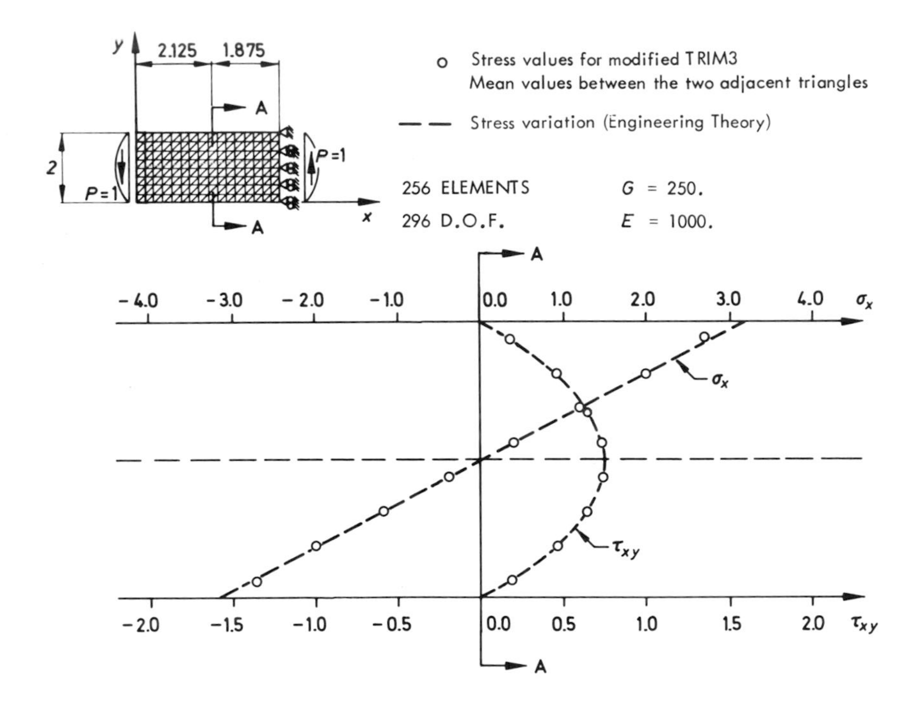

Fig. 8 STRESS VARIATION IN SECTION A - A (λ = 1.)

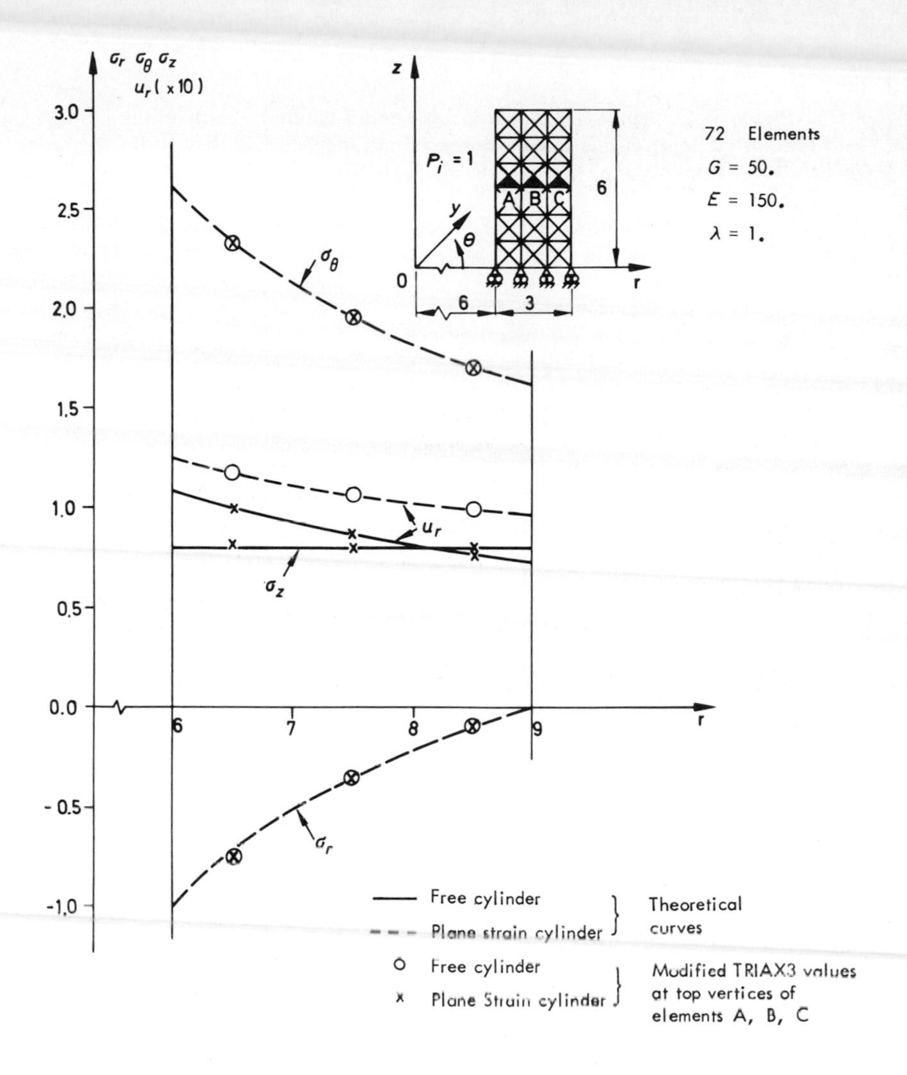

Fig. 9 THICK CYLINDER UNDER INTERNAL PRESSURE (P_i = 1)

REFERENCES

[1] J.H. Argyris, P.C. Dunne, T. Angelopoulos and B. Bichat, Large natural strains and some special difficulties due to non-linearity and incompressibility in finite elements, Comp. Meth. Appl. Mech. Eng. 4 (1974), 220-278.

[2] J.C. Nagtegaal, D.M. Parks and J.R. Rice, On numerically accurate finite element solutions in the fully plastic range, Comp. Meth. Appl. Mech. Eng. 4 (1974), 153-177.

[3] J.H. Argyris, P.C. Dunne, Th.L. Johnsen and M. Müller, Linear systems with a large number of sparse constraints with applications to incompressible materials, Comp. Meth. Mech. Eng. 10 (1977), 105-132.

[4] E.G. Thompson, Average and complete incompressibility in the finite element method, Int. J. Num. Meth. Eng. 9 (1975), 925-932.

[5] D.S. Malkus, A finite element displacement model valid for any value of the compressibility, Int. J. Solids Structures, 12 (1976), 731-738.

[6] I. Fried, Finite element analysis of incompressible material by residual energy balancing, Int. J. Solids Structures, 10 (1974), 993-1002.

[7] R. Temam, Some finite elements in fluid flow, AGARD Lecture Series No. 86 on Computational Fluid Dynamics, AGARD (1977), 6-1 to 6-14.

[8] J.T. Oden, Finite elements in non-linear continua (McGraw Hill, New York, 1972).

[9] M. Crouzeix and O.A. Raviart, Conforming and non-conforming finite element methods for solving the stationary Stokes equations, R.A.I.R.O., R.3, (1973) 33-76.

[10] J.H. Argyris and P.C. Dunne, A simple theory of geometrical stiffness with applications to beam and shell problems, Lecture Notes in Physics, Vol. 58, Ed. R. Glowinski and J.L. Lions, (Springer-Verlag, Berlin, Heidelberg, New York, 1976).

COMPUTATION OF EDDY CURRENTS ON A SURFACE IN $\mathbb{R}^3$ BY FINITE ELEMENT METHODS

by

J.C. NEDELEC

Ecole Polytechnique - Centre de Mathématiques Appliquées

and

J.C. VERITE

E.D.F. - Direction Etudes et Recherches

ABSTRACT

We study eddy currents on a thin conductor surface in $\mathbb{R}^3$. We suppose that these currents result from a periodic alternative time-varying excitation (such as a difference of potential or imposed currents).

In Section 1, we shall write the current equations by using the variable surface density of the current which is a tangential complex vector to Γ. We prove that these integro-differential equations admit a unique solution.

In Section 2, we introduce an approximate problem built by finite elements, and we prove some error estimates.

In Section 3, we present the numerical results obtained by our method. The studied problem was the case of some big conductors of an electric power station of Electricité de France. Finally, we have computed the resulting forces on the conductors.

1 - THE PHYSICAL PROBLEM AND ITS VARIATIONAL FORMULATION.

We study currents in a thin conductor assimilated with a surface Γ in $\mathbb{R}^3$. We suppose this surface to be homeomorphic to a cylinder, i.e. to be composed of a unique layer with two holes. $\partial\Gamma_1$ and $\partial\Gamma_2$ will denote the closed curves of $\mathbb{R}^3$ which are the boundaries of these holes.

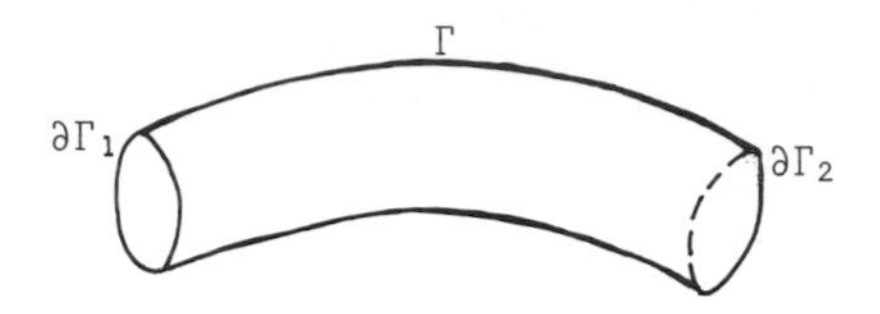

We study the case of alternative currents of period ω . Then the unknowns of the problem are, on the one hand, the surface current J , which is a tangential complex vector, and on the other hand, the scalar potential V , which is a complex function defined in the space $\mathbb{R}^3$.

If we neglect some effects, we may assume the current J is conservative on the surface Γ , which we can express by the equation

(1.1) $$\operatorname{div} J = 0 .$$

We shall define here the operator div on the surface Γ .

In the space C^3 , the Maxwell equation can be written in the following form

(1.2) $$\frac{1}{\sigma} J(x) + \frac{i\omega\mu}{c^2} \int_\Omega \frac{J(x)}{|x-y|} dy = -\operatorname{grad} V ,$$

where σ denotes the conductor surface resistivity, μ , the magnetic permeability, and c , the light speed.

We have two kinds of boundary conditions at the border of the holes : either the potential V is imposed on $\partial\Omega$ (and we can easily see that there intervenes only the potential difference between $\partial\Gamma_1$ and $\partial\Gamma_2$), or the current flux is imposed (i.e. $J.n$, where n is the normal to $\partial\Gamma$ in the tangential plane to Γ).

REMARK 1 : We could consider the case of a pluri-layer surface with more than two holes and with mixt boundary conditions of both types. We have chose the simplest case only to simplify the exposure.

We shall now define the equations (1.1) and (1.2). First, let us give the way to define the surface Γ . It is defined by a finite number p of maps which are bijective mappings Φ_ℓ of an open set θ_ℓ of $\mathbb{R}^2$ into $\mathbb{R}^3$:

$$\xi = (\xi_1, \xi_2) \in \theta_\ell \to \Phi_\ell (\xi_1, \xi_2) = \begin{cases} x_\ell(\xi_1, \xi_2) \\ y_\ell(\xi_1, \xi_2) \\ z_\ell(\xi_1, \xi_2) \end{cases}$$

such that the Φ_ℓ are "sufficiently differentiable", and such that the tangential vectors to the image $\Phi_\ell(\theta_\ell)$, let them be

$$e_1 = \frac{\partial\Phi_\ell}{\partial\xi_1} \quad , \qquad e_2 = \frac{\partial\Phi_\ell}{\partial\xi_2} \quad ,$$

are independent. The images $\Phi_\ell(\theta_\ell)$ cover the surface Γ, and the application Φ_ℓ is a bijection from θ_ℓ onto the image $\Phi_\ell(\theta_\ell)$. The mappings $\Phi_\ell^{-1}\circ\Phi_k$ are bijective and differentiable from the set $\Phi_k^{-1}(\Phi_\ell(\theta_\ell)\cap\Phi_k(\theta_k))$ onto $\Phi_\ell^{-1}(\Phi_\ell(\theta_\ell)\cap\Phi_k(\theta_k))$.

Then, the boundaries $\partial\Gamma_1$ and $\partial\Gamma_2$ are curves of the surface Γ, whose images are regular curves of $\mathbb{R}^2$.

Now, in order to build a triangulation, we suppose we can find, in each domain θ_ℓ, a closed polyhedrion D_ℓ such that

$$\bigcup_{\ell=1}^{P} \Phi_\ell(D_\ell) = \Gamma$$

$$\Phi_\ell(D_\ell) \cap \Phi_k(D_k) \;;\; \ell \neq k \;;\; \text{is a curve of } \Gamma .$$

We shall denote by Γ_ℓ the closed set $\Phi_\ell(D_\ell)$.

The vectors e_1, e_2, allow us to define the first fundamental form of the surface by

$$g_{ij} = e_i \cdot e_j \quad , \quad 1 \leqslant i,j \leqslant 2 \; ,$$

(where $\cdot$ denotes the euclidean scalar product of $\mathbb{R}^3$ and also the hermitian scalar product of $\mathbb{C}^3$).

The inverse matrix will be denoted by g^{ij}. The vectors

$$e^i = g^{ij}\, e_j \quad \text{(repeated index summation convention)}$$

constitute the dual base (e_1, e_2) in the tangential plane.

Let us recall that the area element $d\gamma$ on the surface Γ is

$$d\gamma = \sqrt{g}\; d\xi_1\; d\xi_2 \; ,$$

with

$$g = g_{11}\; g_{22} - (g_{12})^2 = \det(g_{ij})$$

($d\xi_1\, d\xi_2$ is the area element in $\mathbb{R}^2$).

A complex vector J, tangential to Γ, will be located either by its complex coordinates in the base e_1, e_2, denoted by J^1, J^2 (contra-variant coordinates) or by its coordinates in the dual base (co-variant coordinates), denoted by J_1, J_2. Therefore we have

$$J_i = g_{ij}\; J^j \; ,$$
$$J^i = g^{ij}\; J_j \; .$$

The hermitian product of both vectors J and K, defined at the same point, will be

$$J \,.\, K = g_{ij} \, J^i \, \overline{K}^j = J^i \, \overline{K}_i = J_i \, \overline{K}^i = g^{ij} \, J_i \, \overline{K}_j \,.$$

We shall denote the hermitian norm of J by

$$|J| = (J \,.\, J)^{\frac{1}{2}} \,.$$

A function defined on Γ will be said C^k differentiable on Γ, if its images $V \circ \Phi_\ell$ are C^k differentiable functions of θ_ℓ onto $\mathbb{R}^3$, for every map Φ_ℓ, $\ell=1, -, p$.

We define the gradient of a C^1 function V, defined on Γ, as the vector X in the tangent plane, with co-variant coordinates

$$X_i = \frac{\partial V}{\partial \xi^i} \,.$$

The vortex of a C^1 function φ (denoted by $\overrightarrow{\text{curl}}\ \varphi$), defined on Γ, will be the vector X in the tangent plane, with contra-variant coordinates

$$X^1 = -\frac{1}{\sqrt{g}} \frac{\partial \varphi}{\partial \xi^2} \,, \quad X^2 = \frac{1}{\sqrt{g}} \frac{\partial \varphi}{\partial \xi^1} \,.$$

A vector J will be said C^k differentiable when the images $J_1 \circ \Phi_\ell$ and $J_2 \circ \Phi_\ell$ are C^k functions from θ_i into $\mathbb{R}^3$. The divergence of a C^1 vector J in the tangent plane will be the scalar function

$$\text{div } J = \frac{1}{\sqrt{g}} \frac{\partial}{\partial \xi^i} (\sqrt{g} \; J^i).$$

The vortex of a C^1 vector X will be the scalar function

$$\text{curl } X = \frac{1}{\sqrt{g}} \left(\frac{\partial X_2}{\partial \xi^1} - \frac{\partial X_1}{\partial \xi^2} \right) .$$

Then, we can easily verify that

$$|\text{grad}\, \varphi|^2 = g^{ij} \frac{\partial \varphi}{\partial \xi^i} \frac{\partial \overline{\varphi}}{\partial \xi^j} = |\overrightarrow{\text{curl}}\, \varphi|^2 \,.$$

We also define the Laplace operator, which is the following scalar function associated to a C^1 function φ defined on the surface

$$\Delta \varphi = \text{div grad } \varphi = \text{curl } \overrightarrow{\text{curl}} \ \varphi = \frac{1}{\sqrt{g}} \frac{\partial}{\partial \xi^i} \left(\sqrt{g} \,.\, g^{ij} \frac{\partial \varphi}{\partial \xi^j} \right) .$$

We define the space $L^2(\Gamma)$ as the space of functions defined (almost everywhere) on Γ such that

$$L^2(\Gamma) = \left\{ f \mid f \circ \Phi_\ell \in L^2(\theta_\ell) \ , \text{ for } \ell = 1, \ , p \right\}.$$

The space $L^2(\Gamma)$ is then a Hilbert space with the hermitian product

$$(\varphi, \psi) = \int_\Gamma \varphi \ \overline{\psi} \ d\gamma \ ,$$

and the associated norm.

We define the space of vectors in $(L^2(\Gamma))^2$ as the space of vectors such that each one of its coordinates (co-variant or contra-variant) is in $L^2(\theta_\ell)$, for all $\ell = 1, -, p$. This space is also a Hilbert space with the hermitian product (and the associated norm)

$$(J, K) = \int_\Gamma J \, . \, K \ d\gamma \ .$$

We define now the space $H^1(\Gamma)$ (which is again a Hilbert space)

$$H^1(\Gamma) = \left\{ \varphi \in L^2(\Gamma), \ \text{grad } \varphi \in (L^2(\Gamma))^2 \right\}.$$

The scalar product will be

$$((\varphi, \psi)) = (\varphi, \psi)_{L^2(\Gamma)} + (\text{grad } \varphi \ , \text{ grad } \psi)_{(L^2(\Gamma))^2}.$$

The space of vectors in $(H^1(\Gamma))^2$ will be the Hilbert space

$$(H^1(\Gamma))^2 = \left\{ J \in (L^2(\Gamma))^2; \ J_1 \in H^1(\theta_\ell), \ J_2 \in H^1(\theta_\ell), \ell = 1, -, p \right\},$$

with the norm

$$\| J \|^2 = \sum_{\ell=1}^{p} \left\{ |J_1|^2_{H^1(\theta_\ell)} + |J_2|^2_{H^1 \ \theta_\ell)} \right\}.$$

The space $H^2(\Gamma)$ will be the Hilbert space of functions φ such that $\text{grad } \varphi \in (H^1(\Gamma))^2$. We refer to LIONS-MAGENES [2] for additional properties of these Hilbert spaces.

We have the following properties.

<u>*PROPOSITION 1.1*</u> *: Let* J *be a tangential vector to* Γ, *and* φ *be a function defined on* Γ, *such that* $J \in (H^1(\Gamma))^2$, *and* $\varphi \in H^1(\Gamma)$ *; then we have*

$$(1.3) \qquad \int_\Gamma \overline{\varphi} \ \text{div } J \ d\gamma + \int_\Gamma J \, . \, \text{grad } \varphi \ d\gamma = \int_{\partial\Gamma} \overline{\varphi} \ J \, . \, n \ ds \ ;$$

$$(1.4) \qquad \int_\Gamma J \, . \, \vec{\text{curl}} \ d\gamma + \int_\Gamma \overline{\varphi} \ \text{curl } J \ d\gamma = \int_{\partial\Gamma} \overline{\varphi} \ J \, . \, t \ ds \ ,$$

where n *is the exterior normal to* $\partial\Gamma$ *in the tangential plane to* Γ *, and* t *is the oriented tangent to* $\partial\Gamma$.

PROPOSITION 1.2 : Let J *and* X *be vectors tangent to* Γ *. Then we have the following equivalences*

$$(1.5) \qquad \operatorname{div} J = 0 \ \textit{and} \ J \in (L^2(\Gamma))^2 \Longleftrightarrow J = \overrightarrow{\operatorname{curl}}\, \varphi + \lambda J_0 \ , \ \lambda \in C$$

with $\varphi \in H^1(\Gamma)$ *and* $J_0 \in (L^2(\Gamma))^2$, $\operatorname{div} J_0 = 0$ *and* $\int_{\partial\Gamma_1} J_0 . n \, ds \neq 0$;

$$(1.6) \qquad \operatorname{curl} X = 0 \ \textit{and} \ X \in (L^2(\Gamma))^2 \Longleftrightarrow X = \operatorname{grad} V + \nu X_0 \ , \ \nu \in C$$

with $V \in H^1(\Gamma)$ *and* $X_0 \in (L^2(\Gamma))^2$, $\operatorname{curl} X_0 = 0$ *and* $\int_{\partial\Gamma_2} X_0 . t \, ds = 1$.

(φ *and* V *are unique modulo a constant if* J_0 *and* X_0 *are given).*

REMARK 2 : If we impose $J . n = 0$ on a part of the boundary $\partial\Gamma_1$, for example, we obtain an analogous proposition by choosing $\varphi \in H^1$ and $\varphi|_{\partial\Gamma_1} = 0$.

We denote by P_x the projection onto the tangent plane to Γ at the point x of vectors in C^3 . The equation (1.2) is an equation in C^3 because the integral

$$\int_\Gamma \frac{J(y)}{|x-y|} \, dy$$

is a complex vector in C^3 , and grad V is the gradient in C^3 of a complex function.

By projection, we obtain

$$(1.7) \qquad \frac{1}{\sigma} J(x) + \frac{i\omega\mu}{c^2} P_x \int_\Gamma \frac{J(y)}{|x-y|} \, d\gamma_y = - P_x \operatorname{grad} V .$$

But we can easily verify that P_x grad V is the gradient on Γ (in the tangent plane) of the restriction, to the surface Γ , of the function V .

Now, the unknown functions are only defined on Γ .

We can give a variational formulation of the equation (1.2). Let H be the following Hilbert space

$$H = \left\{ K \in (L^2(\Gamma))^2 \ ; \ K \text{ tangent to } \Gamma \ , \ \operatorname{div} K = 0 \right\} .$$

By multiplying by K the equation in the tangent plane, and using (1.3), we can see that the current J will be the solution of (x and J denoting the current points of Γ)

$$(1.8)\quad \begin{cases} \text{Find } J \in H \text{ , such that} \\ \displaystyle\int_\Gamma \frac{1}{\sigma} J \,.\, K \, d\gamma + \frac{i\omega\mu}{c^2} \int_\Gamma\int_\Gamma \frac{J(x).K(y)}{|x-y|} \, d\gamma_x \, d\gamma_y = -\int_{\partial\Gamma} V_0 \, n \,.\, K \, ds, \; \forall \, K \in H. \end{cases}$$

The function V_0 is the imposed potential on $\partial\Gamma$. If we choose J_0 real and regular, with

$$\operatorname{div} J_0 = 0 \quad \text{and} \quad \int_{\partial\Gamma_1} J_0 \,.\, n \, ds = 1 \, ,$$

then, we have the equivalent formulation (from Proposition 1.2)

$$(1.9)\quad \begin{cases} \displaystyle\int_\Gamma \frac{1}{\sigma}(\overrightarrow{\text{curl}}\,\varphi+\lambda J_0).(\overrightarrow{\text{curl}}\,\psi+\nu J_0)\,d\gamma + \frac{i\omega\mu}{c^2}\int_\Gamma\int_\Gamma \frac{(\overrightarrow{\text{curl}}\,\varphi+\lambda J_0)(x).(\overrightarrow{\text{curl}}\,\psi+\nu J_0)(y)}{|x-y|}\,d\gamma_x \, d\gamma_y \\ \displaystyle\qquad = -\int_{\partial\Gamma} V_0 \, n \,.\, (\overrightarrow{\text{curl}}\,\psi+\nu J_0)\,ds, \; \forall \, \psi \in H^1(\Gamma), \; \nu \in \mathbb{C} \,. \end{cases}$$

<u>*THEOREM 1.1*</u> *: If* $V_0 \in H^{\frac{1}{2}}(\partial\Gamma)$*, the problem (1.7) (and therefore, the problem (1.8)) admits a unique solution in* $(L^2(\Gamma))^2$*, and moreover, if* $V_0 \in H^{\frac{3}{2}}(\partial\Gamma)$*, then the solution is in* $(H^1(\Gamma))^2$ *(and therefore* φ *is in* $H^2(\Gamma)$ *if* $J_0 \in (H^1(\Gamma))^2$*).*

(For the definition of $H^{\frac{1}{2}}(\partial\Gamma)$, see LIONS-MAGENES [7]).

2 - FINITE ELEMENT APPROXIMATION.

In order to approximate the solution of the problem (1.8), we shall first introduce a surface Γ_h , and then a space H_h of vectors defined on Γ_h and approximating Γ and H respectively.

For each polyhedrion D_ℓ, $\ell=1, , p$, we construct a triangulation $\mathcal{T}_{h\ell}$ (by triangles the maximal diameters of which are denoted by h and verifying the classical angular condition $\theta \geqslant \theta_0 > 0$, θ_0 independent from h). Then, let us consider the affine interpolate $\Phi_{\ell h}$ of the application Φ_ℓ , defined by

$$\Phi_{\ell h}(a) = \Phi_\ell(a) \text{ , for all vertex } a \text{ of } \mathcal{T}_{h\ell} \text{ ;}$$
$$\Phi_{\ell h} \quad \text{is affine on each triangle of } \mathcal{T}_{h\ell} \text{ .}$$

F_T will denote the restriction of the application $\Phi_{\ell h}$ to a triangle T of $\mathcal{T}_{h\ell}$.

Then, the image of D_ℓ by $\Phi_{\ell h}$ is (at least for h small enough) a continuous surface $\Gamma_{\ell h}$, constituted by plane triangles, the vertices of which are points $\Phi_\ell(a)$, $a \in \mathcal{S}$.

We suppose the triangulation $\mathcal{T}_{h\ell}$ to be compatible within them : the vertices of the triangulation $\Gamma_{\ell h}$ and Γ_{kh} ($\ell \neq k$), located on the curves $\Phi_\ell(D_\ell) \cap \Phi_k(D_k)$, are common. Then, the $\Gamma_{\ell h}$ union gives a surface Γ_h homeomorphic to Γ , and whose maps are the applications $\Phi_{\ell h}$, bijective for h small enough.

The approximate currents will be constant complex vectors on each triangle of Γ_h , tangent to this triangle, which moreover verify the flux equality condition on a common side of two triangles T_1 and T_2 (the normal to this side in T_1 and T_2 being n_1 and n_2 respectively)

$$J_1 \cdot n_1 + J_2 \cdot n_2 = 0 \text{ .} \tag{2.1}$$

H_h will denote this vectorial space.

Then the approximate problem is

$$\int_{\Gamma_h} \frac{1}{\sigma} J_h \cdot K_h \, d\gamma + \frac{i\omega\mu}{c^2} \int_{\Gamma_h}\int_{\Gamma_h} \frac{J_h(x) \cdot K_h(y)}{|x-y|} \, d\gamma_x \, d\gamma_y = -\int_{\partial\Gamma_h} V_{0h} \, n \cdot K_h \, d\gamma \text{ , } \forall K_h \in H_h \text{ .} \tag{2.2}$$

On the surface Γ_h , we can consider the following operator, equivalent to those defined on Γ . For a function $\varphi \in C^0(\Gamma) \left(\bigcup_{T \in \mathcal{T}_h} C^1(T) \right)$, for each face T of Γ_h , we can define

$$\operatorname{grad} \varphi|_T = \begin{cases} \dfrac{\partial\varphi}{\partial\eta^1} \\ \dfrac{\partial\varphi}{\partial\eta^2} \end{cases} \quad \text{for } (\eta^1, \eta^2) \text{ a local orthonormal system of coordinates on } T \text{ ;}$$

$$\vec{\operatorname{curl}} \, \varphi|_T = \begin{cases} -\dfrac{\partial\varphi}{\partial\eta^1} \\ \dfrac{\partial\varphi}{\partial\eta^2} \end{cases}$$

We can also use the local moving system of coordinates, associated to the maps $\Phi_{\ell h}$. Then we have, on each triangle of Γ_h , the definitions of grad φ and $\vec{\text{curl}}\ \varphi$, by the same formulas as for Γ .

Let J_{0h} be in the space H_h such that

$$\int_{\partial\Gamma_1} J_{0h} \cdot n \ ds = 1 .$$

Then we have

PROPOSITION 2.1 :

$$J_h \in H_h \Longleftrightarrow J_h = \lambda J_{0h} + \vec{\text{curl}}\ \varphi_h$$

$$\varphi_h \in \Phi_h$$

$$\Phi_h = \left\{ \varphi_h \in C^0(\Gamma_h) , \ \varphi_h|_T \in \mathbb{P}_1 \ ; \ \forall \ T \subset \Gamma_h \right\}$$

($\mathbb{P}_1$ denotes the space of polynomials of degree 1 in the local coordinates (η^1,η^2)).

From Proposition 2.1, we have an equivalent formulation of the problem (2.2) which is

$$(2.3)\begin{cases} \displaystyle\int_{\Gamma_h} \frac{1}{\sigma}(\vec{\text{curl}}\ \varphi_h+\lambda J_{0h}).(\vec{\text{curl}}\ \psi_h+\nu J_{0h})d + \frac{i\omega\mu}{c^2}\int_{\Gamma_h}\int_{\Gamma_h} \frac{(\vec{\text{curl}}\ \varphi_h+\lambda J_{0h})(x).(\vec{\text{curl}}\ \psi_h+\nu J_{0h})(y)}{|x - y|} d\gamma_x d\gamma_y \\ \displaystyle = - \int_{\partial\Gamma_h} V_{0h} \ n.(\vec{\text{curl}}\ \psi_h+\nu J_{0h})ds \ , \forall \ \nu \in C, \ \varphi_h \in \Phi_h . \end{cases}$$

In order to study the error of approximation, we shall compare, with the vector J , a vector $\tilde{J}_h$ binded with J_h in the following way : a vector J_h of the space H_h will be located by its contra-variant coordinates in the moving coordinates system, associated to the different maps $\Phi_{h\ell}$. Then, to the vector field $J_{h\ell}|_{\Gamma_{h\ell}}$, we can associate the vector field $\hat{J}_{h\ell}$, defined on D_ℓ by co-variant coordinates ($\sqrt{g_h}$ is the area element on Γ_h)

$$(2.4)\begin{cases} \hat{J}_{h\ell_1}|_T = J_h^1 \ \sqrt{g_h} \\ \hat{J}_{h\ell_2}|_T = J_h^2 \ \sqrt{g_h} \ . \end{cases}$$

Then, the property (2.1) allows us to verify that $\hat{J}_{h\ell}$ is a vector field of divergence zero in $L^2(D_\ell)$. (If T_1 and T_2 are two adjacent triangles of D_ℓ , and if $\vec{n}$ is the normal to the common side, then we also have

$$(J_h|_{T_1} - J_h|_{T_2}) \cdot n = 0).$$

Finally, we define the vector field $\tilde{J}_h$ on Γ , as the vector field the contra-variant coordinates of which, in the axes associated to the map Φ_ℓ , are

$$(2.5)\qquad \begin{cases} \tilde{J}_h^1 = \dfrac{1}{\sqrt{g}} \hat{J}_{h_1} , \\ \tilde{J}_h^2 = \dfrac{1}{\sqrt{g}} \hat{J}_{h_2} . \end{cases}$$

Then, we verify that $\tilde{J}_h$ is a vector field of divergence zero.

<u>*THEOREM 2.1*</u> *: Suppose* $J_0 \in (H^1(\Gamma))^2$ *and* $J_{0h} \in H_h$ *are known such that*

$$(2.6)\qquad |J_0 - \tilde{J}_{0h}|_{(L^2(\Gamma))^2} \leqslant C\ h ;$$

then, we have the following error bound

$$(2.7)\qquad |J - \tilde{J}_h|_{(L^2(\Gamma))^2} \leqslant C \left[h\ |J|_{(H^1(\Gamma))^2} + |V_0 - \tilde{V}_{0h}|_{H^{\frac{1}{2}}(\partial\Gamma)} \right]$$

where

$$\tilde{V}_{0h}(x) = V_{0h}(x_h) , \quad \Phi_\ell^{-1}(x) = \Phi_{\ell h}^{-1}(x_h) .$$

3 APPLICATION TO THE ALTERNATOR OUTPUT CONDUCTORS.

3.1 - DESCRIPTION OF THE DEVICE.

It is constituted by the conductors connecting alternator and transformer in a power station, with their sheaths. These conductors and sheaths are hollow cylinders in aluminium, which are submitted to great mechanical constraints. In order to calculate these constraints (and heatings due to Joule effects), it is necessary to know the repartition of the eddy currents in this set of metallic shells submitted to electromagnetic interactions.

Figure 1 shows such a set of conductors.

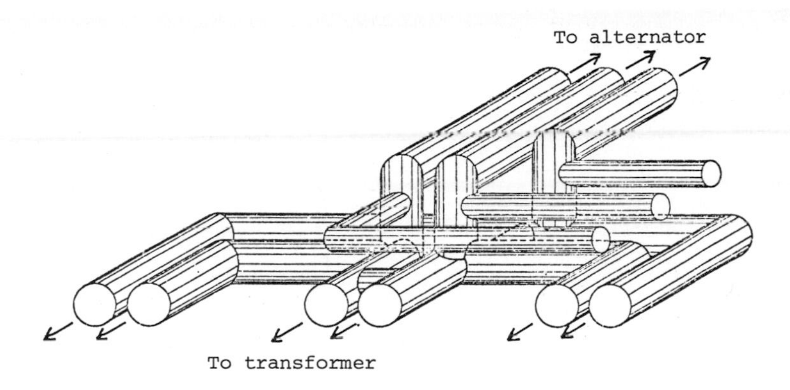

Figure 1

Two types of simplified zone have been studied : a knee (Figure 2) and a junction (Figure 3).

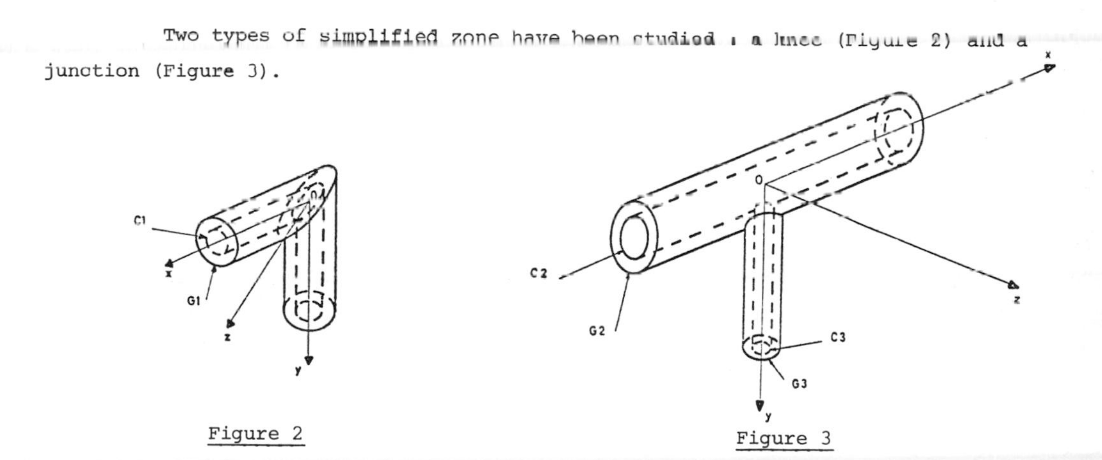

Figure 2

Figure 3

The physical values correspond to a 700 MW alternator of Cordemais, near Nantes (France):

	Main pipes	junction pipes
Radius of the conductor	0,425 m	0,25 m
Radius of the sheath	0,71 m	0,40 m
Thickness of the conductor	0,01 m	0,01 m
Thickness of the sheath	0,02 m	0,02 m
Applied current in the conductor	2.10^4 A	8.10^2 A

The cylinders are thin so that we calculate surfacic densities of current.

3.2 - DESCRIPTION OF THE PROGRAM.

3.2.1 - General description.

The main steps of the program are the following ones :

- Reading of geometric and electrical data.
- Constitution of the mesh, building of the tables setting vertices A_i and triangles T_j .
- Construction of the basic functions w_i .
- Calculation of the terms of the matrix.
- Solution of the system $\rightarrow \alpha_i$.
- Calculation of the density of the current J for each triangle.
- Graphical output of the current lines and current vectors.
- Calculation of the resultant strength on each triangle.
- Calculation of heatings.

3.2.2 - Construction of the triangular mesh.

The vertices are defined as the intersections between the cylinders and two sets of planes.

Figure 4 shows an example of such a mesh near a knee for a semi-cylinder developped in the plane.

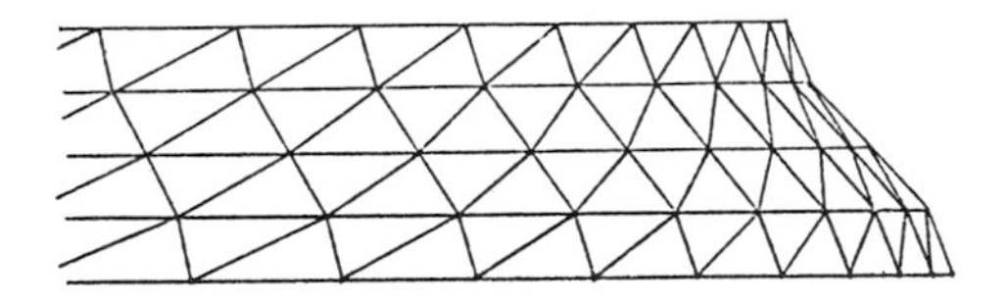

Figure 4

The following tables are given

- COORD (N,3) gives the coordinates of each vertex.
- ITRISO (NTR,3) gives the vertices of each triangle.
- ISOTR (N,9) gives the number of the triangles of each vertex.
- IPOINT (N) indicates if a vertex is on the conductor or the sheath.

3.2.3 - Calculation of the elements of the matrix.

The main difficulty is to calculate the integrals

$$\int_{\Omega}\int_{\Omega} \frac{w_i(x).w_j(y)}{|x - y|} dS_x \, dS_y ,$$

i.e. to calculate the integrals

$$I = \int_{T_i}\int_{T_j} \frac{dS_{x_i} . dS_{y_j}}{|x_i - y_j|} ,$$

where T_i and T_j are two triangles.

If T_i and T_j are far from each other, we calculate $I = \frac{S_i \cdot S_j}{d}$, where S_i, S_j are the surfaces of the triangles and d the distance between the centres of gravity.

If T_i and T_j are adjacent, we use twice a 15th degree, 64 points numerical formula ([8]).

If $i = j$, we calculate the first integral analytically and for the second one, we use a 5th degree, 7 points formula ([8]).

3.2.4 - Solution of the system.

The matrix is full, symmetrical and complex. The ordre of the system is equal to the total number of vertices.

In the case of a unique conductor, we use a direct method, the matrix being decomposed on the form LDL^T .

In the real case with three phases, we have considered that there were three conductors, each of them associated with its sheath, translated from each other. Then, we use a block Gauss-Seidel method, each block corresponding to one conductor and being directly solved in the form LDL^T .

3.2.5 - Calculation of strengths.

The resultant force developed on each triangle is calculated. Constraints are not calculated here.

The force F_{ij} developed on a triangle T_i , crossed by a current J_i , due to a current J_j crossing a triangle T_j , can be written

$$F_{ij} = \frac{\mu_0}{4\pi} J_i \wedge \left(J_j \wedge \int_{T_i}\int_{T_j} \frac{\overrightarrow{M_j M_i}}{|M_j M_i|^3} dS_{M_j} dS_{M_i} \right) .$$

When M_i tends to M_j , the integral has a finite part and cannot be calculated. Happily it can be shown that $F_{ii} = 0$. As to the case of two adjacent triangles, tests have shown that a numerical formula was sufficient. In every case, a 5th degree, 7 points formula has been chosen.

3.2.6 - Storage requirements and CPU time.

The necessary time to build the mesh and the geometric tables is less than one minute in each case. The following table shows that the necessary time to solve the system is short compared with the time for building the matrix and calculating forces. Contrary to the results of the next paragraph, we have not considered symmetry in the following table, where N is the total number of vertices and NTR the total number of triangles. The CPU time is given for an IBM 370/168.

Operation	Nb of conductors	Nb of elements	St. requi.	CPU time
Constitution	1	N = 90 - NTR = 144	72 K bytes	2 mn 20 s.
of the	1	N = 274 - NTR = 496	261 K	12 mn
matrix	3	N = 90 - NTR = 144	80 K	3 mn
	3	N = 274 - NTR = 496	269 K	18 mn
Resolution	1	N = 90 - NTR = 144	93 K	20 s.
of the	1	N = 274 - NTR = 496	497 K	2 mn
system	3	N = 274 - NTR = 496	518 K	7 mn
Calculation	1	N = 274 - NTR = 496	150 K	14 mn 30 s.
of forces	3	N = 274 - NTR = 496	245 K	18 mn

3.3 - RESULTS.

Results concerning a knee are give here. The plane of the knee is considered to be a plane of symmetry, and the distance between the ends of the cylinders and the center of the knee is 12 m, the other sizes have previously been given. Including symmetry, results have been obtained with 224 vertices and 416 triangles for one side of the knee, i.e. 432 vertices and 832 triangles for the complete knee.

3.3.1 - Current density.

Figure 5 represents one side of the knee developed in the plane, and shows the current lines on the sheath, taken at the phase 0 of the applied current. The condition at the extremity on the left of the Figure is $\vec{J} \cdot \vec{n} = 0$, i.e. the sheath is considered to be truncated. Figure 6 shows the current lines at the phase $\pi/2$ of the applied current.

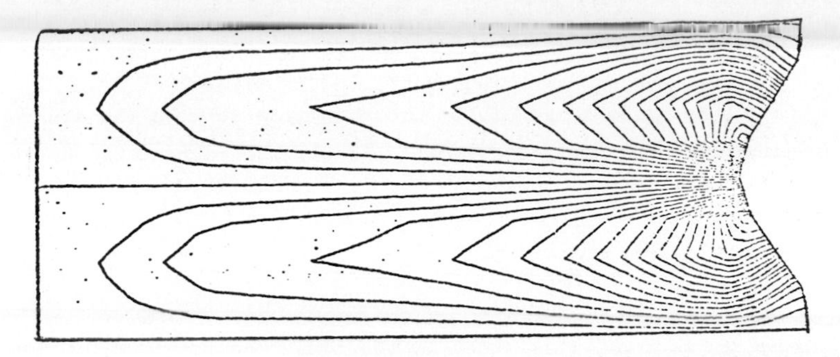

Figure 5

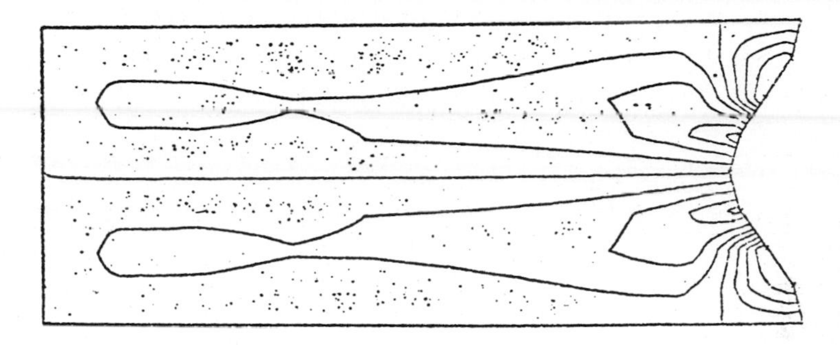

Figure 6

The same results have been obtained by taking V = constant (scalar potential) at the extremity, except a disturbance close to this extremity. The more important are the eddy currents, the more tightened are the current lines. We can see they are very important inside the knee, and they are almost in phase with the applied current. On the conductor, the eddy currents have the same behaviour, but are more confined close to the knee. The maximum density of eddy current is 75 % of the density of the applied current.

3.3.2 - Forces.

The resultant on each triangle has been calculated, just as the total resultant on each side of the knee.

In the case of a unique conductor with its sheath, the total resultant on the conductor is very small. The resultant on the sheath, about 200 Newtons, tends alternatively to open and shut the knee at a frequency of 100 Hertzs.

In the case of three conductors, the resultant is almost the same as in the previous case on the sheath, but in addition, there is a force, 800 Newtons for the conductor in the middle, and 400 Newtons for the other ones, perpendicular to the place of the knee, and tending to move alternatively further and closer the conductors from each other one.

These forces seem to be small, but they grow up with the square of the applied current and in the case of a short-circuit, the current can be ten times superior to the nominal value.

3.4 - CONCLUSIONS.

As far as we know, this study is the first one for this problem. It must be noticed that, even in simple geometry, we do not know any analytical solution, so that the only possible test of our results are direct measurements (some experiments are presently going on). It is important to notice that the electromagnetic influence decreases rather slowly ; so that, in general geometric situations, we cannot neglect any influence. However, in all the cases treated here, the sheath has an efficient shielding-effect, and that allows us to neglect some influence. This last point has been checked by our program.

BIBLIOGRAPHY

[1] DJAOUA, M., *Méthodes d'éléments finis pour la résolution d'un problème intérieur dans* $\mathbb{R}^3$, Rapport Interne du Centre de Mathématiques Appliquées de l'Ecole Polytechnique, n° 3, 1976.

[2] LIONS, J.L., MAGENES, E., *Problèmes aux limites non homogènes et applications*, T.1, Dunod, Paris, 1968.

[3] NEDELEC, J.C., *Curved finite element methods for the solution of singular integral equations on surfaces in* $\mathbb{R}^3$, Comp. Meth. Appl. Mech. Eng., 8 (1976), 61-81.

[4] NEDELEC, J.C., SABRIE, J.L., VERITE, J.C., *Calculation of eddy currents in a conductor and its sheath by a finite element method*, Proc., Conference on the Computation of Magnetic Fields, Oxford, april 1977.

[5] NEDELEC, J.C., *Computation of eddy currents on a surface in* $\mathbb{R}^3$ *by finite element methods*, to appear in SIAM J. on Numerical Analysis.

[6] PLANCHARD, J., VERITE, J.C., *Calcul des forces d'origine électro-magnétique sur des coques métalliques parcourues par des courants de Foucault - Etude Théorique*, Rapport Interne HI 2434/02, E.D.F., mai 1977.

[7] STOLL, R.L., *The analysis of eddy currents*, Clarendon Press, Oxford, 1974.

[8] STROUD, H.A., *Approximate calculation of multiple integrals*, Prentice Hall, 1971.

[9] VERITE, J.C., *Calcul de la répartition des courants de Foucault dans les gaines coaxiales de sortie des alternateurs. I - Partie théorique - II - Application au cas d'un coude*, Rapport Interne HI 2204/02, E.D.F. , septembre 1976.

[10] VERITE, J.C., *Calcul de la répartition des courants de Foucault dans les gaines coaxiales de sortie des alternateurs. Résultats concernant un coude*, Rapport Interne HI 2217/02, E.D.F., septembre 1976.

[11] VERITE, J.C., *Calcul de la répartition des courants de Foucault dans les gaines coaxiales de sortie des alternateurs. Résultats concernant un embranchement*, Rapport Interne HI 2421/02, E.D.F., avril 1977.

[12] VERITE, J.C., *Calcul des forces d'origine électro-magnétique dans les gaines coaxiales de sortie des alternateurs. Résultats concernant un coude et un embranchement*, Rapport Interne HI 2453/02, E.D.F., juin 1977.

THE CLUB MODULEF

A library of subroutines for finite element analysis

A. PERRONNET
Laboratoire d'Analyse Numérique
Université Pierre et Marie Curie
PARIS
IRIA-LABORIA
Domaine de Voluceau
78150 LE CHESNAY

SUMMARY :
The goal of this lecture is to introduce the club MODULEF. MODULEF gathers together people from industries and members of universities in order to conceive and realise a library of modules for finite element analysis. The purpose of the library, the methods, and a list of members are described in the first paragraph.
The present achievements which are already realized and those which have to be prepared are the object of the second paragraph.
The third paragraph contains briefly the software standards and the levels from which one may use the library.
The conclusion of this lecture is drawn in the fourth paragraph.

1. - THE DESCRIPTION OF THE CLUB

1.1. Purpose of its existence

Nowadays industry and research must take into account :

. costly software developments to cope with specific, real situation
- mesh generation, large linear systems, visual projection... -

. expensive packages non commercially available
- ASKA (1), GENESYS (2), MARC-ORSINOR (3), NASTRAN (4), SESAM 69 (5),...

. specific purpose packages - structure mechanics -

. their large size - 300 000 cards for ASKA, NASTRAN -

and industries must also face the following options :

. the development of small and limited sets of programs usually mutually redundant regardless of any generality

. their dependence on foreign softwares. Indeed in many instances these are the only ones able to integrate independently created modules into a cohe-

rent application without dramatic loss of performance.
Furthermore researcher's highly theoretical approach to problems find difficult the engineer's practical requirements.
For all these reasons the LABORIA decided to organise a club to join together people interested in scientific computation from university as well as from industry.

1.2. Goals

It's goals are :

- Get finite element people together and allow them to exchange their point of view
- Bring together all institutions interested in the same subject so as to set up standards
- Test the different methods, compare results and eventually find the most efficient solving method and actually program it in the form of one or more modules
- Develop a library of modular programs on finite elements based on results from functional analysis and numerical methods.
- Release periodically to each member the latest version of this library
- Establish programming standards in order to make the whole coherent, i.e. describe and write programs according to the same rules and with the same meanings and insure maximum portability
- Efficiently dispose of machine resources : main memory (M C), secondary storage (MS) and special terminals (Benson, Tektronix,...)
- Make reference material homogeneous in order to improve efficiency.

1.3. The club's life

1.3.1. Rules and regulations (6)

In order to respect the large diversity of the members the rules apply the following rights and restrictions :

- Full independence of each member
- Efficiency requirements usual for all lucrative operations
- Free circulation for research and development purposes with the club but prohibition of any commercial use of such programs or parts of them without prior permission from the authors.

1.3.2. Means

The library of modules is established with the contributions of each members. Essentially the club's ressources are those made available by its members. In order to make available some basic modules, IRIA may order those modules to interested members. Essentially the club is a co-operative association dedicated to develop programs in the field of finite elements.

1.3.3. The Structure of the club (6)

The regulation divides the club into two committees and members. The "Comité Directeur" manages the club. It accepts or rejects membership applications, establishes head lines in development and research and administers the subsidies.
The "Comité de Coordination" is in charge of implementing decisions reached at the "Comité Directeur's" level, supervises the programming work, checks for the appropriate use of accepted coding standards, improves these standards as soon as this proves necessary, supervises the draft of user booklets and insures delivery of such information to members, reports to members of the lastest achievements.
It is not responsible with the maintenance of the library.
Finally it submits possible errors to the one who produces the modules and brings out the corrections to its members.

ORGANISM	REPRESENTATIVE
AEROSPATIALE Division Systèmes balistiques et spaciaux	MM. BRETON
AVIONS MARCEL DASSAULT/BREGUET AVIATION	PERIAUX - NAVES
CCSA	CAMPEL - DAMIENS
CEBTP	ABSI
ECOLE CENTRALE DE LYON	MAITRE
ECOLE POLYTECHNIQUE LAN	NEDELEC
EDF	FEINGOLD
FACULTE DES SCIENCES D'ORLEANS	Mme BOUJOT
HEBREW UNIVERSITY OF JERUSALEM	MM. BERCOVIER
INSA LYON	LEMAIRE
INSTITUT FUR STATIK UND DYNAMIK DER LUFT STUTTGART	ARGYRIS
LABORIA-IRIA	GLOWINSKI
LAN BORDEAUX	HAUGAZEAU
LAN GRENOBLE	CHATELIN
LAN NICE	CEA
LAN ORSAY	TEMAM
LAN PARIS VI	RAVIART
LAN PAU	ARCANGELI
LAN PAVIE	MAGENES
MESSIER HISPANO	MASCLET
SNCF Direction du matériel	CLAVIER
THOMSON - CSF	RAULT
UER Mathématiques Pures et Appliquées de LILLE	POUZET
UT COMPIEGNE	TOUZOT

1.3.4. Working methods

General Assembly of all members will be organized every nine months. Every new module will then be presented by his author. The program as well as his explanations will be freely distributed to members.
A working plan for the next six months will be established at the same time. Each member points out what module he is interested to develop. Naturally some kind of coordination will be needed at this point, as several members will work on the same module. The next three months will allow the "Comité de Coordination" to gather the modules, their written documents and verify their standards.

2. - ACHIEVEMENTS

First of all a short overview of the terms employed might prove usefull (cf. (7)).

- A DATA STRUCTURE (SD) is a set of arrays to which are associated the means to manage their physical support (main memory and/or secondary storage).
These arrays are the result of an important computational step and serve as input for several other modules as different options may be taken into account at this point. Examples : mesh description, assembled stiffness matrix ...

- A MODULE is a set of subroutines. It implies distinct tasks which can be performed independently of how other tasks are performed. A MODULE transforms input data structures (SDE) in output data structures (SDS). So modules are "interconnected" by the means of the data structure.

2.1. Existing modules

They are made up from eight logical sets

2.1.1. Modules for the mesh-generation

They are briefly described in the following illustrations :

▭ a module
input cards
○ variables or arrays
a data structure

The data structure MEFI describes a bidimensional mesh, NOPO a general one (bi or tridimensional)

COMEFI : generates a bidimensional mesh from input data cards

QUACOU : builds a mesh with triangles and quadrilaterals originating a curve 'quadrilaterals'cf. (8)

TRIAUT : the input data for this triangulation is the boundary of the domain cf. (9)

ALTERN : builds the mesh of the alternator's stator and rotor cf. (10)

AFFLOC : subdivides the mesh in the viscinity of certain vertices cf. (11)

RETRIA : subdivides in N^2 elements the triangles and/or the quadrilaterals of a data mesh cf. (9)

SYMDR : provides a symmetry in respect with a straight line

TRAROT : performs the translation or rotation of the mesh

RECOL : forms a mesh from two adjacent meshes

ADPOIN : adds the nodes which are not vertices

RENCMK : renumbers the nodes and eventually the elements using the algorithm described in (12)

RENEL : reorders the elements in order to solve the linear system by the Gauss frontal method (13)

TRABE1 : provides a plotting of the mesh

VISU1 : displays the mesh and allows conversational modifications of vertices coordinates, elements type, and points displacement on a 2250 IBM or a TEKTRONIX visual projection unit.

LANEF : a preprocessor which allows chaining of mesh generation in an oriented language. The amount of data is thus diminished and easier to handle cf. (14)

MEFNOP : restructures the data structure MEFI (corresponding to a bidimensional mesh) in a more general one NOPO cf (15)
The data structure MEFI will soon pass away.

TN2D3D : transforms a bidimensional mesh 2D, a tridimensional one, 3D with the help of a subroutine F2D3D(X,Y,XX,YY,ZZ) written by the user XX = FX(X,Y), YY = FY(X,Y), ZZ = FZ(X,Y) - cf. (16)-

COLIBR : divides an element in a subfamily of elements. An element may be : a straight line, a triangle, a quadrilateral , a tetrahedral, a pentahedron, a hexahedron (cf. (17))

RECOLN : forms a mesh from two adjacent meshes

The following examples illustrate some possible uses of these modules :

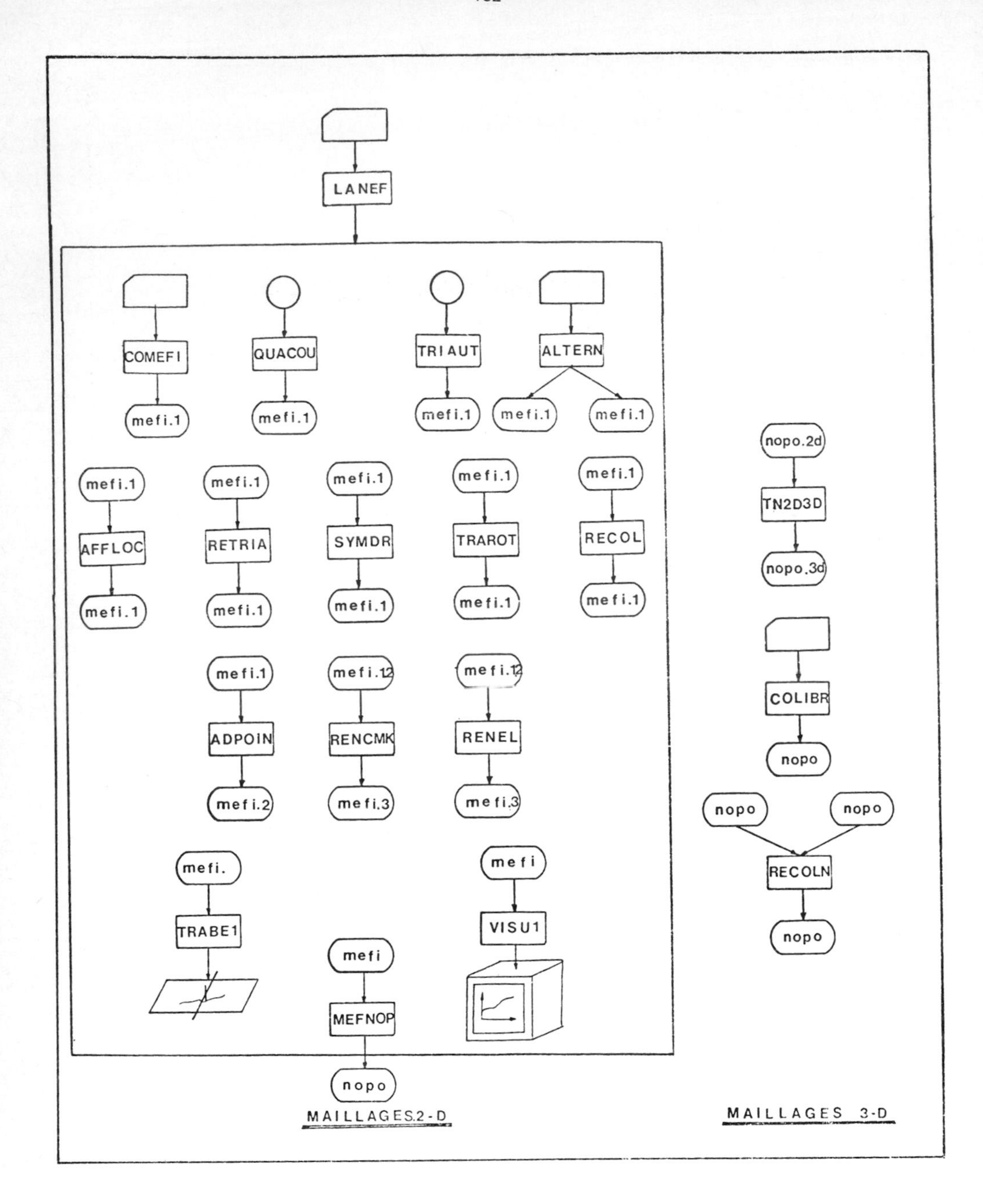

Modules for the mesh generation

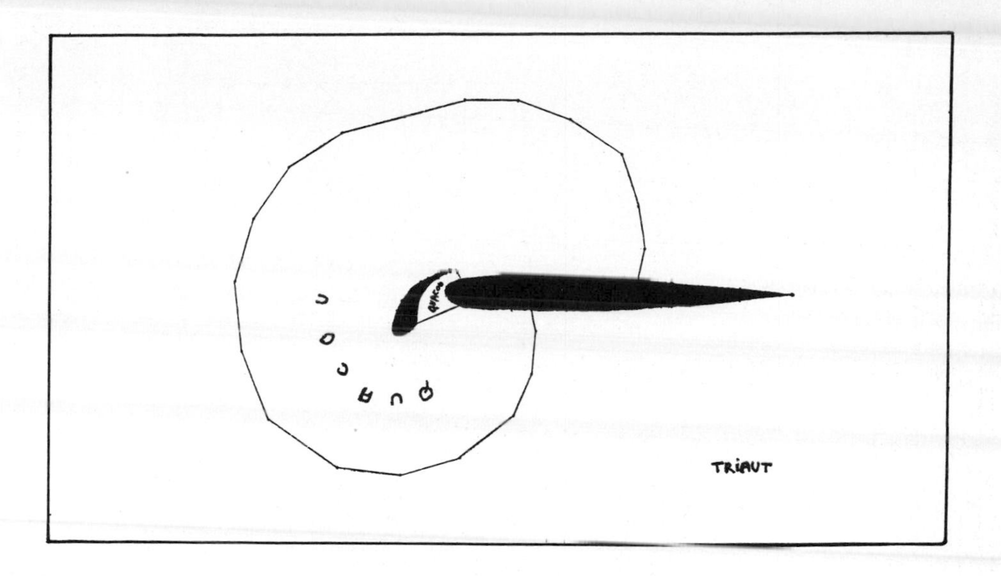

USED MODULES : QUACOU + QUACOU + TRIAUT

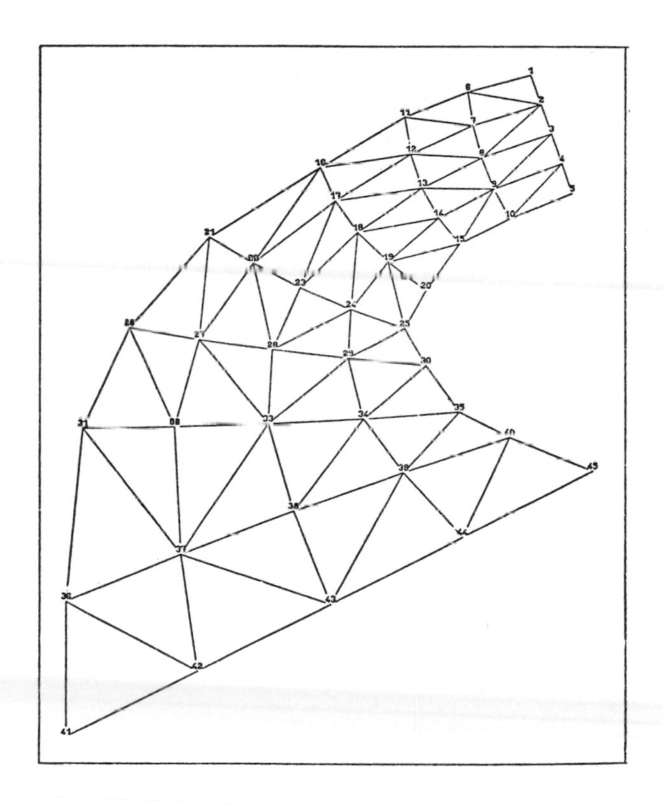

FRONT PART OF THE WING

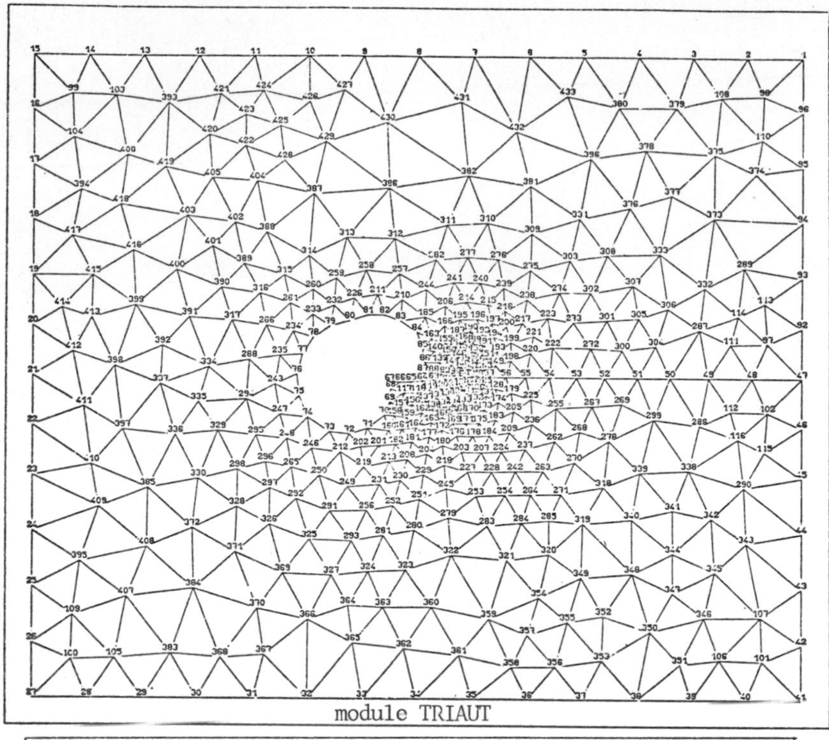

module TRIAUT

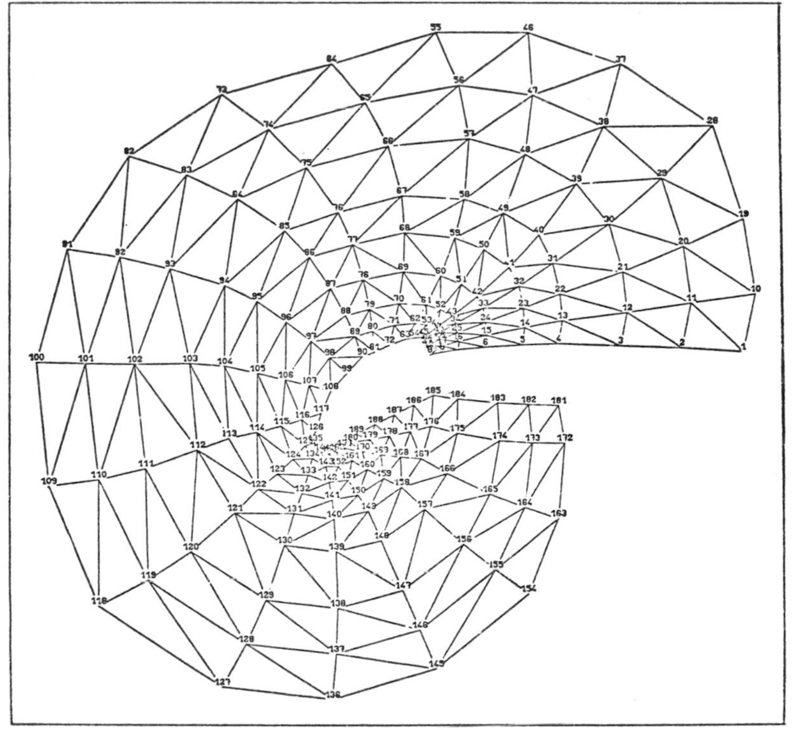

module QUACOU

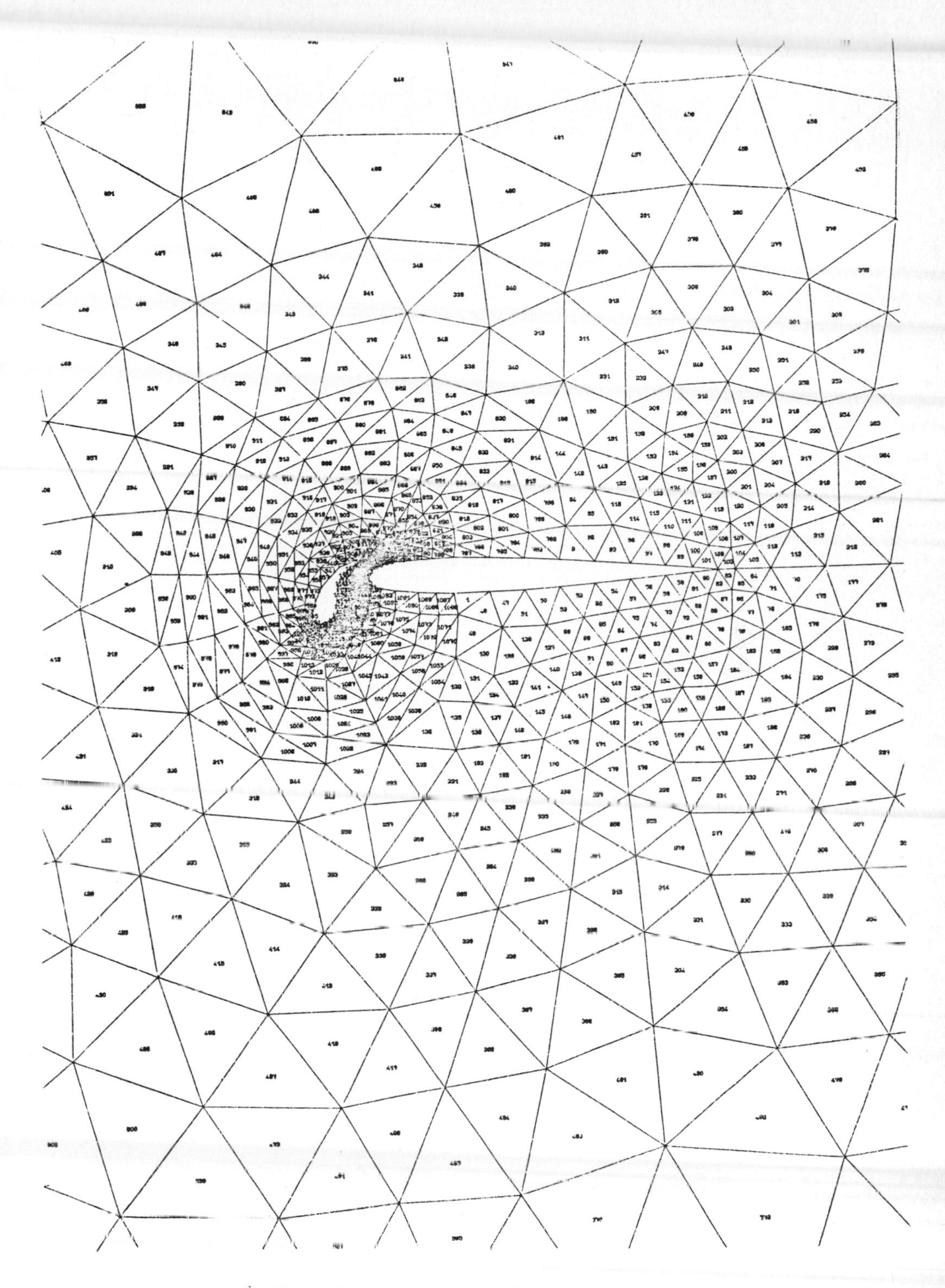

QUACOU + QUACOU + TRIAUT + RECOL + ADPOIN + RENCMK

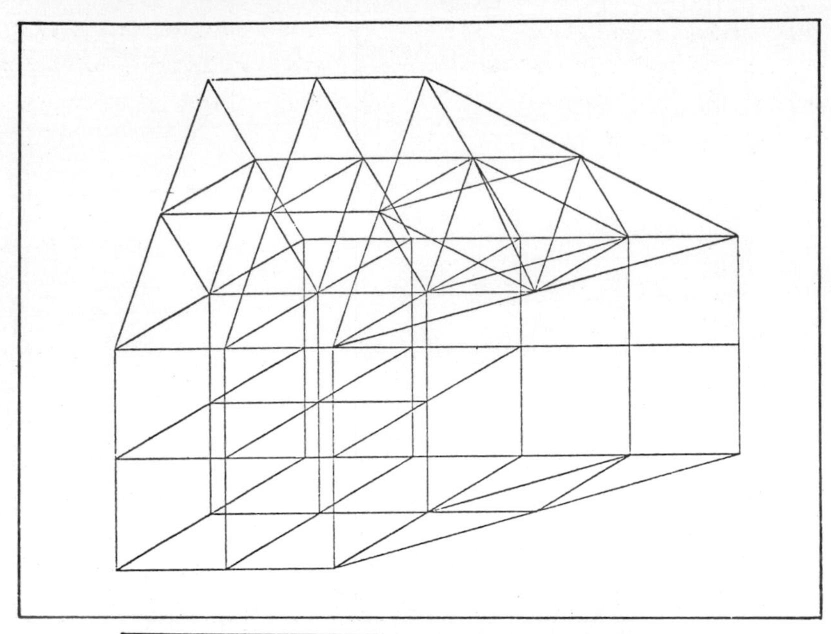

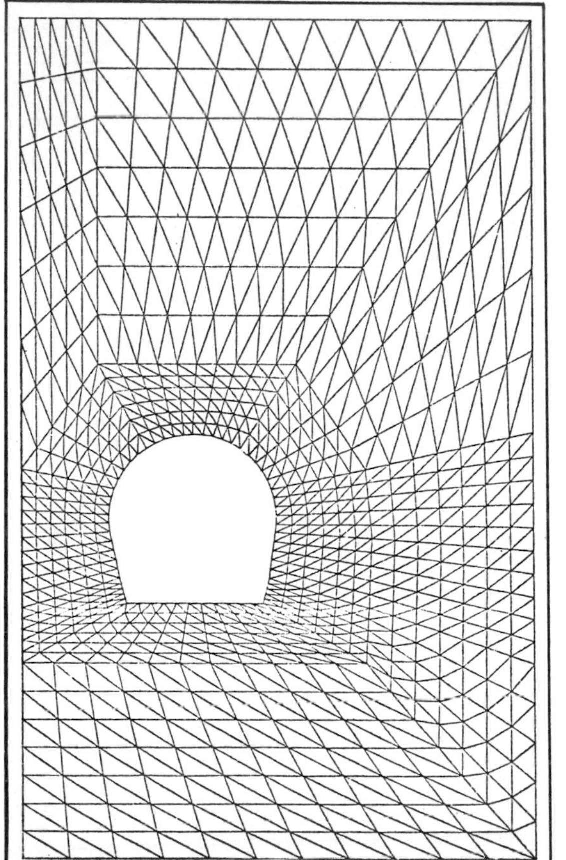

module

COLIBR

A NUMBER OF MATERIAL also called NUMBER OF SUB-DOMAIN is associated to each element. In addition to that, a number called REFERENCE NUMBER may be assigned to the sides, the edges, the vertices. Further the reference number allows the following :

- to project the points on the surface number n. (The user gives the equation or other characterisation of this surface in a subroutine).
- to describe only once by reference number the heat wave, the boundary conditions instead of doing it for each side, edge or node. Finally the input data set is minimized.

2.1.2. Modules for the choice of the interpolation, building of the arrays associated to elements.

According to the variety of applications (frame structures, heat problems, fluid mechanics, electrical engineering), to the increase and the diversity of finite elements, an important formal information was done.
For instance the three components of the displacement, the flow's velocity, the temperature are all called variational unknown variables.
In each node at least one of them has degrees of freedom (d.l.). They are described by their mnemonics. (VN-value at the node, DX - X derivative...). At the same time, some arrays are able to comment the signification of every variational unknown variable...
So all the characteristics of points, nodes, finite element's type are easily found in the data-structure MAIL and allow to do automatically operations such as : changing a basis visual projection of one part of the mesh, print in detail the data set cf. (18).

COMACO : extends to the description of the mesh the description of the interpolation (19)

THELAS : builds the associated arrays : elementary matrices and right members of a mesh using a "FINITE ELEMENT" library for a thermic problem - THER - or an elasticity one - ELAS cf. (20) - The material's properties are found in the data structure MILI and the efforts or heat quantity in the data structure FORC

The user can make a choice between a finite element existing in the library and one of his own elements.
If for the moment (end 1977) their amount is not important, essentially one or two degrees polynomials on triangles, quadrilaterals, tetrahedrals, hexahedrons... a special effort will be done in order to program mathematical's study results of F. BREZZI (21), J.M. THOMAS (22), P.A. RAVIART (23), M. BERNADOU-P.G. CIARLET (24), B. MERCIER-M. JEAN-FRANCOIS, B. DEROUBAIX (25)
The previous finite elements perform satisfactorily with regard to the computation time but the interpolation, the type of problem and the numerical integration formula are rigid. Unlike these ones the POLBAS module points out whether the elements

$\hat{\Sigma}$ are unisolvent - cf. (26) - using user's cards desbribing :

- the nodes and the associated degrees of freedom, the points and their features,
- the interpolation formula for each variational unknown variable on the reference triangle and/or quadrilaterals $\hat{\Sigma}$
- the interpolation formula for each component of the mapping F : $\hat{\Sigma} \to \Sigma$ current element
- the spaces of polynomials $\hat{P}$ used on $\hat{\Sigma}$
- the variational formulation of the problem : $\int_{\Omega} D^{\alpha}u_i D^{\beta}v_j d\Omega$, $\int_{\Gamma} D^{\alpha'}u_i D^{\beta'}v_j d\Gamma$
- the integration formulae to be used

The FORM module computes the elementary matrix and the elementary right members - cf. (27),(28) -
Therefore those two modules allow quick check up of new elements and to choose and store the best solution in the library ELAS or THER.

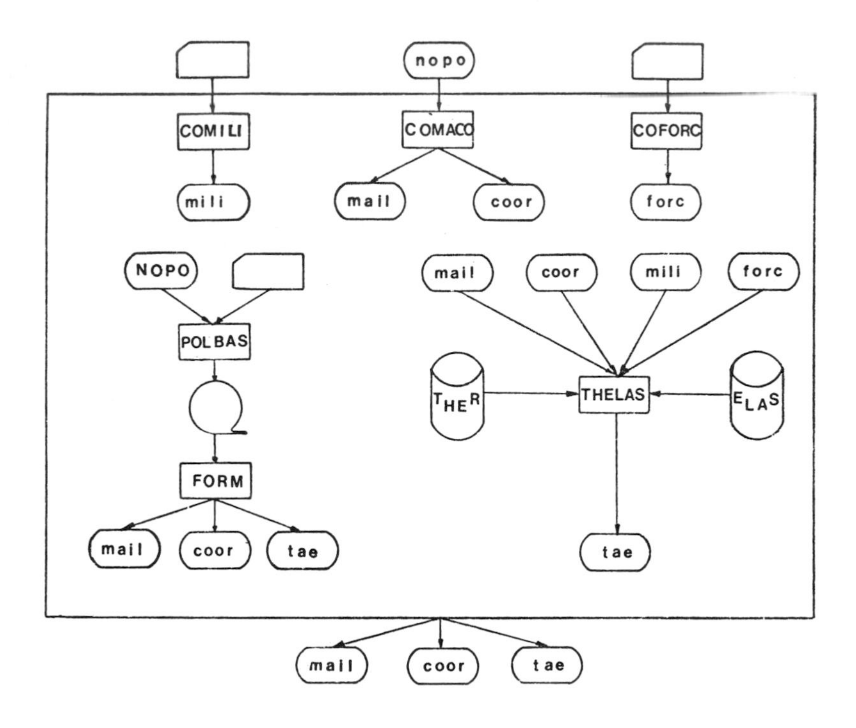

Choice of the interpolation

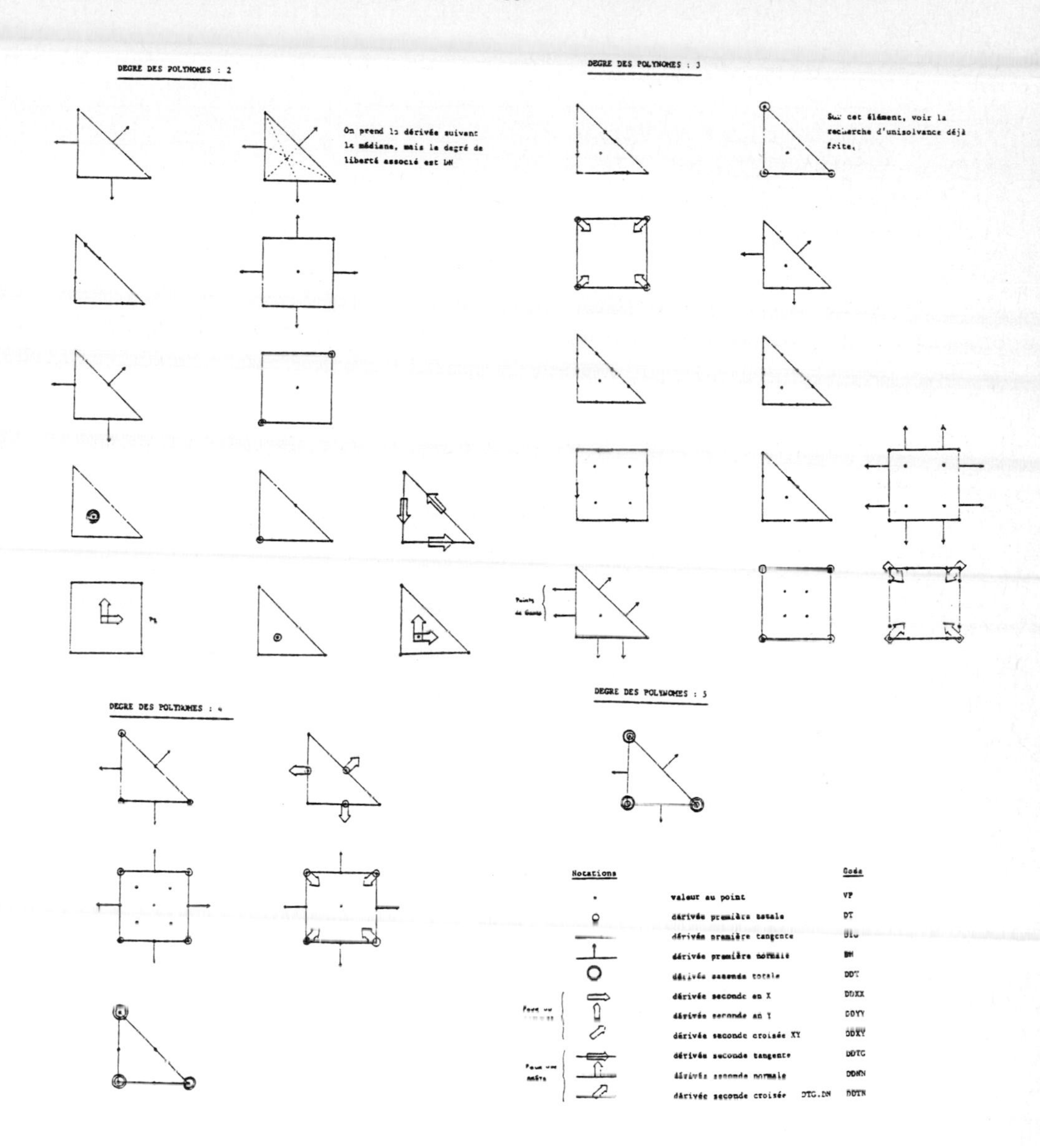
DEGRE DES POLYNOMES : 2
On prend la dérivée suivant la médiane, mais le degré de liberté associé est DN
DEGRE DES POLYNOMES : 3
Sur cet élément, voir la recherche d'unisolvance déjà faite,
DEGRE DES POLYNOMES : 4
DEGRE DES POLYNOMES : 5
Notations
Code
valeur au point
VP
dérivée première totale
DT
dérivée première tangente
dérivée première normale
dérivée seconde totale
DDT
dérivée seconde en X
DDXX
dérivée seconde en Y
DDYY
dérivée seconde croisée XY
DDXY
dérivée seconde tangente
DDTG
dérivée seconde normale
DDNN
dérivée seconde croisée DTG.DN
DDTN

Examples

2.1.3. Linear system solving modules

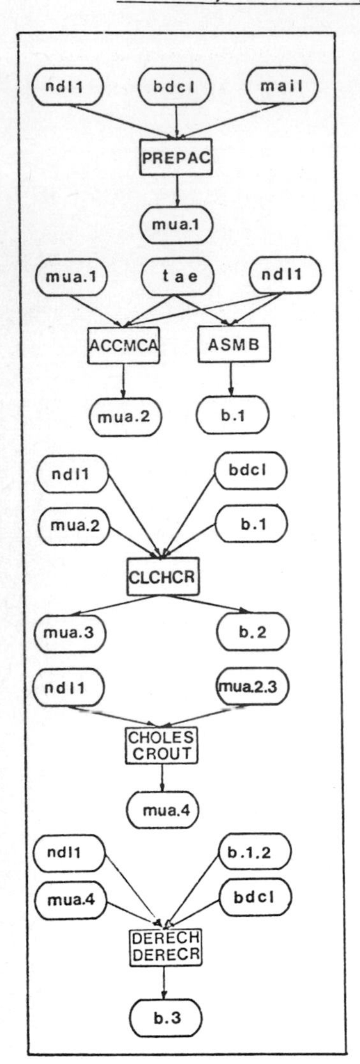

$A = L^tL$ or LD^tL or L.U.
in main memory SKYLINE storage - cf. (9) -

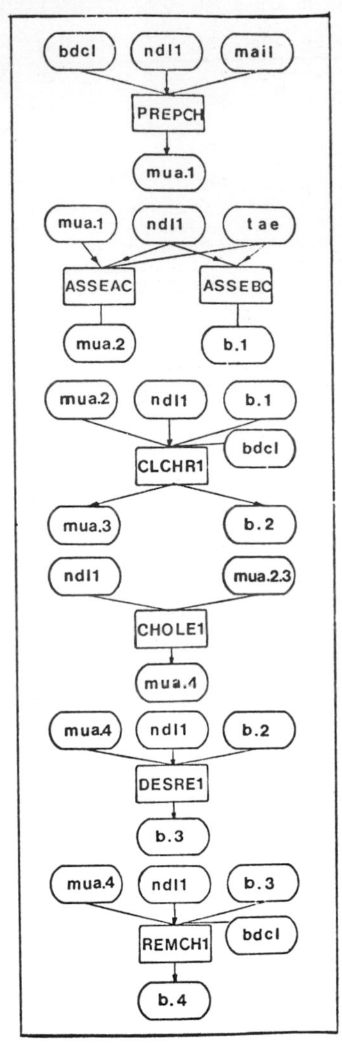

$A = L^tL$ or LD^tL
on secondary storage using the direct access SKYLINE storage

A = L.U
a symmetric or non-symmetric wave frontal storage - cf. (13)

The data structure BDCL desbribes the degrees of freedom to assign. The COBDCL module creates BDCL from data cards.
The module CONDL1 builds, if necessary, the array containing the number of the last degree of freedom of each node.
The preparatory work for each method establishes the amount of main memory to be used for the storage and the partition in blocks of the matrix and of the right

member if needed.
Each step - assembly, taking into account the boundary conditions, factorisation, solving the linear systems with a triangular matrix (upward or downward) mades up a single module whenever it is possible. For instance if several systems with the same matrix are to be solved this matrix needs being factorized only once.
The blocking conditions for the boundary conditions described in the data structure BDCL may be linear combinations between degrees of freedom (except for the FRONT module).

2.1.4. Modules which allow to compute the flow, the initial deformations, the stress

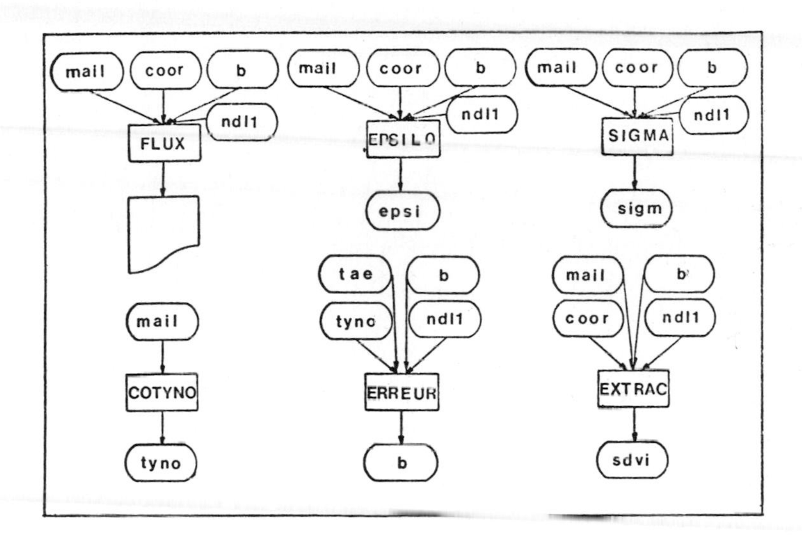

FLUX : computes the flow of heat on the boundaries
EPSILO : computes the initial thermic deformations
SIGMA : computes the stress of continuum medium
COTYNO : the data structure TYNO contains in detail the signification of each degree of freedom already computed
ERREUR : estimates ($\sum_T [A_T] \{B\} - \sum_T \{b_T\}$, (where $[A_T]$ denotes the elementary matrix, $\{b_T\}$ the elementary right member, {B} the computed solution) and stores the results in the data structure B
EXTRAC : selects out of the data structure B the chosen values for a display - S.D. SDVI - .

2.1.5. Modules for the result's visualisation

TRASIG : plots the principal stresses submitted to a continuum bidimensional medium
CRBNIV : draws the countour lines of the chosen solution on a full bidimensional do-

main or on selected parts of it - (cf. (29) -

ZEGFXY : displays a function z = F(X,Y) as a sequence of cross sections, removing the hidden sides - cf. (30),(31) -

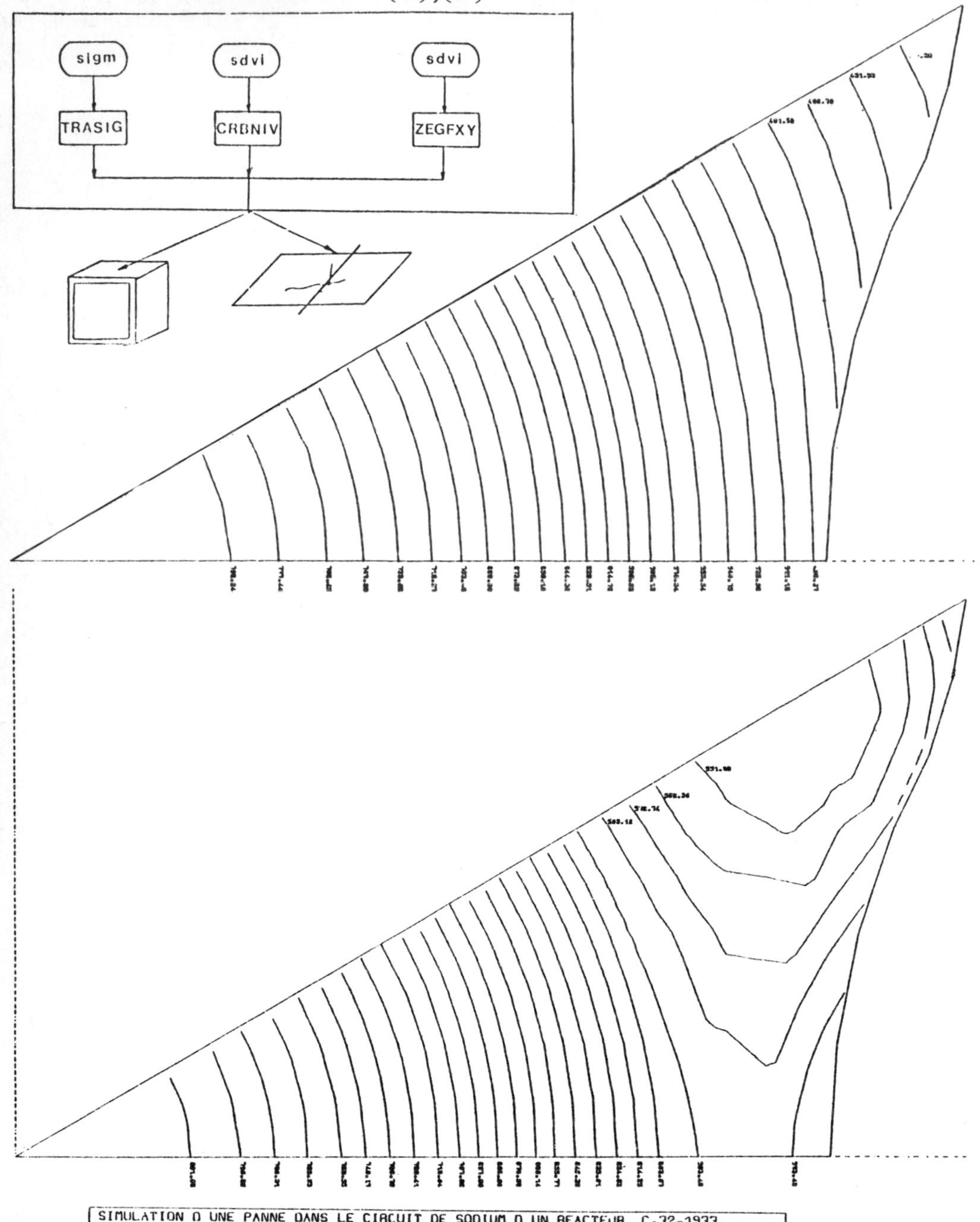

2.1.6. Modules describing the generalized eigenvalues and vectors

The aim is to calculate for [K] (resp. [M]), the stiffness matrix (resp. the matrix mass of the structure), the lowest eigenvalues and corresponding eigenvectors such that $([K]-\lambda[M])\ \{x\} = 0$ - cf. (35), (36) -

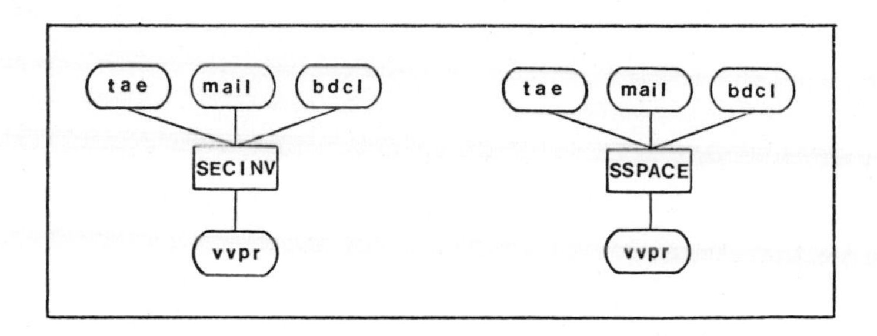

SECINV : the secant method and the inverse iteration with translation - cf. (32)-(33)-
SSPACE : the subspace iteration method (the simultaneous iteration method) - cf. (32)-(33) -

2.1.7. The alternator - cf. (34) -

The potential vector A of the magnetic inductivity generated in an alternator is evaluated from the current j and material's permeability.
The meshes of the rotor and stator are obtained from the ALTERN module - cf. (2.1.1)-
The problem is to find :

$$A \in H_0^1(\Omega) \text{ solution of}$$

$$\int_\Omega \nu(x,A) \text{ grad } A \text{ grad } v \, d\Omega = \int_\Omega jv \, d\Omega \quad \forall v \in H_0^1(\Omega)$$

$$A|_\Gamma = 0$$

with $\nu = \begin{vmatrix} \dfrac{1}{4\pi\ 10^{-7}} \text{ MKSA in air} \\ \alpha + (1-\alpha)\dfrac{(|\text{grad } A|^8)}{|\text{grad } A|^8 + \beta} \end{vmatrix}$

$\alpha = 4{,}5\ 10^{-4}$ in rotor-iron
$\beta = 2{,}2\ 10^4$
$\alpha = 3\ 10^{-4}$ in stator-iron
$\beta = 1{,}6\ 10^4$

by EGSN algorithm - cf. (34) -

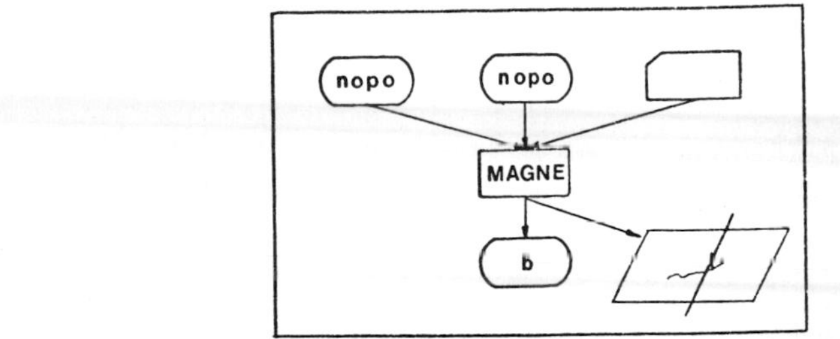

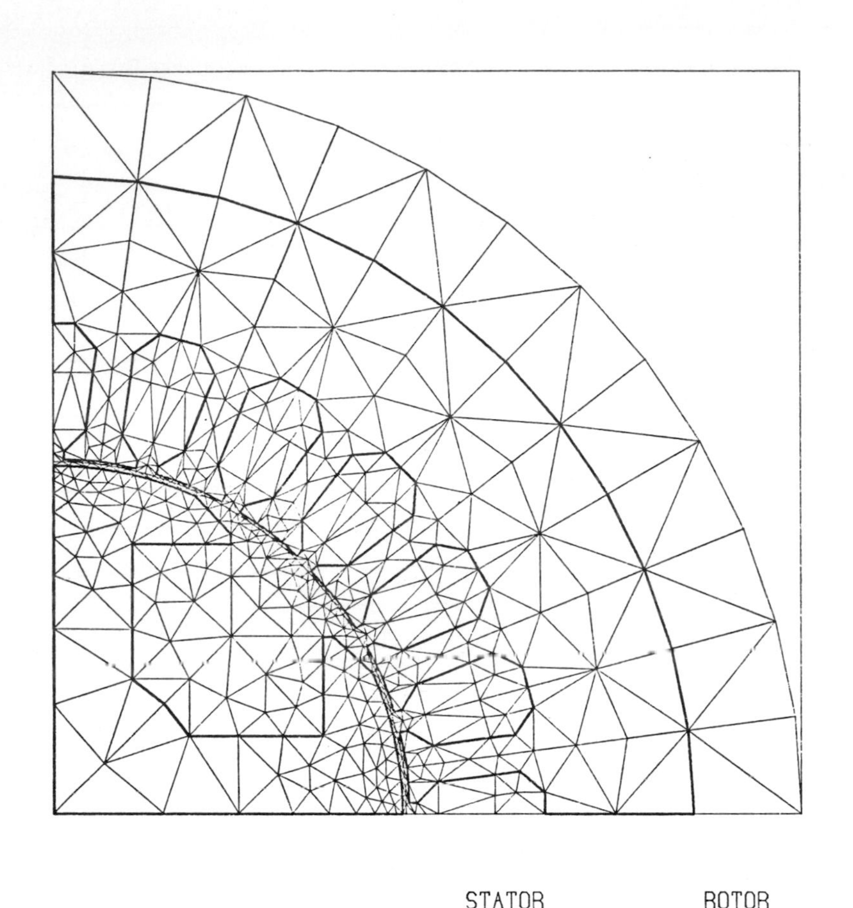

	STATOR	ROTOR
NOMBRE D ELEMENTS	504	336
NOMBRE D ELEMENTS FER	228	150
NOMBRE DE NOEUDS	297	209

LABORIA- A.MARROCCO A.PERRONNET

MESH

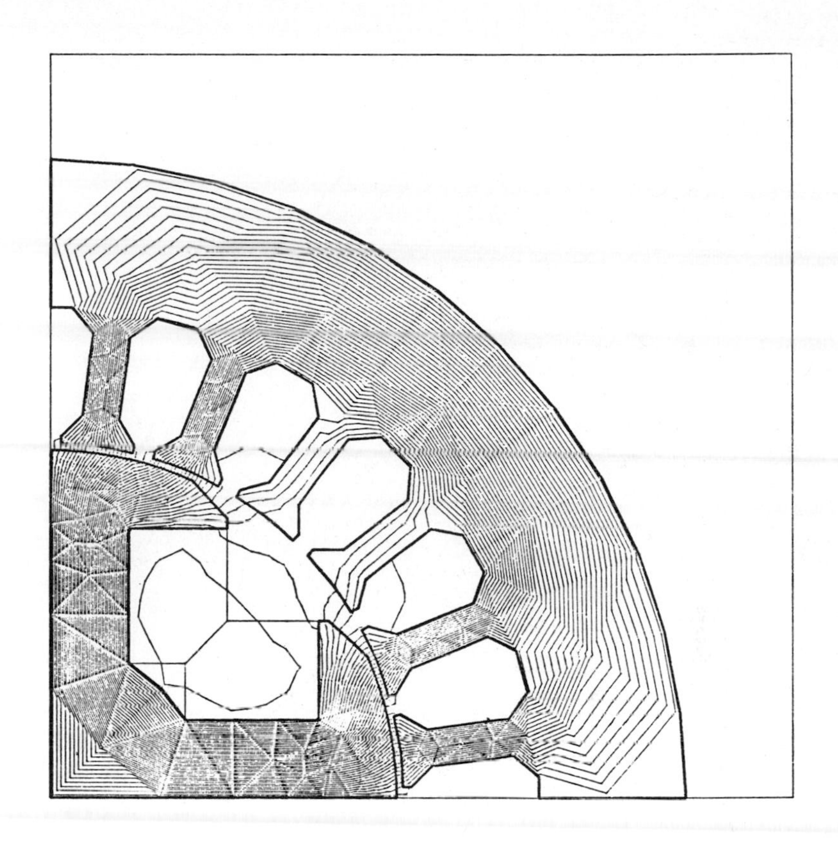

PAS ENTRE 2 LIGNES 0.500E-3 POS. 0
VALEURS EXTREMES DE LA FONCTION -0.163E-4 0.220E-1
COURANT ROTOR
3 DENSITE MKSA 0.100E+8
4 DENSITE MKSA 0.100E+8

LABORIA- A.MARROCCO A.PERRONNET

MAGNETIC INDUCTIVITY

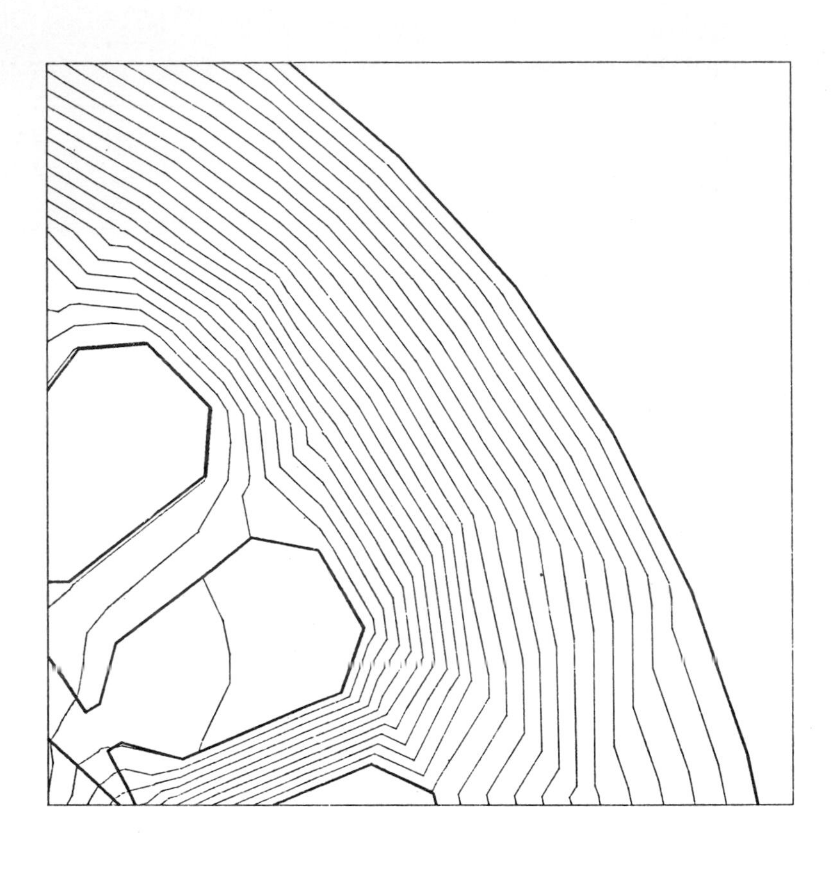

PAS ENTRE 2 LIGNES 0.916E-3 POS. 0

VALEURS EXTREMES DE LA FONCTION -0.163E-4 0.220E-1

COURANT ROTOR

3 DENSITE MKSA 0.100E+8

4 DENSITE MKSA 0.100E+8

LABORIA- A.MARROCCO A.PERRONNET

ZOOM

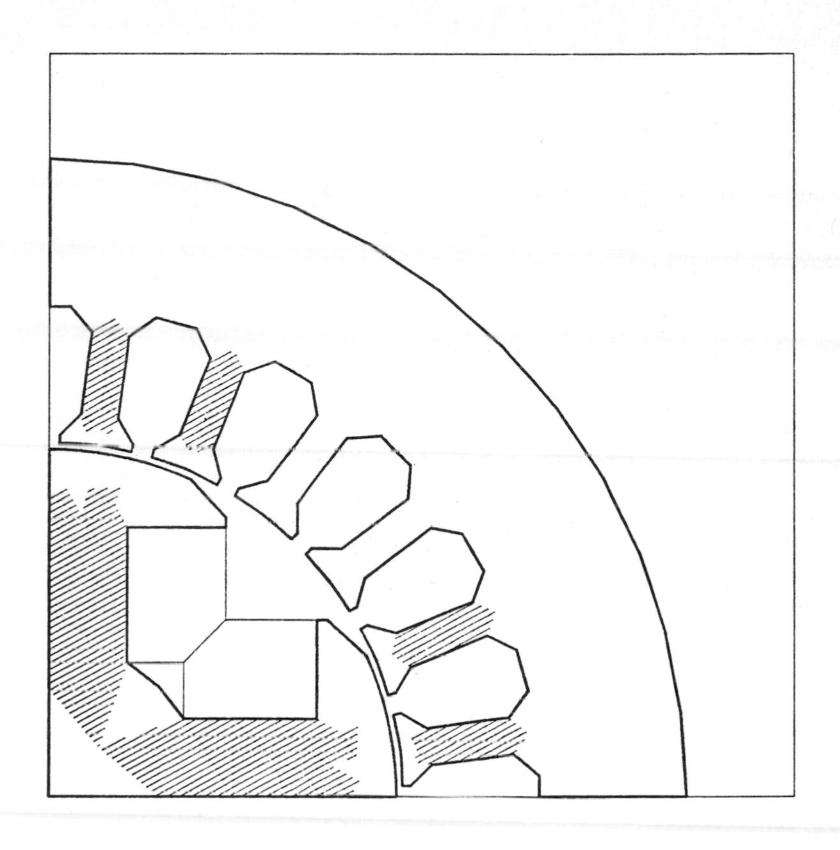

ZONE OU L INDUCTION EST SUPERIEURE A 1.60

LABORIA- A.MARROCCO A.PERRONNET

2.1.8. Modules solving the bidimensional Navier-Stokes formulae for a viscous incompressible fluid

Let $\vec{u}$,p be the unknown variables of the Navier-Stokes problem :

$$- \nu\Delta\vec{u} + \sum_{i=1}^{2} u_i \frac{\partial\vec{u}}{\partial x_i} + \text{grad } p = \vec{f} \text{ in } \Omega$$

$$\text{div } \vec{u} = 0 \text{ in } \Omega$$

$$\vec{u} = \vec{g} \text{ on } \Gamma .$$

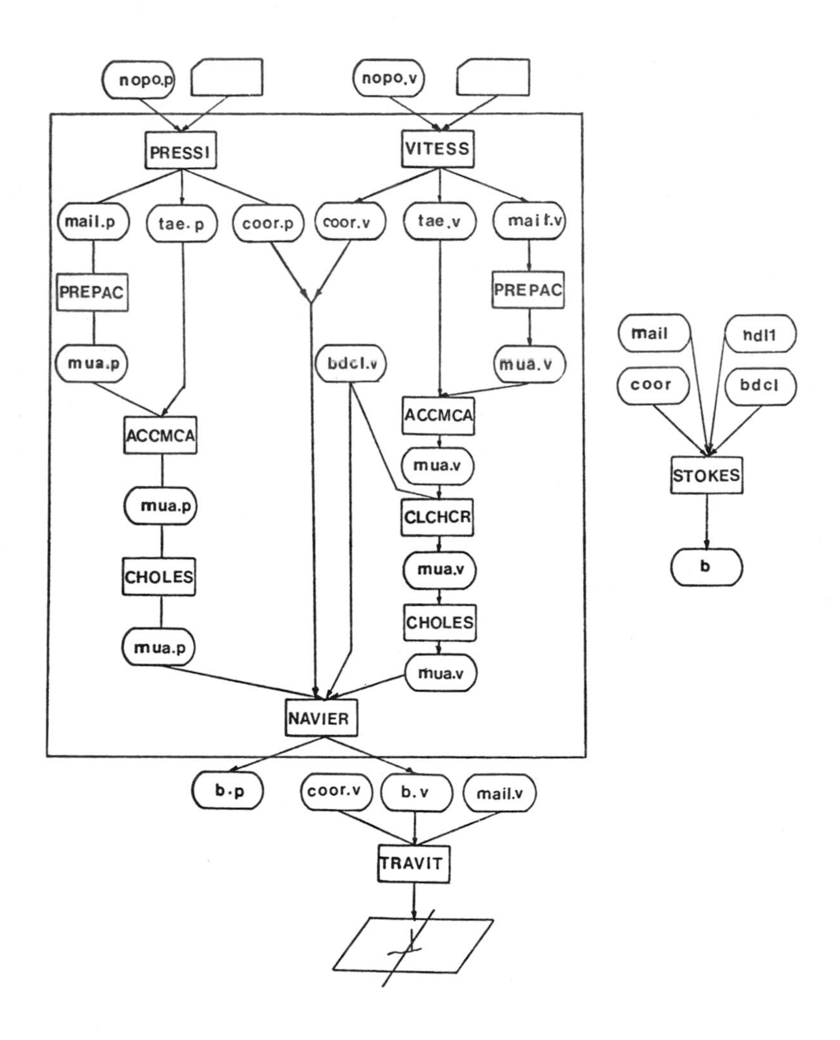

Two modules solving the bidimensional NAVIER-STOKES equations

-n if it resides on secondary storage unit N, in direct access mode.

OUTPUT DATA STRUCTURE (S.D.S.) : 0 no need for the data structure to be saved on secondary storage

n if the data structure saved is to be performed in sequential access on unit n

-n if the data structure saved is to be performed in direct access on unit n.

. if needed, provides input data subroutines such as a surface equation, an effort depending of the coordinates...
. assigns libraries in order to help solving unsatisfied external references
. if needed, segments the program into several overlays
. assigns secondary storage to support the data structures
. supplies data on punched cards in the format described in the user's manual.

Therefore anyone who knowns FORTRAN and knows how to assign files is able to set up a sequence of existing modules and run it in one or several passes, automatically saving on secondary storage relevant intermediate results as well as final results.

3.2. Programming standards for modules

The calling sequence for each module is easy to code and provides all the necessary saved options, though these advantages are paid by a sensible programming burden. It is advisable that users get acquainted with the following :
. the statements of a portable FORTRAN subset
. using machine specific extension of the standard FORTRAN subset in special purpose subroutines - character string use, direct access on secondary storage - cf.(43) -
. portable dynamic storage of arrays
. the DATA STRUCTURE notion, management of its physical support
. the notion of MODULE and its structure :
   - to set the input data structure's arrays into memory
   - to define the output data structure's arrays
   - to perform the numeric algorithm - as standard FORTRAN subroutine
   - to save the ouptut data structures.

3.3. Standards for inserting a module into a library

Each new subroutine must be inserted into a particular library.
The user must ensure that its name, the names of newly introduced, dynamically managed arrays do not already exist , equally that the software standards were respected. Subroutines checking these standards will soon be implemented.

2.2 Modules under development

Immediate aims :

- generation of the nodes non vertices of a mesh described by the data structure NOPO.
- display of a tridimensional mesh
- renumbering method for nodes and elements having NOPO as input data structure - cf. (37) -
- finite element programming in thermodynamics and elasticity - cf. (21) to (24) -
- QR method for solving $([K] - \lambda[M])\ \{x\} = 0$
- solving of a potential problem using an integral singular equation.

In the next future :

- solving the linear systems :
  - Cholesky with hyper matrix storage - cf. (38) -
  - relaxation method
  - conjugate gradient method (39)
- other methods for solving the Navier-Stokes equations (40)
- computing step by step transient processes
- computation of egeinvectors of $[K]\{x\} + [B]\{\dot{x}\} + [M]\{\ddot{x}\} = 0$
- expanding LANEF call capabilities to any module.

In the long run

- programming of the method of substructures
- development of a programming language as described in (41)
- development of conversional, interactive versions of the modules.

3. - SOFTWARE STANDARDS OF THE CLUB

Three programming levels are to be distinguished - cf. (42) -

3.1. Standard call procedures

The purpose is to link up modules within the library. The user must write a main program in FORTRAN which :

- defines a working array M
- initializes calling parameters for each module - considered as a subroutine - according to the data structure support number conventions :

  INPUT DATA STRUCTURE (S.D.E.) : 0 if in main memory

  n if it resides on secondary storage unit N, in sequential access mode

Moroover the library choice is important due to some peculiar features of linkage editors. Then the FORTRAN source code is compiled, edited and stored on secondary storage within its library.
Changing from a computer installation to another leads to record the non-portable subroutines, gathered into a specific library. At the moment there are such subroutines.

4. - CONCLUSION

The present modules lay the foundation of the Finite Elements programming. The calling sequences are designed to be as simple as possible. Programming standards are significantly improving FORTRAN's characteristics. Machine dependent coding is reduced and isolated so that changes are easy to make. This experiment of building from the cooperation of research laboratories or industry a comprehensive software on finite elements is started in a satisfactory manner.

BIBLIOGRAPHY

(1) J.H. ARGYRIS, ASKA, Institut für Statik und Dynamik der Luft, University of Stuttgart.

(2) GENESIS Limited, Pennine House, Lemington Street, Loughborough, LEICS, England.

(3) MARC ORDISOR, Marc Analysis Research Corporation, Providence, Rhode Island.

(4) R.M. Mac NEAL, Some organizational aspects of NASTRAN, Nuclear engineering and design (29), 1974, 254-265, North-Holland Publ.

(5) D. EGELAND, P.O. ARALDSEN, SESAM 69 a general purpose finite element method program, Computers and Structures, Vol. 4, 41-68, Pergamon Press, 1974.

(6) Dossier d'adhésion au Club MODULEF, Réglement.

(7) A. PERRONNET, Gestion dynamique des tableaux - Structures de données - Modules, IRIA, Dec. 1976.

(8) W.A. COOK, Body oriented coordinates for generating... Int. Jour. for numerical methods, Vol. 8, 27-43 (1974)

(9) A. GEORGE, Computer implementation of the finite element method, STAN CS-71, 208 Feb. 1974, Stanford University, Comp. Science Department.

(10) A. PERRONNET, Module ALTERN maillage d'un alternateur - MODULEF - IRIA, Octobre 1977.

(11) O. KOUTCHMY, P. JOLY, A. PERRONNET, Les modules de maillage de MEFISTO-LAN 189 - Université Paris 6 (76005)

(12) CUTHILL-MacKEE, Reducing the bandwith of sparse symmetric matrices. Proc. 24th Nat. Conf. Assoc. Comp. Mech. (1969) ACM Publication.

(13) B.M. IRONS, A frontal solution program for finite element analysis, Int. Jour.

num. methods Engineering, 2 (1970), 5-32.

(14) M. ITALIANI, G. SERAZZI, Un apprecio generalizzato per alcuni linguaggi, "Probelm Oriented", LAN Pavia, 1976.

(15) P. LAUG, le module MEFNOP, Club MODULEF, IRIA, Octobre 1977.

(16) P. LAUG, le module TN2D3D, Club MODULEF, IRIA, Octobre 1977.

(17) PIERROT, A. PERRONNET, VAZEILLES, Génération de maillages tridimensionnels par subdivision d'éléments grossiers, Module COLIBR, MODULEF, Octobre 1977.

(18) A. PERRONNET, Description des structures de données, MODULEF-IRIA, Octobre 1977.

(19) D. LEROY, le module COMACO, Club MODULEF, IRIA, Octobre 1977.

(20) M. ROTTMAN, le module THELAS, Club MODULEF, IRIA, Octobre 1977.

(21) L.D. MARINI, Implementation of hybrid finite element methods and associated numerical problems, LAN Pavia, 1976, N° 136.

(22) J.M. THOMAS, Thèse doctorat d'Etat, Analyse des méthodes d'éléments finis hybrides et mixtes, 27 Mai 1977, LAN 189, Paris, Université Pierre et Marie Curie.

(23) P.A. RAVIART, P.A. THOMAS, Primal Hybrid Finite Element Methods... Math. of Comput. 31, (Avril 77).
A mixed finite element method for 2nd Symposium, Rome (dec. 1975).

(24) BERNADOU-DUCATEL, Méthodes conformes d'éléments finis pour des problèmes elliptiques du 4ème ordre avec intégration numérique, IRIA, Mai 76.

(25) B. MERCIER-M. JEAN-FRANCOIS-B. DOURBAIX, Résolution numérique du problème de la flexion élasto-plastique des plaques minces e.f. mixte HERMANN-JOHNSON, Ecole Polytechnique, Palaiseau.

(26) P.G. CIARLET-P.A. RAVIART, The combined effect of curved boundaries and numerical integration, Mathematical Foundations of the f.e.m. (Aziz ed.) 409-474 Academic Press, New York (1972).

(27) P. JOLY, Le module POLBAS, LAN 189 Université Pièrre et Marie Curie, Paris, (1977)

(28) M. LEGENDRE, Le module FORM, LAN 189, Université Pierre et Marie Curie, Paris, (1977)

(29) D. KOUTCHMY, Le module CRBNIV, LAN 189, Université Pierre et Marie Curie, Paris, (1977)

(30) M. WILLIAMSON, Hidden line plotting program Algorithm 420., Communication of the ACM (Feb. 1972), Vol. 15, N° 2.

(31) O. KOUTCHMY, Le Module ZEGFXY, LAN 189, Université Pierre et Marie Curie, Paris, (1977).

(32) K.J. BATHE, Solution methods for large generalized eigenvalue problems, Berkeley, California, Nov. 1971.

(33) BENAZET-BROCHARD-GOURDIN, Analyse dynamique et module par la m.e.f., LAN 189, Paris, (Jan. 1977)

(34) A. MARROCCO, Etude numérique du champ magnétique dans un alternateur, Comp. methods in Applied Mech. and Eng., 3, (1974), North-Holland, p. 55-85.
A. MARROCCO, Le module MAGNE, Club MODULEF, IRIA, Octobre 1977.

(35) MANTEL, POIRIER, PERIAUX, Le module NAVIER AMD/BA, 78, Quai Carnot, 92 St Cloud.

(36) M. BERCOVIER, Thèse de doctorat d'Etat, Univ. de Rouen (1976).

(37) G. AKHRAS-G. DHATT, An automatic node relabelling scheme for minimizing a matrix Int. Journal for numerical methods in engineering, Vol. 10, 787-797 (1976).

(38) G.V. FUCHS-ROY-SCHREM, Hyper matrix solution of large sets of symmetric... Comp. Meth. in Applied Mech. Eng. (72), 197-216.

(39) O. AXELSSON, A class of iterative methods for finite element equations 75.03R Chalmers University, Göteborg.

(40) P.A. RAVIART, A mixed finite element method for the Navier Stokes equations, LAN 189, Paris, 77010.

(41) E. SCHREM, From program systems to programming systems for finite element analysis, ISD report N° 214, Stuttgart, 1977.

(42) A. PERRONNET, Normes d'appel et de programmation des modules du club MODULEF IRIA, Octobre 1977.

(43) P. LAUG, Gestion des fichiers en accès direct (MEFDIR), Manuel d'utilisation Club MODULEF, IRIA, Juin 1977.

FINITE ELEMENT ANALYSIS FOR STRESS INTENSITY FACTORS

Yoshiyuki Yamamoto

Dept. of Naval Architecture, Univ. of Tokyo
Bunkyo-ku, Tokyo 113 Japan

Yoichi Sumi

Dept. of Naval Architecture, Yokohama National University
Hodogaya-ku, Yokohama 240 Japan

SUMMARY The finite element method is developed on the basis of the concept of superposition of solutions for stress intensity factor analyses. The method is discussed in a general form for boundary value problems with geometric singularities, and numerical results are shown for two- and three-dimensional problems.

INTRODUCTION

Linear fracture mechanics has been developed on the basis of the concept of the stress intensity factor, which is analyzed by solving the boundary value problem for a cracked body.[1-3] Boundary value problems of this kind cannot be treated by a conventional numerical method. Motz[4] and Woods[5] attacked this problem for a two-dimensional potential problem with the finite difference method. Recently, this problem has been developed mainly in relation to the stress intensity factor analysis by various methods including the finite element method.

The stress intensity factor is determined from singular terms in the complete solution, and therefore, its analysis can be performed with a conventional numerical method if results obtained are manipulated with the aid of physical and mathematical relations.[6-9] This approach is convenient for practical applications, but it may give less accurate results.

For accurate analyses of the stress intensity factor, the finite element method seems to be promising because of its versatility. Characteristics of singular solutions at the crack front can be investigated analytically, and singular solutions obtained can be used as shape functions for elements near the crack front. The shape function can also be expressed as a sum of such singular terms and conventional functions, and this method have been applied mainly to two-dimensional problems. The present authors proposed a method on the basis of the concept of superposition of analytical and finite-element solutions for the stress concentration problems,[10] and have applied it to stress intensity factor analyses.[11-16] In the present paper, this method will be discussed in a general form, and some numerical results of the stress intensity factor analysis will be presented.

BOUNDARY VALUE PROBLEM FOR A CRACKED DOMAIN

Consider a self-adjoint boundary value problem defined in a two-dimensional domain D by a set of linear equations

$$-L[u] = b \qquad \text{in } D \tag{1}$$

$$B[u] = f \qquad \text{on } \partial D_1 \tag{2}$$

$$u = v \qquad \text{on } \partial D_2 \tag{3}$$

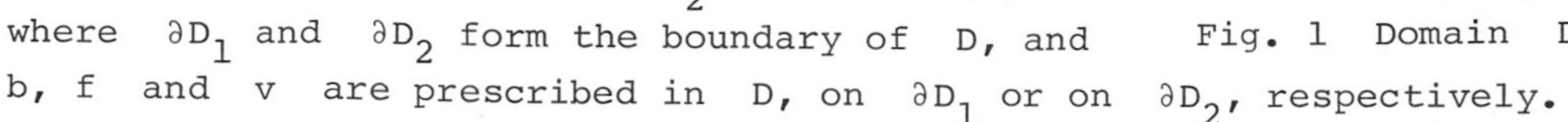

Fig. 1 Domain D

where ∂D_1 and ∂D_2 form the boundary of D, and b, f and v are prescribed in D, on ∂D_1 or on ∂D_2, respectively.

First, singularities caused by b and/or f will be considered by assuming that ∂D_1 has no singularities. The singularity of solutions is determined by b and/or f, and can be expressed by a singular function u_s. Assume the solution of the original boundary value problem as

$$u = u_s + u_r \tag{4}$$

where u_r is the residual solution determined by

$$\left.\begin{aligned} -L[u_r] &= b + L[u_s] && \text{in } D \\ B[u_r] &= f - B[u_s] && \text{on } \partial D_1 \\ u_r &= v - u_s && \text{on } \partial D_2 \end{aligned}\right\} \tag{5}$$

By the assumption, the right-hand sides of the first two equations of (5) have no singularities; therefore, u_r is regular everywhere in D, and can be determined by a conventional numerical method by disregarding singularities caused by b and/or f. This idea can be applied to the case where ∂D_1 has singular points.

Before entering discussion of singularities caused by geometry, the boundary value problem will be rewritten in a weak form. Let δu be an arbitrary regular function satisfying the condition

$$\delta u = 0 \qquad \text{on } \partial D_2 \tag{6}$$

Since the boundary value problem under consideration is self-adjoint, the inner product of $-L[u]$ and δu in D can be expressed as

$$(-L[u],\ \delta u) = \langle u,\ \delta u\rangle - ((B[u],\ \delta u)) \tag{7}$$

Here $\langle u, \delta u\rangle$ is a symmetric bilinear form such that

$$\langle u, \delta u\rangle = \langle \delta u, u\rangle\ ,\quad \langle u_1 + u_2, \delta u\rangle = \langle u_1, \delta u\rangle + \langle u_2, \delta u\rangle$$

In general, it can be written in the form

$$\langle u,\ \delta u\rangle = \int_D \varepsilon[u]^t\, C\, \varepsilon[\delta u]\, dD \tag{8}$$

where ε is a linear vector operator, t indicates the transpose of an appropriate quantity, and C is a symmetric tensor function. In most of physical problems, C is non-negative definite. The corresponding quadratic form is twice as much as an energy function, and its variation can be written as

$$\delta\langle u, u\rangle = 2\langle u, \delta u\rangle$$

The inner product $((B[u], \delta u))$ is defined on ∂D_1. Introducing Eqs.(1)

and (2) into Eq. (7) leads to the Galerkin-type equation

$$\langle u, \delta u\rangle - (b, \delta u) - ((f, \delta u)) = 0 \tag{9}$$

or equivalently the variational problem

$$1/2.\langle u, u\rangle - (b, u) - ((f, u)) = \text{minimum} \tag{10}$$

with the subsidiary condition given by Eq. (3). The finite element method can be formulated on the basis of Eq. (9) or (10). It should be noticed that the solution of the boundary value problem defined by Eqs. (1)--(3) is uniquely determined within trivial solutions, which make $\varepsilon[u]$ vanish everywhere. Singularities of the solution u is characterized by $\varepsilon[u]$ or $C\varepsilon[u]$.

Assume that the boundary ∂D_1 has a singular point P, although b and f are regular everywhere. Let [D] be a neighborhood of P, and let $[\partial D_1]$ be the corresponding part of ∂D_1. Introduce the polar coordinates (r, θ) with the origin P. Then the characteristics of the singular solution u_s can be investigated by introducing $u_s = r^\lambda g(\theta)$ into the homogeneous equations

$$\left.\begin{aligned} -L[u_s] &= 0 && \text{in } [D] \\ B[u_s] &= 0 && \text{on } [\partial D_1] \end{aligned}\right\} \tag{11}$$

Then λ and $g(\theta)$ can be determined as an eigenvalue and the corresponding eigenfunction. The solution u_s of Eq. (11) can be expressed in the form

$$u_s = \sum_{j=1}^{N} m_j u_{sj} + \sum_{j=N+1}^{\infty} m_j u_j \tag{12}$$

where u_{sj}, $(j = 1,\ldots, N)$, are singular solutions, and u_j, $(j = N+1,\ldots\ldots)$, are nonsingular. A regular solution u_a of the non-homogeneous equations

$$\left.\begin{aligned} -L[u_a] &= b && \text{in } [D] \\ B[u_a] &= f && \text{on } [\partial D_1] \end{aligned}\right\} \tag{13}$$

can, in general, be obtained. It will be assumed that the domain of definition for u_{sj}'s and u_a can be extended throughout D. Moreover, the singular solutions u_{sj}'s can also be determined as analytic solutions of similar problems.[11]

Now determine the solutions u_o and u_{rj}, $(j = 1,\ldots, N)$, by solving numerically the boundary value problems

$$\left.\begin{aligned} -L[u_o] &= b + L[u_a] && \text{in } D \\ B[u_o] &= f - B[u_a] && \text{on } \partial D_1 \\ u_o &= v - u_a && \text{on } \partial D_2 \end{aligned}\right\} \tag{14}$$

and

$$\left.\begin{aligned} -L[u_{rj}] &= L[u_{sj}] && \text{in } D \\ B[u_{rj}] &= -B[u_{sj}] && \text{on } \partial D_1 \\ u_{rj} &= -u_{sj} && \text{on } \partial D_2 \end{aligned}\right\} j = 1,\ldots, N \tag{15}$$

These calculations can be performed by any of conventional numerical methods, such as the conventional finite element method. The solution of the original boundary value problem can then be written in the form

$$u = u_a + u_o + \sum_{j=1}^{N} m_j (u_{sj} + u_{rj}) \tag{16}$$

where m_j's are coefficients to be determined. Assuming that m_j, u_o, and u_{rj} are unknown quantities, and introducing u from Eq. (16) into Eq. (10) lead to the following equation and the natural conditions equivalent to Eqs. (14) and (15):

$$\begin{aligned} &\langle u_a + u_o + \sum_{k=1}^{N} m_k (u_{sk} + u_{rk}),\ u_{sj} + u_{rj} \rangle \\ &- (b,\ u_{sj} + u_{rj}) - ((f,\ u_{sj} + u_{rj})) = 0 \qquad j = 1, \ldots, N \end{aligned} \tag{17}$$

Since u_j, $(j = N+1, \ldots)$, are regular, $\varepsilon[u_j]$ is finite at the point P, and that for j of higher order vanishes at this point. Let $(C\varepsilon[u])_\mu$ be the μ-component of $C\varepsilon[u]$. By the analysis of Eq. (11), it can be seen that

$$(C\varepsilon[u_j])_{\mu_k} = 0 \qquad \text{at P; } k = 1, \ldots, M;\ j = N+1, \ldots \tag{18}$$

with respect to the coordinates chosen properly, like the x-and y-axes in Fig. 1. If M is larger than or equal to N, the unknown m_j, $(j = 1, \ldots, N)$, can be determined by using Eq. (18):

$$(C\varepsilon[u_a + u_o + \sum_{j=1}^{N} m_j u_{rj}])_{\mu_K} = 0 \qquad \text{at P; } k = 1, \ldots, N$$

or

$$\sum_{j=1}^{N} m_j (C\varepsilon[u_{rj}])_{\mu_k} + (C\varepsilon[u_a + u_o])_{\mu_k} = 0 \quad \text{at P; } k = 1, \ldots, N \tag{19}$$

Eq. (19) can be regarded as a kind of collocation equations.

From the numerical point of view, evaluations of $C\varepsilon[u]$ at P may yield significant numerical errors; it is then preferable to evaluate it at some collocation points other than P in the neighborhood [D]. In this case, the relation corresponding to Eq. (18) does not hold exactly. This difficulty can be resolved by moving some of the regular terms u_j's to u_{sj}'s; the newly added ones of u_{sj}'s should be such that

$$C\varepsilon[u_{sj}] = O(r^{\alpha}) \qquad 0 < \alpha < 1 \tag{20}$$

Then Eq. (18) holds approximately in the neighborhood of P; errors expected are, at most, of order r.

Eq. (15) can be rewritten as

$$\left.\begin{aligned} -L[u_s + u_r] &= 0 && \text{in } D \\ B[u_s + u_r] &= 0 && \text{on } \partial D_1 \\ u_s + u_r &= 0 && \text{on } \partial D_2 \end{aligned}\right\} \tag{21}$$

This means that $u_{sj} + u_{rj}$ vanishes within trivial solutions if Eq.(15) is solved rigorously. In reality, u_{rj} is a solution by a conventional numerical method; therefore, there holds the relation

$$u_{sj} + u_{rj} = 0 \qquad \text{other than} \quad [D];\ j = 1,\dots, N \tag{22}$$

although u_{rj} fails to be a good approximation of $-u_{sj}$ in [D]. It follows from Eq. (8) that

$$\left.\begin{aligned} b + L[u_a] &= 0 \qquad \text{in} \quad [D] \\ f - B[u_a] &= 0 \qquad \text{on} \quad [\partial D_1] \end{aligned}\right\}$$

Then there holds the approximate relation

$$(b + L[u_a],\ u_{sj} + u_{rj}) + ((f - B[u_a],\ u_{sj} + u_{rj})) = 0 \tag{23}$$

Introducing $u = \sum_{k=1}^{N} m_k u_{sk} + u_a$ and $\delta u = u_{sj} + u_{rj}$ into the equality given by Eq. (7) leads to

$$\begin{aligned}(-L[u_a],\ u_{sj} + u_{rj}) = &< \sum_{k=1}^{N} m_k u_{sk} + u_a,\ u_{sj} + u_{rj} > \\ &- ((B[u_a],\ u_{sj} + u_{rj}))\end{aligned} \tag{24}$$

From Eqs. (17) and (24), it follows that

$$\begin{aligned}&< \sum_{k=1}^{N} m_k u_{rk} + u_o,\ u_{sj} + u_{rj} > - (b + L[u_a],\ u_{sj} + u_{rj}) \\ &\qquad - ((f - B[u_a],\ u_{sj} + u_{rj})) = 0 \\ &\qquad j = 1,\dots, N\end{aligned} \tag{25}$$

Now, m_j's can be determined from Eq. (25), assuming that the solution u is given in the form of (16) with the finite element solutions u_o and u_{rj}'s and introducing Eq. (10). Pian-Tong-Luk[17] and Yagawa et al.[18] proposed the formulation, which is equivalent to Eqs. (14), (15) and (25), on the basis of the variational problem given by Eq. (10). In this formulation, fairly complicated integrals should be evaluated for determining the inner products. With the use of Eq. (23), Eq. (25) can be simplified in the form

$$\sum_{k=1}^{N} m_k < u_{rk},\ u_{sj} + u_{rj} > + < u_o,\ u_{sj} + u_{rj} > = 0 \qquad j = 1,\dots, N \tag{26}$$

Eq. (26) can be rewritten as

$$\int_{[D]} \left(\sum_{k=1}^{N} m_k \varepsilon[u_{rk}] + \varepsilon[u_o]\right)^t C \varepsilon[u_{sj} + u_{rj}]\, dD = 0 \qquad j = 1,\dots, N \tag{27}$$

By virtue of Eq. (21), $\varepsilon[u_{sj} + u_{rj}]$ is significant in a small neighborhood [D*], ($\subset$ [D]), of P, and can be evaluated approximately by

$$\varepsilon[u_{sj} + u_{rj}] = \varepsilon[u_{sj}] \qquad \text{in} \quad [D^*]$$

It then follows that

$$\int_{[D]} \left(\sum_{k=1}^{N} m_k \varepsilon[u_{rk}] + \varepsilon[u_o]\right)^t C \varepsilon[u_{sj}]\, dD = 0 \qquad j = 1,\dots, N \tag{28}$$

or approximately

$$\sum_{A_i \subset [D^*]} \left(\sum_{k=1}^{N} m_k \varepsilon[u_{sk}] + \varepsilon[u_o]\right)^t C \varepsilon[u_{sj}]\, A_i = 0 \qquad j = 1,\dots, N \tag{29}$$

where A_i is the area of the finite element i in [D*], and its coefficients are evaluated at the representative point of the respective element. In general, $\varepsilon[u_{sj}]$, $(j = 1,\dots, N)$, have significant compoments, say $(\varepsilon[u_{sj}])_{\mu_k}$, $(k=1,\ldots,M)$, and then Eq. (29) can be simplified as

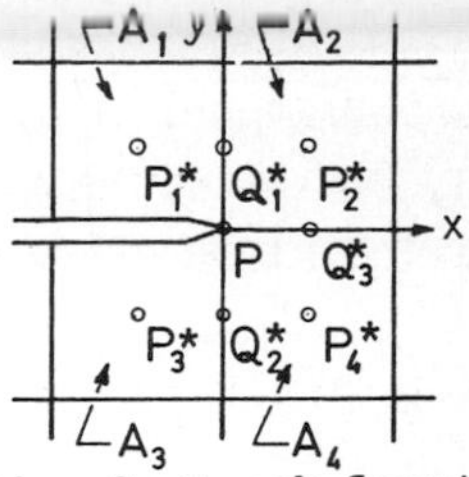

$$(\sum_{j=1}^{N} m_j C\varepsilon[u_{rj}] + C\varepsilon[u_o])_{\mu_k} = 0 \quad k=1,\ldots,M \tag{30}$$

Evaluating this equation at a certain number of collocation points P_i^*'s chosen properly in $[D^*]$ gives m_j's. Eq. (19) and Eq. (30) are equivalent from the practical point of view.

Fig. 2 Crack front

The linear theory of elasticity is described by the following equations corresponding to Eqs. (1)--(3):

$$\left.\begin{array}{lll} -\nabla\sigma = b & \text{in} & D \\ \sigma n = f & \text{on} & \partial D_1 \\ u = v & \text{on} & \partial D_2 \end{array}\right\} \tag{31}$$

where σ and u are the stress or displacement, n is the outward unit normal vector, and b, f and v are the prescribed force or displacement. The stress is expressed in terms of the displacement by

$$\sigma = C\varepsilon[u] \tag{32}$$

where ε is a linear operator expressing the strain caused by u, and C is the elastic rigidity tensor. Therefore, $1/2.\langle u, u\rangle$ is the elastic strain energy caused by the displacement u, and Eq. (10) corresponds to the principle of minimum strain energy.

TWO-DIMENSIONAL PROBLEMS

According to Williams[19], the stress σ is expressed near a crack front in the following form with reference to the coordinate shown in Fig. 2:

$$\sigma = \sum_{j=1,2} K_j\sigma_{sj} + \sum_{j=3,4} m_j\sigma_{sj} + \sigma_a + \sigma^* + O(r) \tag{33}$$

Here the singular terms of stresses are given by

$$\left.\begin{array}{l} \left.\begin{array}{l}\sigma_{s1x}\\ \sigma_{s1y}\end{array}\right\} = \cos(\theta/2)\,[1 \mp \sin(\theta/2)\sin(3\theta/2)]/\sqrt{2\pi r} \\ \tau_{s1} = \sin(\theta/2)\cos(\theta/2)\cos(3\theta/2)/\sqrt{2\pi r} \end{array}\right\} \tag{34}$$

$$\left.\begin{array}{l} \sigma_{s2x} = -\sin(\theta/2)\,[2+\cos(\theta/2)\cos(3\theta/2)]/\sqrt{2\pi r} \\ \sigma_{s2y} = \sin(\theta/2)\cos(\theta/2)\cos(3\theta/2)/\sqrt{2\pi r} \\ \tau_{s2} = \cos(\theta/2)\,[1-\sin(\theta/2)\sin(3\theta/2)]/\sqrt{2\pi r} \end{array}\right\} \tag{35}$$

$$\left.\begin{array}{l} \left.\begin{array}{l}\sigma_{sjx}\\ \sigma_{sjy}\end{array}\right\} = \sqrt{r}\,[\frac{5}{4}(2+F_j\{\begin{array}{c}\sin^2\theta\\ \cos^2\theta\end{array} \pm \frac{3}{2}\sin(2\theta)\frac{dF_j}{d\theta} + \begin{array}{c}\cos^2\theta\\ \sin^2\theta\end{array}\}\frac{d^2F_j}{d\theta} \\ \tau_{sj} = \sqrt{r}\,[-\frac{5}{8}\sin(2\theta)F_j - \frac{3}{2}\cos(2\theta)\frac{dF_j}{d\theta} + \frac{1}{2}\sin(2\theta)\frac{d^2F_j}{d\theta} \end{array}\right\} \quad j=3,4 \tag{36}$$

where

$$F_3 = \cos\frac{\theta}{2} - \frac{1}{5}\sin\frac{5\theta}{2}, \qquad F_4 = \sin\frac{5\theta}{2} - \sin\frac{\theta}{2} \tag{37}$$

The coefficients K_1 and K_2 are called the stress intensity factors of the opening or shear mode, respectively. As for the constant stress σ^*, the components σ^*_y and τ^* vanish. The stresses σ_{s3} and σ_{s4} are of order $\sqrt{r}$, and cannot be expressed by conventional shape functions; therefore, it will be treated as a singular term as described before.

In view of the property of σ^*, Eq. (19) can be written in the following form:

$$\left.\begin{aligned} \sum_{j=1,2} K_j\sigma_{rjy} + \sum_{j=3,4} m_j\sigma_{rjy} + \sigma_{ay} + \sigma_{oy} = 0 \\ \sum_{j=1,2} K_j\tau_{rj} + \sum_{j=3,4} m_j\tau_{rj} + \tau_a + \tau_o = 0 \end{aligned}\right\} \quad (38)$$

These relations should be evaluated at more than two collocation points, which are so chosen that the resulting equations for determining K_j's and m_j's are independent of each other. For example, the points P_1^* (or P_3^*) and P_2^* (or P_4^*) in Fig. 2 can be used for this purpose. Manipulating the equations evaluated at P_1^*, P_2^*, P_3^*, and P_4^* gives better results. The points Q_1^*(or Q_2^*) and Q_3^* can be used for the same purpose.

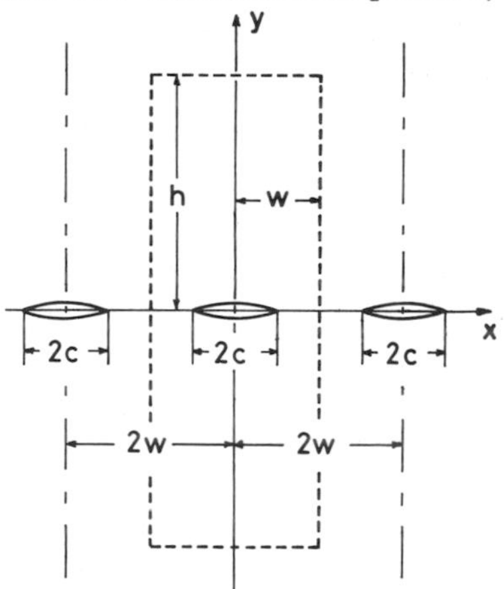

Fig. 3 Collinear Cracks

As an example, consider a stretched infinite plate with collinear cracks of length 2c shown in Fig. 3. This problem was solved analytically by Westergaard with Airy's stress function.[20] In view of periodicity of stress distributions, one bay shown in Fig. 3 will be analyzed with the meshes shown in Fig. 4; the domain to be analyzed is subjected to the edge

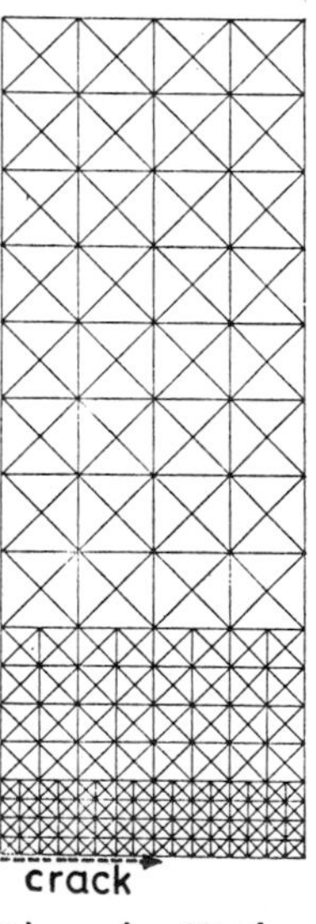

Fig. 4 Mesh Subdivision

Table 1. Infinite Plate with Collinear Cracks

Solution	σ_{s1} & σ_{s2}	σ_{s1}	Exact solution
c/w	$K_1/[\sigma_o/\sqrt{(\pi c)}]$		
1/16	1.0357	1.1189	1.0016
2/16	1.0161	1.0607	1.0065
3/16	1.0197	1.0501	1.0149
4/16	1.0300	1.0534	1.0270
6/16	1.0663	1.0830	1.0651
8/16	1.1288	1.1430	1.1284
10/16	1.2346	1.2489	1.2347
12/16	1.4309	1.4496	1.4315
13/16	1.6061	1.6319	1.6072
14/16	1.9111	1.9567	1.9125
15/16	2.5989	2.7498	2.6258

stress derived from Westergaard's solution. In view of symmetry of stress distributions, the stress intensity factor K_2 vanishes, and the stress intensity factor K_1 can be determined by the first equation of (38) alone. Results obtained are shown in Table 1 together with analytical solutions, and it shows good agreement. It can be seen that the mesh size near the crack front should be less than $c/3$ to obtain satisfactory accuracy, and that the second term u_{s2} in singular displacements becomes significant only when c/w is very small or nearly equal to unity, where $2w$ is the pitch of cracks.

CURVED CRACK FRONT LINES

Three-dimensional singular solutions will be investigated for a curved crack shown in Fig. 5. Without loss of generality, it can be assumed that the crack surface lies in a plane Π near a point P on the crack front line, which forms a closed smooth curve. Introduce the coordinates (x, y, z) with P as the origin, such that the x, y-plane lies in the plane Π and the y-axis is normal to it. Let ξ be the arc length measured along the crack front line from the point P. Then the locally orthogonal coordinates (r, θ, ξ) can be used conveniently for the analysis near the point P, and the relation between the two coordinate systems is given approximately by

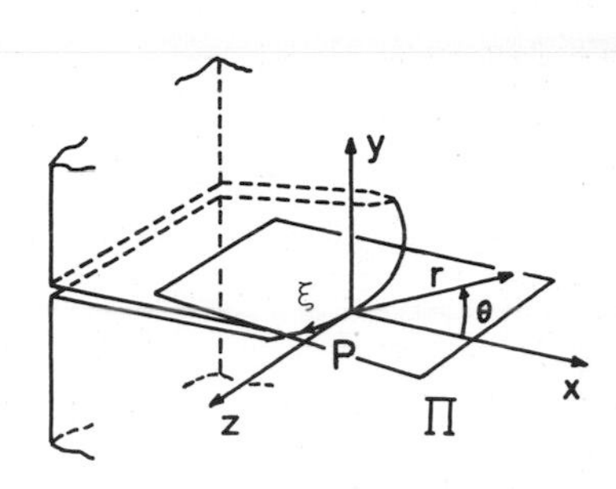

Fig. 5 Crack Front Line

$$\left.\begin{aligned} x &= r\cos\theta - 0.5(\xi/a)^2(\pm a + r\cos\theta) \\ y &= r\sin\theta \\ z &= \xi \pm r(\xi/a)\cos \end{aligned}\right\} \qquad (39)$$

Here a, $(a > 0)$, is the radius of curvature of the crack front line at the point P, and the center of curvature locates at the point $(\mp a, 0, 0)$.

Displacements can be expressed in terms of the displacement potential h_j, $(j = 1, 2, 3)$:[21]

$$\left.\begin{aligned} u_x &= (1-2\nu)\frac{\partial h_1}{\partial x} + y\frac{\partial^2 h_1}{\partial x \partial y} - 2(1-\nu)\frac{\partial h_2}{\partial y} + y\frac{\partial}{\partial x}\left(\frac{\partial h_2}{\partial x} + \frac{\partial h_3}{\partial z}\right) \\ u_y &= -2(1-\nu)\frac{\partial h_1}{\partial y} + y\frac{\partial^2 h_1}{\partial y^2} - (1-2\nu)\left(\frac{\partial h_2}{\partial x} + \frac{\partial h_3}{\partial z}\right) + y\frac{\partial}{\partial y}\left(\frac{\partial h_2}{\partial x} + \frac{\partial h_3}{\partial z}\right) \\ u_z &= (1-2\nu)\frac{\partial h_1}{\partial z} + y\frac{\partial^2 h_1}{\partial y \partial z} - 2(1-\nu)\frac{\partial h_3}{\partial y} + y\frac{\partial}{\partial z}\left(\frac{\partial h_2}{\partial x} + \frac{\partial h_3}{\partial z}\right) \end{aligned}\right\} \qquad (40)$$

Singular terms derived from h_1, h_2 and h_3 correspond to the openning, in-plane-shear, or out-of-plane-shear mode, respectively. In the absence of body forces, h_j's satisfy the condition

$$\Delta h_j = 0 \qquad j = 1, 2, 3 \tag{41}$$

A typical solution of this equation is given by

$$h_{j\lambda} = H_{j\lambda}(\xi)\, r^{\lambda} \left[\begin{matrix}\sin\\ \cos\end{matrix}\lambda\theta \mp (r/4a)\begin{matrix}\sin\\ \cos\end{matrix}(\lambda - 1)\theta + O(r^2)\right] \tag{42}$$

where $H_{j\lambda}(\xi)$ is a function of ξ. The first two terms on the right-hand side of Eq. (42) satisfy the equation of equilibrium asymptotically as $r \to 0$. As for λ, the finiteness of strain energy requires that it must be larger than unity, and it will be determined by the stress-free condition on the crack surfaces:

$$\left.\frac{\partial^2 h_j}{\partial \theta^2}\right|_{\text{at}\ \ \theta = \pm\pi} = 0 \tag{43}$$

which gives

$$\lambda = 2, 3, \ldots \qquad \text{or} \qquad \lambda = 3/2, 5/2, \ldots \tag{44}$$

Stresses of order $1/\sqrt{r}$ or $\sqrt{r}$ are derived from the following expression, respectively:

$$\left.\begin{aligned} h_{j(3/2)} &= r^{3/2}[\cos(3\theta/2) \mp (r/4a)\cos(\theta/2)] \\ h_{j(5/2)} &= r^{5/2}\cos(5\theta/2) \end{aligned}\right\} \tag{45}$$

The second term in $h_{j(3/2)}$ expresses the effect of curvature of the crack front line; these expressions become identical with two-dimensional ones as a tends to infinity. These terms can be used as singular solutions for the present method.

As a simple example, consider a long round bar with a circumferential crack shown in Fig. 6. The cylindrical coordinates (ρ, ϕ, z) will be introduced so that the z-axis coincides with the axis of the bar. When the bar is twisted by the moment T, the stress around the crack front is expressed in the form

$$\begin{Bmatrix}\tau_{\rho\phi}\\ \tau_{\phi z}\end{Bmatrix} = \frac{K_3}{\sqrt{2\pi r}}\begin{Bmatrix}-\sin\theta/2\\ \cos\theta/2\end{Bmatrix} + O(1) \tag{46}$$

where K_3 is the stress intensity factor for the out-of-plane-shear mode. The singular terms of the displacement can be derived from $h_{3(3/2)}$ and $h_{3(5/2)}$ given by Eq. (45):

$$\left.\begin{aligned} u_{\phi 1} &= (1-\nu)r^{1/2}[3\sin\theta/2 \\ &\quad - (r/2a)(\sin 3\theta/2 + (3/2)\sin\theta/2)] \\ u_{\phi 2} &= (1-\nu)r^{3/2}\sin 3\theta/2 \end{aligned}\right\} \tag{47}$$

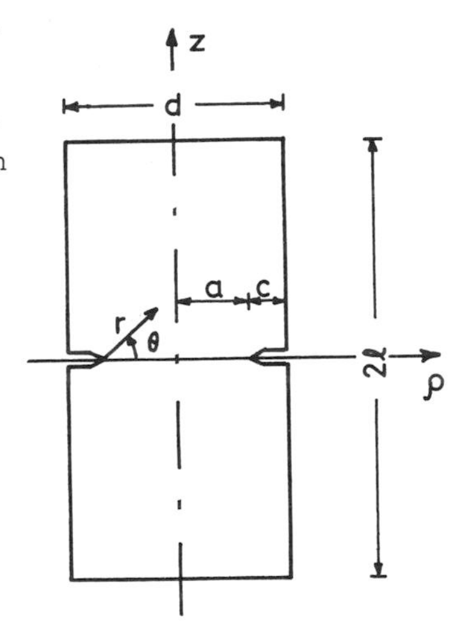

Fig. 6 Cracked Round Bar (analyzed with meshes shown in Fig. 5)

Numerical results obtained are shown in Fig. 7, in which K_3 is expressed in a non-dimensional form. An asymptotic approximation formula was proposed by Benthem and Koiter,[22] and its validity is proved by the present results. Moreover, the second analytical expression becomes significant only in the case of small crack length.

When the bar is stretched by the force P, the stress components σ_ρ, σ_z, $\tau_{\rho z}$ around the crack front are similar to those for the two-dimensional case given by Eq. (33). The values K_1 obtained with the use of $h_{1(3/2)}$ are shown also in Fig. 7 with the asymptotic approximation by Benthem and Koiter.[22]

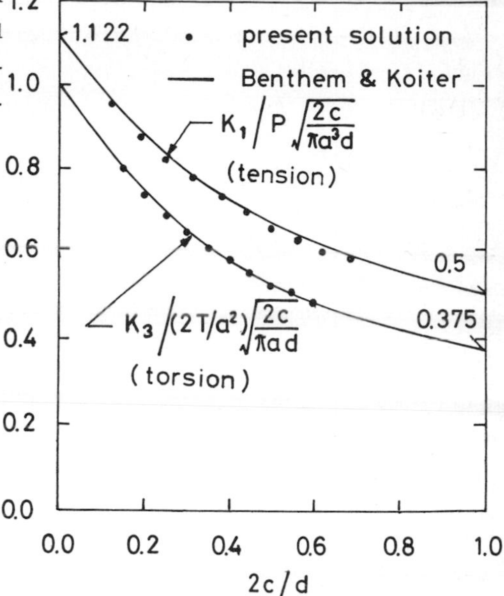

Fig. 7 Stress intensity factor for a bar with a circumferential crack

THROUGH CRACK IN A PLATE

Singular solutions near the terminating point of the crack front line were obtained by Benthem[23,24] and Bazant-Estenssoro[25] by the eigenvalue problem method. It is possible to introduce such singular solutions into analyses; in reality, however, the layer which is affected by this term is very thin, and therefore, this effect will be disregarded at the first stage of the following analysis.

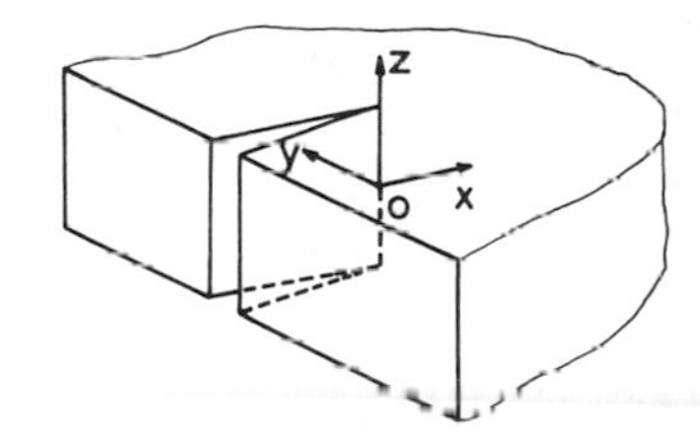

Fig. 8 Through crack

Consider a plate with a through crack, which is subjected to in-plane forces, and is shown in Fig. 8. It will be assumed that the crack front line penetrates through the thickness normally to the plate surfaces. In this case, deformations lack the out-of-plane mode, and therefore, it may be assumed that the singular terms for u_ξ or u_z vanish. The singular terms for (u_x, u_y) given in the preceding section are the same as two-dimensional ones. The corresponding stresses for the plane strain state may not satisfy the equilibrium conditions in the z-direction, but this difficulty can be resolved by introducing the induced body force and surface traction properly. According to this line of thought, Sih[26] investigated the distribution of the stress intensity factor of the present problem, and it is found that the stress intensity factor vanishes at the plate surfaces. This fact is confirmed by Benthem and Bazant-Estenssoro. The stress around the crack front line is expressed in the form

$$\left.\begin{array}{l} \left.\begin{array}{l}\sigma_x\\ \sigma_y\end{array}\right\} = K_1(z)(2\pi r)^{-1/2}\cos\theta/2\,(1 \mp \sin\theta/2\sin 3\theta/2) + O(1)\\ \sigma_z = 2K_1(z)(2\pi r)^{-1/2}\cos\theta/2 + O(1)\\ \tau_{xy} = K_1(z)\nu(2\pi r)^{-1/2}\sin\theta/2\cos\theta/2\cos 3\theta/2 + O(1)\\ \tau_{yz} = 0\\ \tau_{zx} = 0 \end{array}\right\} \quad (48)$$

Let t be the thickness of the plate. The plate thickness will be subdivided into $2(N-1)$ layers by the planes

$$z = z_1 (= 0),\quad \pm z_1, \ldots,\quad \pm z_N (= \pm t/2) \quad (49)$$

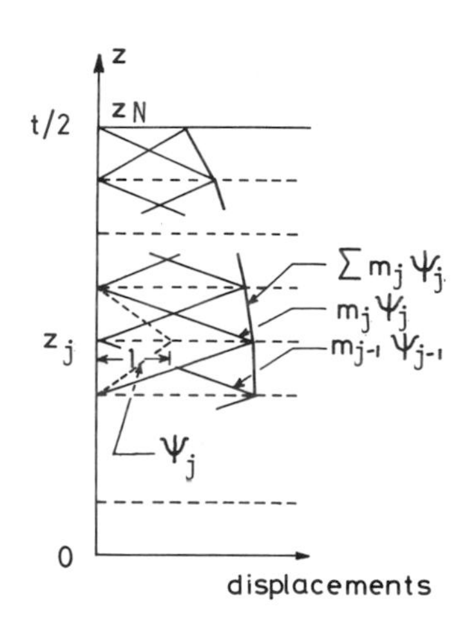

Fig. 9 Shape function

Displacements and stresses will be assumed to be linear or constant in the z-direction in each layer. Therefore, the analytical expression of the displacement which gives stress singularities is expressed by a linear combination of the following terms in the upper half of the thickness:

$$\left.\begin{array}{l} \left.\begin{array}{l}u_{sxj}\\ u_{syj}\end{array}\right\} = (1/G)\,r^{1/2}\,(1 - 2\nu + \sin^2\theta/2)\\ \qquad \cdot\left\{\begin{array}{l}\cos\\ \sin\end{array}\right.\theta/2 \cdot \psi_j(z)\\ u_{szj} = 0 \end{array}\right\} \quad (50)$$

where G is the shear modulus, and ψ_j is defined by

$$\begin{array}{lll} \psi_j(z) = (z - z_{j-1})/(z_j - z_{j-1}) & z_{j-1} < z \leqq z_j & \\ \quad = (z_{j+1} - z)/(z_{j+1} - z_j) & z_j < z < z_{j+1} & (51)\\ \quad = 0 & \text{otherwise} & \end{array}$$

The corresponding expressions for the stress become as

$$\left.\begin{array}{l} \left.\begin{array}{l}\sigma_{sxj}\\ \sigma_{syj}\end{array}\right\} = r^{-1/2}\cos\theta/2\,(1 \mp \sin\theta/2\sin 3\theta/2)\,\psi_j(z)\\ \sigma_{szj} = 2\nu r^{-1/2}\cos\theta/2 \cdot \psi_j(z)\\ \tau_{sxyj} = r^{-1/2}\sin\theta/2\cos\theta/2\cos 3\theta/2 \cdot \psi_j(z)\\ \left.\begin{array}{l}\tau_{syzj}\\ \tau_{szxj}\end{array}\right\} = r^{1/2}\left\{\begin{array}{l}\sin\\ \cos\end{array}\right.\theta/2\,(1 - 2\nu + \sin^2\theta/2)\dfrac{d\psi_j(z)}{dz} \end{array}\right\} \quad (52)$$

These stresses are in equilibrium with the induced body force $b_{sj} = -\nabla\sigma_{sj}$

and the surface traction $\{\pm\tau_{szxN}, \pm\tau_{syzN}, \pm\sigma_{szN}\}$ on the plate surface $z = \pm t/2$. The induced body force is expressed by

$$\left.\begin{aligned} \left.\begin{matrix} b_{sxj} \\ b_{syj} \end{matrix}\right\} &= -\sqrt{r}\left\{\begin{matrix}\cos\\ \sin\end{matrix}\theta/2\,(1 - 2\nu + \sin^2\theta/2)\frac{d^2\psi_j(z)}{dz^2}\right. \\ b_{szj} &= -1/\sqrt{r}\cdot\cos\theta/2\,\frac{d\psi_j(z)}{dz} \end{aligned}\right\} \tag{53}$$

The x-and y-components of the induced body force vanish almost everywhere, and they can be regarded as the internal surface forces. It is obvious that the stress intensity factor is expressed in the form

$$K_1 = \sqrt{\pi}\sum_{j=1}^{N} m_j\psi_j(z) \tag{54}$$

In the above discussion, the second singular terms are disregarded, but it is easy to introduce them.

As an example, consider a slit infinite plate stretched by the uniform stress $\bar{\sigma}$ at infinity (cf. Fig. 10). The small domain around the slit will be analyzed under the edge stress determined by the two dimensional theory of Westergaard.[20] According to the general procedure, the finite element solutions u_o, u_{rj}; σ_o, σ_{rj} can be determined. Calculations were performed with the aid of the general purpose program SOLID SAP with eight-point-three-dimensional elements.[27] The induced body forces and surface tractions can be expressed as nodal forces by numerical integrations. The mesh subdivision used is shown in Fig. 10, and one eighth of the domain is analyzed by the symmetry of geometry. Results obtained are shown in Figs. 11 and 12. If Poisson's ratio vanishes, the three-dimensional stress intensity factor coincides with the two-dimensional value $\bar{\sigma}\sqrt{\pi c}$ as can be seen in Fig. 12. The three-dimensional effects become significant with the increase of Poisson's ratio.

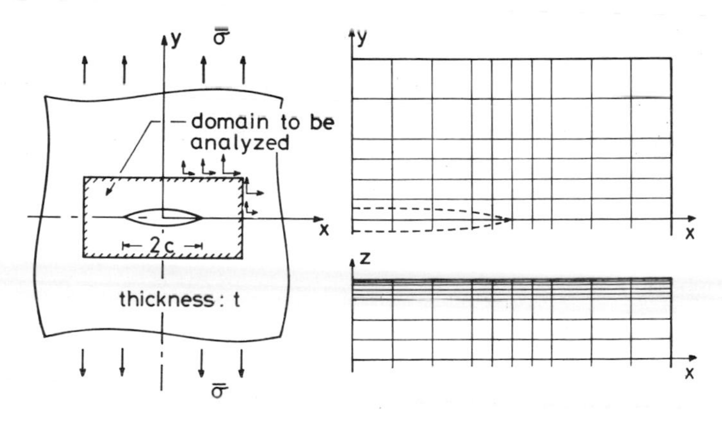

Fig. 10 Infinite plate with a slit

Detailed distributions of the stress intensity factor near the plate surface can be investigated with the aid of Benthem's theory.[23, 24] According to his theory, the stress intensity factor near the plate surface can be represented assymptotically in the following form:

$$K_1 = a_1\zeta^{0.0477} + a_2\zeta^{0.718} + a_3\zeta^{1.181} + \dots \qquad \nu = 0.3$$
$$\qquad\qquad\qquad\qquad\qquad\qquad\qquad\qquad\qquad (55)$$
$$= a_1\zeta^{0.125} + a_2\zeta^{0.573} + a_3\zeta^{1.347} + \dots \qquad \nu = 0.45$$

where $a_1, a_2, a_3, \dots$ are coefficients to be determined, and ζ is given by

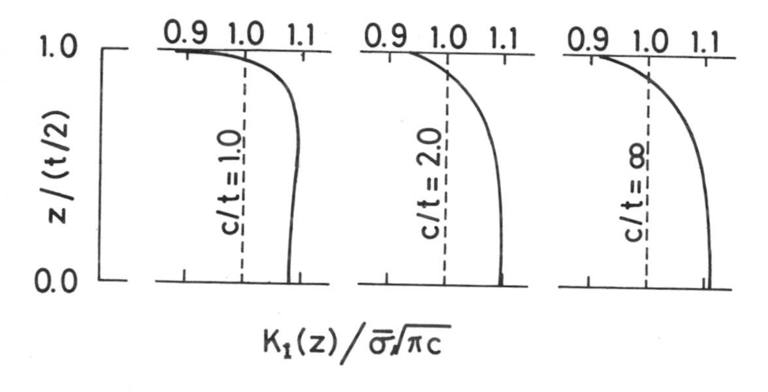

Fig. 11 Stress intensity factor in a slit plate (effect of slit width, $\nu = 0.45$)

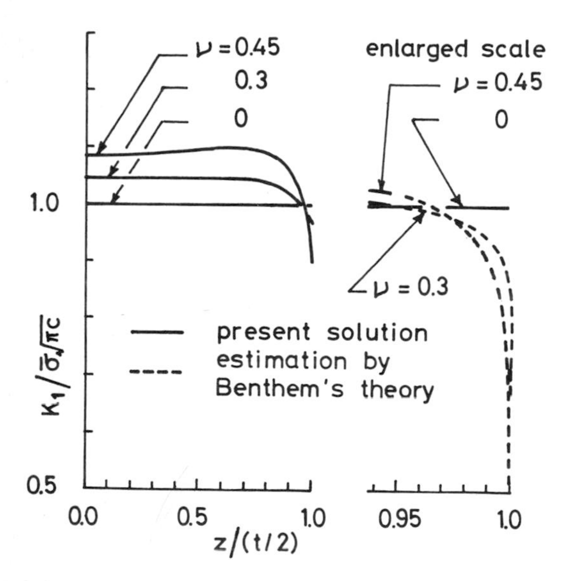

Fig. 12 Stress intensity factor in a slit plate (effect of Poisson's ratio, $c/t = 1.0$)

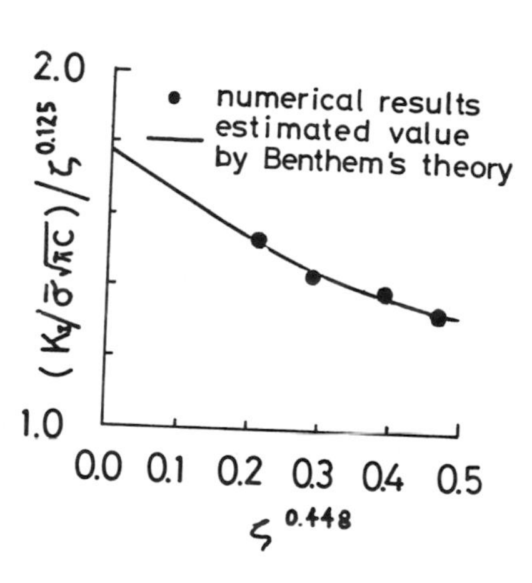

Fig. 13 Determination of coefficients ($c/t = 1.0$)

$$\zeta = 1 - 2z/t$$

Introducing numerical results for the stress intensity factor K_1, and applying the least square method yield the coefficients a_1, a_2 and a_3. Here numerical value of K_1 at the plate surface is disregarded. The resulting curve corresponding to Eq. (55) is almost linear as shown in Fig. 13, which suggests validity of Benthem's theory. Detailed distributions of the stress intensity factor thus obtained are shown in Fig. 12, which shows a significant drop of the stress intensity factor in a very thin layer near the plate surface as discussed before.

CONCLUSION

In the present paper, a numerical method for solving a self-adjoint boundary value problem with geometrical singularities is developed on the basis of the concept of superposition. It is applied to the stress intensity factor analysis of two-and three-dimensional problems, and it gives satisfactory results. Calculations can be performed by a conventional finite element method without complicated integrations.

REFERENCES

1. Irwin, G.R., Fracture. Handbuch der Physik, Bd. 4, Springer, 1958, 551-590.
2. Paris, P.C., & Sih, G.C., Stress Analysis of Cracks. Proc. of the Symposium on Fracture Toughness and its Applications, STP 381, ASTM, 1965, 38-80.
3. Sih, G.C.(ed.), Mechanics of Fracture, Vol. 1, Noordhoff, 1973.
4. Motz, H., The Treatment of Singularities of Partial Differential Equations by Relaxation Methods. Quart. Appl. Mech., Vol. 6, 1953, 371-377.
5. Woods, L.C., The Relaxation Treatment of Singular Points in Poisson's Equation, Q. J. Mech. & Appl. Math., Vol. 6, 1953, 163-185.
6. Wilson, W.K., Finite Element Methods for Elastic Bodies Containing Cracks, in Ref. 3, 484-515.
7. Dixon, J. R., & Pook, L.P., Stress Intensity Factors Calculated Generally by the Finite Element Technique. Nature, Vol.224, 1969, 166-167.
8. Rice, J.R., A Path Independent Integral and the Approximate Analysis of Stress Concentration by Notches and Cracks, J. Appl. Mech., Vol. 35, 1968, 379-386.
9. Miyata, H., Shida, S.,& Kusumoto, S., The Simple Method of Evaluation of Stress Intensity Factor Using the Finite Element Method. Proc. 1974 Symp. on Mech. Behavior of Materials, Vol. 1, Soc. of Material Sci., Japan, 1974, 63-80.
10. Yamamoto, Y., Finite Element Approaches with the Aid of Analytical Solutions, Recent Advances on Matrix Meth. of Struct. Analysis and Design, Univ. of Alabama Press, 1971, 85-103.
11. Yamamoto, Y., & Tokuda, N., Determination of Stress Intensity Factors in Cracked Plates by the Finite Element Method. Int. J. Numer. Meth. in Engng., Vol. 6, 1973, 427-439.
12. Yamamoto, Y., Tokuda, N., & Sumi, Y., Finite Element Treatment of Singularities of Boundary Value Problems and its Application to Analysis of Stress Intensity Factors. Theory and Practice in Finite Element Structural Analysis, Univ. of Tokyo Press, 1973, 75-90.

13. Yamamoto, Y., & Sumi, Y., Stress Intensity Factors of a Twisted-Round-Bar with a Circumferential Crack. Int. J. of Fracture, Vol. 10, 1974, 269-271.
14. Yamamoto, Y., Sumi, Y, & Ao, K., Stress Intensity Factors of Cracks Emanating from Semi-Elliptical Side Notches in Plates. Int. J. of Fracture, Vol. 10, 1974, 593-595.
15. Yamamoto, Y., & Ao, K., Stress Intensity Factors of Cracks in Notched Bend Specimens. Int. J. of Fracture, Vol. 12, 1976, 495-498.
16. Yamamoto, Y., & Sumi, Y., Stress Intensity Factors for Three - Dimensional Cracks. to be published in Int. J. of Fracture.
17. Pian, T.H.H., Tong, P., & Luk, C.H., Elastic Crack Analysis by a Finite Element Hybrid Method. Proc. Third Conf. on Matrix Method of Struct. Mech., AFFDL-TR-71-160, 1971, 690-711.
18. Yagawa, G., Nishioka, T., Ando, Y., & Ogura, N., The Finite-Element Calculation of Stress Intensity Factors Using Superposition, Computational Fracture Mechanics, ASME Special Publicstion, 1975, 21-34.
19. Williams, M.L., On the Stress Distribution at the Base of a Stationary Crack. J. Appl. Mech., Vol. 24, 1957, 109-114.
20. Westergaard, H.M., Bearing Pressures and Cracks. J. Appl. Mech., Vol. 6, 1939, A49-A53.
21. Sih, G.C., & Liebowitz, H., Mathematical Theory of Brittle Fracture. Fracture: An Advanced Treatise, Vo. 2, Academic Press, 1968, 67-190.
22. Benthem, J. P., & Koiter, W.T., Asymptotic Approximations to Crack Problems. in Ref. 3, 131-178.
23. Benthem, J.P., Three-Dimensional State of Stress at the Vertex of a Quarter-Infinite Crack in a Half Space. Rep. No. 563, Laboratory of Engng. Mechanics, Delft Univ. of Technology, 1975.
24. Benthem, J.P., State of Stress at the Vertex of a Quarter-Infinite Crack in a Half-Space. Int. J. of Solids & Struct., Vol. 13, 1977, 479-497.
25. Bazant, Z.P., & Estenssoro, L.F., General Numerical Method for Three-Dimensional Singularities in Cracked or Notched Elastic Solid, Fracture 1977, Vol. 3, ICF4, Waterloo, Canada, 1977, 371-385.
26. Sih, G.C., A Review of the Three-Dimensional Stress Problem for a Cracked Plate. Int. J. of Fracture Mech., Vol. 7, 1971, 39-385.
27. Wilson, E.L., SOLID SAP, UCSESM 71-19, Struct. Engng. Laboratory, Univ. of California, Berkeley, California, 1971.

THE SOMMERFELD (RADIATION) CONDITION ON INFINITE DOMAINS AND ITS MODELLING IN NUMERICAL PROCEDURES

O.C. Zienkiewicz*, D.W. Kelly**, and P. Bettess**

1. INTRODUCTION

In wave and transient dynamic problems it is insufficient to prescribe zero displacement (or velocity) conditions at large distances from the disturbance as is the custom in problems of statics. It is in addition necessary to ensure that only outgoing waves are present so that all energy is radiated outward. Without such a condition an infinite energy buildup can occur in the domain in the absence of internal damping and spurious results will be obtained. Indeed the energy radiation condition must and will ensure that under all circumstances the infinite system is damped. The conditions imposed are known as those if finiteness and radiation (viz p.216 Fung 1965).

The problem is of obvious practical importance in the computation of

- . acoustic (pressure wave) and electromagnetic wave propagation problems
- . surface wave studies of harbours or obstacles in an infinite sea

and . foundation-structure interaction in earthquakes and related studies.

In 'discrete' numerical computation such as conducted by finite difference/finite element methods it is usually necessary to curtail the infinite boundary and replace it by another placed at a finite distance from the disturbance - and here it will in general be impossible to impose both the zero displacement and the radiation condition. The latter will always be found to necessarily take precedence.

---

* Professor of Civil Engineering, University College of Swansea

** Lecturer, Department of Civil Engineering, University College of Swansea

An exception to the above is provided by boundary type solutions which can satisfy both the infinity conditions exactly. Here the trial functions are solutions to the governing equations on infinite or semi-infinite domains and the imposition of near field boundaries is the only discretization required.

After discussing the preliminaries in the next sections we shall present in this paper the possibilities open for dealing numerically with

- periodic response of linear problems with nonhomogeneous inner regions
- transient response of problems where a possible nonlinearity of the inner region exists.

An extensive bibliography has also been appended to the paper.

2. BASIC FORMULATION. WAVE EQUATIONS AND INFINITY CONDITIONS. SIMPLE BOUNDARY DAMPERS

2.1. The scalar wave problem - compression waves

A typical example here is that of acoustic and surface waves in fluids and many features of the general problems are conveniently illustrated on this. Similar treatment is relevant to electro-magnetic waves etc.

The general equation governing can be written in terms of the unknown pressure and is

$$\frac{\partial^2 p}{\partial t^2} + \underset{\sim}{\nabla}^T c_p^2 \underset{\sim}{\nabla} p + \alpha \frac{\partial p}{\partial t} = 0 \tag{1}$$

where c_p is the local compression wave velocity given by

$$c_p = \sqrt{\frac{K}{\rho}} \qquad \text{(K-bulk modulus, } \rho \text{ -density)}$$

α - stands for a distributed damping parameter.

In general we shall seek either periodic or transient solutions of (1) with appropriate boundary conditions, for example

$$\frac{\partial p}{\partial n} = 0 \qquad \text{on rigid boundaries} \qquad (1a)$$

and $p = 0$ or

$$\frac{\partial p}{\partial z} + \frac{1}{g}\frac{\partial^2 p}{\partial t^2} = 0 \qquad \text{on the free surfaces.} \qquad (1b)$$

In exterior domains a homogeneous situation is admitted and here equation (1) reduces (with neglect of damping) to *

$$\frac{\partial^2 p}{\partial t^2} + c_p^2 \nabla^2 p = 0 \qquad (2)$$

Equation (2) has a general wave solution of the form

$$p = F(x - c_p t) + f(x + c_p t) \qquad (3)$$

which represents plane compression waves in $\pm x$ directions. In equation (3) x can be any arbitrary direction in the fluid.

If we consider a boundary normal to the direction of propagation of such waves and we want to ensure that only waves in direction of positive x exist (i.e. $f = 0$) then we can write

$$\frac{\partial p}{\partial x} = F'$$

and

$$\frac{\partial p}{\partial t} = -c_p F'$$

where dashes denote differentiation with respect to ($x - c_p t$). Eliminating F' we have (Zienkiewicz and Newton 1969)

$$\frac{\partial p}{\partial x} + \frac{1}{c_p}\frac{\partial p}{\partial t} = 0 \qquad (4)$$

as a necessary radiation condition .

If periodic solutions only are considered, i.e. of the form

$$p = \bar{p} \exp(\iota \omega t)$$

* The situation where damping is present in the exterior region is of some interest but little attention has yet been given to it.

then equation (4) can be written as

$$\frac{\partial \bar{p}}{\partial x} + \frac{i\omega}{c_p} \bar{p} = 0 \qquad (5)$$

In cylindrical or spherical coordinates the wave expression equivalent to (3) can be written down as

$$p = \frac{1}{r^{(n-1)/2}} \left(F(r - c_p t) + f(r + c_p t) \right)$$

where n = 2 for the cylindrical problem (approximately only) and n = 3 for spherical geometry (exactly).

Identical considerations to those used in the derivation of equation (4) will give

$$\frac{\partial p}{\partial r} + \frac{n-1}{2} \frac{1}{r} p = -\frac{1}{c_p} \frac{\partial p}{\partial t} \qquad (4a)$$

Indeed if we take the case of plane waves associated with a single dimension we obtain expression (4) again, hence the above is perfectly general.

Clearly at infinite distances from the disturbance we can equate x with the radial distance r, and the condition of <u>radiation</u> (5) is applicable for all dimensions. To simultaneously ensure <u>finiteness</u>, Sommerfeld (1912) and Rellich (1943) write equation (5) in the form

$$\lim_{r \to \infty} r^m \left(\frac{\partial p}{\partial r} + \frac{i\omega}{c_p} p \right) = 0 \qquad (6)$$

Clearly condition (6) combines an additional requirement to that of equation (4) which we still use in finite analysis.

As the natural boundary condition in problems governed by a weak form of equation (1) concerns $\frac{\partial p}{\partial n}$, the imposition of equation (4) or (4a) is easily incorporated on any <u>finite boundary</u> and presents the most obvious form of a boundary damper which prevents any reflection of waves hitting such a boundary in a normal direction.

Lysmer and Kuhlemeyer (1969) investigated the problem of waves impinging on such a boundary obliquely in the context of elastic media. It is shown that almost all reflection is prevented for angles of incidence of $\pm 60^{\circ}$ from the normal which accounts for the efficiency of such filters.

The use of the more efficient condition, (4a), was first suggested by Newton (1973). In this form dampers are augmented with a 'spring' boundary condition.

As the velocity of compression waves is constant, in accoustic problems we can simply incorporate the transmitting boundary into the finite element approximation at a finite distance from the disturbance and the only approximation made is that concerning normality of wave incidence - a condition certainly met at large distances. Boundaries placed too close will show certain degrees of unwarranted reflection.

Such boundary dampers are not dependent on frequency and therefore can be used in transient and periodic analysis to eliminate compression waves as in Fig. 1a. It is easy to verify that in a purely one-dimensional problem the whole semi-infinite domain can be exactly represented by a single damper.

It is appropriate at this stage to indicate the mode of reflections which occur at boundaries. For a plane wave impinging normally on a boundary, the reflected wave will have the same form as the incident wave so we can write, for example,

$$p = A \exp(i\omega/c)(x+ct) + B \exp(i\omega/c)(x-ct)$$

where A and B are constants. For a rigid boundary at $x=0$, $\frac{\partial p}{\partial x} = 0$ gives

$$A = -B$$

For a free surface at $x=0$, $p=0$ gives

$$A = B$$

The first reflection reinforces the oncoming wave in the vicinity of the boundary at $x=0$, the second cancels it.

2.2. The scalar wave problem - surface waves

If in the fluid problem discussed in the previous section a free surface exists, an additional 'wave' is possible due to the boundary conditions given by equation (1b). Now a general wave in the form of equation (3) is not admitted but for sinusoidal shapes we can write a solution valid in the horizontal direction x only as

$$p = f(kz) \sin k(x \pm c_s t) \tag{7a}$$

where z is measured in the vertical upward direction,

$$k = \omega/c_s = 2\pi/\lambda \qquad (\lambda\text{-wavelength}), \tag{7b}$$

$$c_s^2 = g/k \tanh(kH) \qquad \text{(H-fluid depth)}, \tag{7c}$$

and

$$f(kz) = \cosh(k(z+H))/\cosh(kH) . \tag{7d}$$

Once again the radiation condition equivalent of equation (4) can be re-established by a precisely identical argument as that used before and no reflection of waves given by equation (7) will occur if

$$\frac{\partial p}{\partial x} + \frac{1}{c_s}\frac{\partial p}{\partial t} = 0 \qquad (8)$$

This condition is

(a) in apparent contradiction to that needed to eliminate compression waves and can not be applied simultaneously at least at finite distances,

and (b) is valid only for elimination of sinusoidal waves of a given k (or frequency) as c_s depends on this (7c).

The difficulties presented above make it impracticable to use such 'dampers' in transient analysis due to frequency dependence and indeed seem to make the transmitting (radiation) boundary impossible when both compression and free surface waves occur simultaneously. However if we note that

(a) surface wave effects are only of importance close to the surface and decay rapidly with depth (viz equation (7d)),

(b) their decay with distance from the disturbance is less rapid than that of pressure waves.

and finally,

(c) that pressure waves near the surface and parallel to it must be small,

then the possibility of combined use of both sets of dampers in the manner indicated in Fig. 1b appears practicable. The depth at which transition from one to the other type of boundary condition is made is obviously dependent on decay of the surface wave with depth and therefore in frequency.

For the problem of surface waves alone impinging on a cylinder the radiation condition (8) can be directly applied. Results for the surface elevation on the cylinder for the finite element mesh shown in Fig. 2 have been compared to the analytical solution.

The performance of plane and cylindrical dampers is shown to be very effective viz a viz the more complex alternatives, the cylindrical dampers yielding particularly good results.

2.3. Waves in elastic media

The equations of motion for homogeneous isotropic media are given by

$$\rho \frac{\partial^2 u_j}{\partial t^2} = (\lambda + G) \frac{\partial \bar{\varepsilon}}{\partial x_j} + G \nabla^2 u_{j,i} \qquad j = 1,3 \tag{9}$$

where

$$G = \frac{E}{2(1+\nu)}$$

$$\lambda = \frac{\nu E}{(1+\nu)(1-2\nu)}$$

$\varepsilon_x + \varepsilon_y + \varepsilon_z = \bar{\varepsilon}_{ii}$ is the cubic dilation or volumetric expansion,

and the constants E and ν are Young's modulus and Poisson's ratio respectively.

Two solutions can be found for the equations of motion. For example, differentiating (9) with respect to x_j, $j = 1,3$ and adding the three solutions together leads to the equation.

$$\rho \frac{\partial^2 \bar{\varepsilon}}{\partial t^2} = (\lambda + 2G) \nabla^2 \bar{\varepsilon}$$

or

$$\frac{\partial^2 \bar{\varepsilon}}{\partial t^2} = c_p^2 \nabla^2 \bar{\varepsilon} \tag{10}$$

Equation (10) is exactly the form of the wave equation where

$$c_p = \sqrt{\frac{\lambda + 2G}{\rho}} = \sqrt{\frac{K}{\rho}} \tag{11}$$

The dilatation $\bar{\varepsilon}$ therefore propagates with a velocity c_p. This solution therefore describes the propagation of a dilatational wave.

A second solution gives a pure rotational (equivolume or distortional) wave which propagates with a velocity

$$c_s = \sqrt{\frac{G}{\rho}} \tag{12}$$

Application to the arguments leading to the radiation condition of equation (4) can once again be made to equation (9) for dilatational waves or similarly for distortional waves. Some additional calculation is necessary however to identify the natural boundary conditions which are given now in terms of tractions i.e. not directly in terms of normal gradients. It can be readily shown that these give simply

$$t_n - \frac{K}{c_p}\frac{\partial u_n}{\partial t} = 0 \tag{13}$$

and

$$t_s - \frac{G}{c_s}\frac{\partial u_s}{\partial t} = 0$$

as a relation between tangential(s) and normal(n) directions of tractions and displacements, thus yielding now realistic 'dampers'. Using (11) and (12) we can write

$$t_n = \rho c_p \dot{u}_n$$

and

$$t_s = \rho c_s \dot{u}_s$$

Obviously the damping constants required are ρc_p and ρc_s per unit length respectively.

Once more if no free surface occurs a straight forward physical approximation to radiation conditions is available which is valid for transient as well as periodic problems. If an elastic half-space is considered (i.e. when a free surface exists) a third solution becomes possible corresponding to motion confined to a zone near the boundary of the half-space. This wave was first studied by Lord Rayleigh (1885) and later described in detail by Lamb (1904). If z is the direction normal to the half-space surface then the displacements for a Rayleigh wave travelling with a velocity c_R in the positive x-direction are given by

$$u_x = f(kz) \sin k(x - c_R t)$$

and

$$u_z = g(kz) \cos k(x - c_R t) \tag{14}$$

where k is the wave number defined by

$$k = \omega / c_R$$

The velocity of the Rayleigh wave is $c_R = c_s / \eta$. (15)

The value of η and the functions $f(kz)$ and $g(kz)$ vary with Poisson's ratio. For $\nu = \frac{1}{4}$ they are (Lysmer and Kuhlemeyer 1969)

$$\eta = 1.08766$$

$$f(kz) = A\left[\exp(-0.8475\,kz) - 0.5773\,\exp(-0.3933\,kz)\right]$$

and $$g(kz) = A\left[-0.8475\,\exp(-0.8475\,kz) + 1.4679\,\exp(-0.3933\,kz)\right]$$

in which A is a constant

The wave field generated by a circular footing therefore has the three components shown in Fig. 3. All of the waves encounter an increasingly larger volume as they travel outward and therefore are damped geometrically. In three dimensions the amplitude of the dilatational and distortional waves decrease as $\frac{1}{r}$ (r is the distance from the input source) except along the surface of the half-space where the amplitude decreases as $\frac{1}{r^2}$. The amplitude of the Rayleigh wave decreases as $\frac{1}{\sqrt{r}}$.

For a vertically oscillating, uniformly distributed, circular energy source on the surface of a homogeneous, isotropic, elastic half-space, Miller and Pursey (1955) determined the distribution of total input energy among the three elastic waves to be 67 per cent Rayleigh wave, 26 per cent distortional (shear) wave and 7 per cent compression wave. The fact that two-thirds of the total input energy is transmitted away from a vertically oscillating footing by the Rayleigh wave and that the Rayleigh wave decays much more slowly with distance than the body waves indicate that the Rayleigh wave is of primary concern for foundations on or near the surface of the earth.

As Rayleigh waves move parallel to the surface an appropriate set of radiation boundary conditions relating tractions and displacements on vertical surfaces can be found. The computation is quite complex to achieve such relations and a basically more involved result than that in fluids is obtained (Lysmer and Kuhlemeyer 1969) in the form

$$t_n \equiv t_x \quad \text{and} \quad t_x + \frac{\alpha}{c_R}\frac{\partial u_x}{\partial t} = 0$$

$$t_s \equiv t_y \quad \text{and} \quad t_y + \frac{\beta}{c_R}\frac{\partial u_s}{\partial t} = 0 \qquad \ldots (16)$$

now $\alpha = \alpha(kz)$ and $\beta = \beta(kz)$

and although c_R, the Rayleigh wave velocity, is constant the 'damper' values are frequency dependent and therefore not available for transient problems. The strength of the dampers must now vary with depth (see Fig.4) and careful matching is necessary. In Fig. 4 we have used equations (11), (12) and (14) to write

$$t_n = a(kz)\,\rho c_p\,\dot{u}_x$$

and

$$t_s = b(kz)\,\rho c_s\,\dot{u}_s$$

The damping constants a and b have then been plotted.

All the remarks made previously in the context of the possibility of eliminating simultaneously the reflections of surface and stress waves apply. Once again the use of the dampers eliminating Rayleigh waves near the surface is important (Lysmer and Kuhlemeyer 1969) as these are predominant. The analogy is not complete with fluids as there a very simple constant strength 'damper' suffices (and it is indeed tempting to speculate on the existance of pure gravity waves in solids.)

It should be mentioned here also that the elastic domain is very often layered. Multiple total reflections can then occur in the upper layer and lead to a horizontally polarized shear wave called a Love wave. In addition the reflection of elastic waves is not as simple as that indicated in Section 2.1. as mode conversion occurs at the boundary. Rayleigh wave reflection, for example, also leads to the generation of body waves necessary to satisfy the stress conditions at the boundary. We will however not discuss these matters further here and interested readers should consult appropriate texts in the bibliography.

It is apparent however that procedures for representing the exterior, non-reflecting domain, other than the 'damper' radiation conditions, may be preferable (or at least equally efficient). It is with such possibilities that the next part of the paper is concerned.

3. NUMERICAL TREATMENT OF EXTERIOR DOMAINS BASED ON ANALYTICAL SOLUTIONS

3.1. General

We have noted in the previous section that the direct imposition of the radiation condition by means of a curtailed boundary and a simple, natural condition imposed there is practicable in problems with no free surface for both transient and periodic analysis. As soon as a free surface appears difficulties arise and the boundary conditions appear to be less clear cut. In this section we shall therefore investigate alternative analytical procedures which reduce such limitations.

Such analytical procedures are limited to the frequency domain and it is only with such problems that we shall be here concerned, thus eliminating the possibility of nonlinear transient analysis which will be considered in Section 4.

3. 2. Boundary Solution Procedures

The boundary solution procedures are discretiztion procedures based on trial functions which satisfy the governing equations a priori. The discretization is then confined to the inner surface for a problem defined on an infinite domain. For a homogeneous exterior domain equation (2) reduces to

$$c_p^2 \nabla^2 \bar{p} - \omega^2 \bar{p} = 0$$

or

$$\nabla^2 \bar{p} - k^2 \bar{p} = 0 \qquad (17)$$

where

$$p = \bar{p} \exp(i\omega t)$$

The singular solutions utilized by some boundary solution processes are now solutions of

$$\nabla^2 G(p,q) - k^2 G(p,q) = \delta(p-q) \qquad (18)$$

where

$$\delta(p-q) = 0 \qquad p \neq q \qquad (19)$$

and

$$\int_{\Omega_\infty} \delta(p-q)\, d\Omega = 1 \qquad (20)$$

where Ω_∞ denotes the infinite region in two and three dimensions. The free space Green's functions thus defined are

$$G(p,q) = H_o(kr) \qquad \text{in two dimensions} \qquad (21)$$

and

$$G(p,q) = e^{-kr}/r \qquad \text{in three dimensions}$$

where $H_o(kr)$ is a Hankel function and $r(p,q)$ is the distance between the field point p and the singular point q. Alternative solutions can be proposed but in all cases the extension of the region to infinity is correctly modelled so that the radiation condition is satisfied a priori.

The efficiency of the boundary solution procedures to solve problems on an infinite domain is well supported in the literature. Variational formulations based on the boundary solution procedures have also been proposed by Zienkiewicz et al (1977). These formulations lead to symmetric matrices so that the exterior region 'element' can be linked to an inner mesh of finite elements in a coupled finite element and boundary solution analysis which enables nonhomogeneity to be easily modelled in the finite element region.

We consider again the problem of surface waves impinging on a cylinder using the direct boundary integral procedure to model the outer domain. The surface elevation on the cylinder is superimposed on Fig. 2.

3.3. Infinite elements

3.3.1. The scalar wave equation

Finite elements assume some interpolation function over a finite domain (almost invariably a polynomial) and then process this function using a weighted residual method or a variational statement to get a set of equations approximating the solution of the governing equation in that domain. The infinite elements proposed by Bettess (1975), and Zienkiewicz and Bettess (1975) follow precisely a similar process except that they are based on an infinite domain and must use interpolation functions more appropriate to that domain (for which the basic clue is given by the solutions of the previous section). In principle many interpolation functions are admissable, the criterion they must satisfy being the satisfaction of the radiation condition (6) and simulation of the decay of wave amplitude with increasing distance from the source of the disturbance to the free field solution.

For example, if the conventional shape function for the ith node is written

$$e^{ik\xi} \quad m_j(\xi,\eta)$$

the infinite shape function extending to infinity in the ξ direction can be written

$$N_j(\xi) = e^{(\xi_j - \xi)/L} \; e^{ik\xi} \; M_j(\xi,\eta) \tag{22}$$

In this equation L is an arbitrary distance giving a measure of the severity of the exponential decay. This shape function obviously satisfies automatically the radiation boundary condition (6).

We can demonstrate the accuracy of this approach by again considering waves impinging on a cylinder. The solutions obtained are presented in Fig. 2 .

3.3.2. Layer elements for waves in layered elastic media

The infinite elements described previously can obviously be applied to single wave types propagating in elastic media. However special elements have been developed by Lysmer (1970) and Lysmer and Waas (1972) for solving the wave propagation problem in layered elastic media founded on baserock (Fig.5). Such a system is an accurate model of geotechnical conditions for most seismic problems.

The procedure for generating the layer stiffness matrix is again based on satisfaction of the radiation condition. In a continuous system the displacements of a plane harmonic wave travelling in the x-direction with phase velocity c and the frequency ω can be written in the form

$$\delta_x = u_x(y) \exp i(\omega t - kx) \tag{23}$$

and

$$\delta_y = i\, u_y(y) \exp i(\omega t - kx) \tag{24}$$

A force displacement relationship is then written for an infinite layer following the solution of an eigenvalue problem to find the wave numbers k which can be carried by the discretized model.

Roesset and Ettouney (1977) have shown that these layered elements can be brought very close to the transmitting source and achieve excellent results absorbing essentially all of the transmitted waves. In Fig. 6 we superimpose some of the results given in this reference to indicate firstly the importance of modelling an elastic foundation, and secondly to show the influence of the transmitting boundary.

4. ALTERNATIVE TRUNCATED MODELS FOR TRANSIENT ANALYSIS

4.1. General

We have already discussed in some detail the simplest process of boundary dampers. For surface waves the damping required is frequency dependent so that they do not provide suitable boundary conditions for general transient analysis. The restriction of these procedures and those discussed in Section 3 to the frequency domain then excludes any possibility of nonlinear analysis. Here we shall comment on two alternative approximations.

(a) Replacement of the exterior by a boundary mass, stiffness and damper tuned to minimize the frequency dependence and simulate accurately the compliance of the exterior region,

and (b) a procedure based on superposition of results to cancel boundary reflections which requires the assumption of linear behaviour only in the immediate vicinity of the boundary.

In many nonlinear problems material damping is high and the numerical model of the infinite domain can always be extended until the energy of any reflected wave is dissipated before it reaches the zone of interest. However, as the computational expense for non-linear behaviour demands a minimum number of variables be included in the model, it seems appropriate to discuss the above alternatives which may allow near field truncation.

4.2. Combinations of mass stiffness and damper on the boundary

Pedro (1976) has suggested that the difficulties caused by the frequency dependence of the simple boundary conditions can be minimised if a combination of mass, stiffness and damping is imposed at a finite boundary to simulate the exterior domain. For a harmonic loading we can set

$$\underset{\sim}{u} = \underset{\sim}{\bar{u}}\, e^{i\omega t}$$

in

$$\underset{\sim}{M}\,\ddot{\underset{\sim}{u}} + \underset{\sim}{C}\,\dot{\underset{\sim}{u}} + \underset{\sim}{K}\,\underset{\sim}{u} = \underset{\sim}{\bar{p}}_0\, e^{i\omega t} \tag{25}$$

to give

$$(-\omega^2 \underset{\sim}{M} + i\omega \underset{\sim}{C} + \underset{\sim}{K}) \bar{\underset{\sim}{u}} = \bar{\underset{\sim}{P}} \tag{26}$$

If the time response on the surface of the infinite domain surrounding the truncated model is given by any of the methods in the previous sections as

$$\underset{\sim}{f} = \left[\underset{\sim}{A}(\omega) + i\underset{\sim}{B}(\omega) \right] \bar{\underset{\sim}{u}} \tag{27}$$

we can attempt to identify the terms as

$$-\omega^2 \underset{\sim}{M} + \underset{\sim}{K} = \underset{\sim}{A}(\omega) \tag{28}$$

$$\omega \underset{\sim}{C} = \underset{\sim}{B}(\omega) \tag{29}$$

In general the matrices $\underset{\sim}{M}$, $\underset{\sim}{K}$, and $\underset{\sim}{C}$ will be frequency dependent but with suitable choice this dependence can be minimised in the neighbourhood of frequencies expected in the transient solution. Replacements of a similar kind have been proposed by Richart and Whitman (1967) and by Lysmer (1965) for the case of a rigid footing on an infinite half-space. Here the matrices reduce to scalars and can be simply identified.

4.3. Boundary conditions designed for cancellation of reflected waves by superposition of solutions

The analysis given in Section 2 of a simple plane wave impinging normally on a boundary showed that the reflected wave from a fixed boundary was equal in magnitude but opposite in sign to that from a free boundary. Simple averaging of these two solutions will therefore cancel the reflections and lead to a solution satisfying the radiation condition.

Smith (1974) proves that cancellation will also occur for surface waves, and for the waves in elastic media, if the simple fixed and free conditions are modified to first fixing the normal displacement and then the tangential displacements on the boundary, with other displacements free. The results given by him show excellent cancellation of reflections of dilatational distortional and Rayleigh waves. However his approach suffers from the fact that

- the superposition of complete time histories demands linear behaviour.
- multiple solutions are required when a wave reflects from more than one face (2^n solutions when reflections are required to be eliminated on n surfaces),

and
- certain high order reflections resulting from multiple reflections from opposing surfaces cannot be eliminated.

Research by the present authors encouraged by the initial success of Cundall and Kunar et.al. (1977)uses a technique incorporating the averaging process into the time marching scheme so that the waves are progressively eliminated in the region of the boundary. Multiple reflections are then not possible and linear behaviour is only assumed in the immediate vicinity of the boundary. Reflections are progressively eliminated and numerically stable schemes have been recovered by reinterpreting Smith's fixed condition as a zero acceleration over subsequent time steps, and the free condition as a constant reacting force based on displacements at the beginning of the averaging sequence. The need to take a number of steps before averaging and restarting the process is being investigated but appears to reduce numerical modelling difficulties at the boundary. The recovery of the multiple solutions and the averaging process is restricted to the boundary layer of elements.

To demonstrate that these procedures can meet with some success they have been applied to two problems of linear elasticity. The first considers one dimensional wave propagation produced by a sinusoidally varying load applied to a mesh of only two finite elements. The averaging process was carried out at every tenth step of an explicit time marching scheme, the time interval chosen such that reflected waves would be just leaving the boundary element. In Figure 7 the stress at the indicated Gauss integration point (2 x 2 rule) is superimposed on the exact solution. Results obtained when dampers were used on the transmitting boundary were almost identical and are not given. When the transmitting boundary was left either free or fixed for the complete analysis reflection occurred and the resulting stress history is shown.

The two-dimensional model shown in Fig. 8 was also loaded sinusoidally and boundary conditions applied on the vertical face below the load to prevent a Rayleigh wave propagating down that face. Here dilatational, distortional and, on the free surface, Rayleigh waves propagate away from the load source. Stress histories are compared in the figure for fixed boundaries on the bottom and right hand face, and for the new boundary condition suggested here. These results are not conclusive and further investigation is needed to determine whether this approach will eliminate the reflection of all wave types.

5. CONCLUDING DISCUSSION

Satisfaction of the radiation condition is essential for successful modelling of dynamic problems defined on infinite or sem-infinite domains. We have, in this paper, attempted to review those procedures which have been successfully applied to this type of problem.

Procedures for modelling linear problems appear well established and the references sited give details where only the essential features have been pointed out here. There appears however a need to consolidate the research on procedures allowing a full transient analysis for nonlinear behaviour. Considerable advances have however been made since the work of Zienkiewicz and Newton (1969) and most practical problems on infinite domains can be successfully modelled.

REFERENCES

Ang, A.H.S., and Newmark, N.M., "Development of a Transmitting Boundary for Numerical Wave Motion Calculation," Report DASA 2631 to Defence Atomic Support Agency, Washington D.C., 1971

Banaugh, R.P., and Goldsmith,W., "Diffraction of Steady Acoustic Waves by Surfaces of Arbitrary Shape", J. of the Acoustical Soc. of America, 35, No. 10, 1590-1601, 1963.

Berkhoff, J.G.W., "Linear Wave Propagation Problems and The Finite Element Method",Chapter 13, Finite Elements in Fluids, Vol. 1, Ed. R.H. Gallagher et al, Wiley, 1975

Bettess, P. (1977) "Infinite Elements", Int. J. Num. Meth. Engng, Vol. 11, No. 1, 53-64

Bettess, P., and Zienkiewicz, O.C., "Diffraction and Refraction of Surface Waves Using Finite and Infinite Elements", Int. J. Num. Meth. Engng., Vol.11 (1977)

Castellani, A.; "Boundary Conditions to simulate an Infinite Space", MECCANICA, No. 4, Vol. IX, pp. 199-205, Jour. Italian Association of Theoretical and Applied Mechanics AIMETA, Milano, Italy 1974

Chen, J.H.; "Numerical Boundary Conditions and Computational Modes", Journal of Computational Physics, Vol. 13, pp. 522-535, 1973

Chen, H.S. and Mei, C.C., "Oscillations and Wave Forces in an Offshore Harbor', M.I.T. Report No. 190, August, 1974.

Chen, H.S. and Mei, C.C., "Hybrid element method for water waves" Proc. Modelling Techniques (Modelling 75) San Francisco, Sept.3-5 1975,Vol.1, pages 63-81

Costantino, C.J., "Finite Element Approach to Stress Wave Problems", Jnl. of Eng. Mech. Div. ASCE, Vol.93 No. EM2 (1967).

Courant, R., and Hilbert, D., "Methods of Mathematical Physics Vol.II", Interscience Publishers, 1962, pp 315-318

Craggs, J.W., "On Two-dimensional Waves in an Elastic Half-Space", Proc. Cambridge Phil. Soc. 56 (1960)

Cruse, T.A., "A Direct Formulation and Numerical Solution of the General Transient Elastodynamic Problem, II", J. Math. Anal. Appl. 22, 341-355, 1968

Cruse, T.A., and Rizzo, F.J., "A Direct Formulation and Numerical Solution of the General Transient Elasto-Dynamic Problem I", J. Math. Anal. Appl. 22, 244-259, 1968

Cruse, T.A., and Rizzo, F.J. (Eds), "Boundary-Integral Equation Method: Computational Applications in Applied Mechanics", ASME Proceedings AMD-Vol.II (1975)

Cundall, P.A., Kunar, R., Martī,J. and Carpenter, P. Oral presentation, Dames and Moore Seminar, London, September, 1977.

Dobry, R. and Whitman, R.V., Discussion of the paper "Estimating the Damping of Real Structures " by J.D. Raggett, J. of the Struct. Div., Proc. ASCE, Vol. 102, ST5, May 1976.

Drake, L.A.; "Love and Rayleigh Waves in Horizontally Layered Media", Bull. Seism. Soc. Am., Vol.62, No. 5, pp. 1241-1258, Oct.1972

Drake, L.A.; "Rayleigh Waves in an Alluvial Valley", Natural Physical Science, Vol. 240, pp. 113-114, Dec.1972

Drake, L.A.; and Mal, A.K., "Love and Rayleigh Waves in the San Fernando Valley", Bull. Seism. Soc. Am., Vol. 62, No. 6, pp. 1673-1690, Dec. 1972

Dungar, R., and Eldred, P.J.L., "The Dynamic Response of Gravity Platforms," Int. Jnl. Num. Meth. in Engng. (to be published)

Ewing, W.M., Jardetsky, W.S. and Press, F., "Elastic Waves in Layered Media", McGraw-Hill, New York, 1957.

Fung, Y.C., Foundations of Solid Mechanics, Prentice-Hall, 1965.

Graff, K.F., "Wave motion in Elastic solids', Clarendon Press Oxford, 1975.

Hadala, P.F.; "Evaluation of a Transmitting Boundary for a Two-Dimensional Wave Propagation Computer Code", Technical Report S-71-16, U.S. Army Engineer Waterways Experiment Station, Vicksburg, Mississippi, December 1971.

Hogben, N. "Fluid Loading on Offshore Structures, A State of Art Appraisal: Wave Loads". Maritime Technology Monograph No. 1, Royal Institute of Naval Architects

Hogben, N., and Standing, R.G., "Wave Loads on Large Bodies", Proc. International Symposium on Dynamics of Marine Vehicles and Structures in Waves. Inst. of Mech. Eng. London, 1974.

Hwang, R.N. Lysmer, J. and Berger, E., "A Simplified Three-Dimensional Soil-Structure Interaction Study", Proc. Second ASCE Speciality Conference on Structural Design of Nuclear Plant Facilities, New Orleans, December, 1975.

Hwang, L.S. and Tuck, E.O., "On the Oscillations of Harbors of Arbitrary Shape", J. of Fluid Mechanics, Vol.42, 447-464, 1970

Kabori, T., "Dynamic Response of Rectangular Foundations on an Elastic Half-Space", Proc. Japan National Symposium on Earthquake Engineering, Tokyo, 1967.

Kausel, E.; "Forced Vibrations of Circular Foundations on Layered Media", Research Report R74-11, Department of Civil Engineering, Massachusetts Institute of Technology, Cambridge, Massachusetts, January, 1974.

Kausel, E., Roesset, J.M. and Waas, G., "Dynamic Analysis of Footings on layered Media", J. of the Eng. Mech. Div., Proc. ASCE, EM5, Vol. 101, 677-693, 1975

Lamb, H., "On the propagation of tremors over the surface of an elastic solid", Philosophical Transactions of the Royal Society, London Ser. A, Vol. 203 (1904) pp 1-42

Liang, V.C., "Dynamic Response of Structures in Layered Soils", Research Report R74-10, Department of Civil Engineering, Massachusetts Institute of Technology, Cambridge, Massachusetts, January, 1974

Lindman, E.L.,"Free-Space" Boundary Conditions for the Time Dependant Wave Equation", J. of Computational Physics, 18, (66-78), 1975

Lysmer, J., "Lumped mass method for Rayleigh Waves", Bulletin of the Seismological Society of America Vol. 60, No.1, pages 89-104, February, 1970

Lysmer, J., and Drake, L.A., "The Propagation of Love Waves across Non-Horizontally Layered Structures", Bull, Seism. Soc. Am., Vol.61, pp. 1233-1256, 1971

Lysmer, J. and Drake, L.A., "A Finite Element Method for Seismology" Chapter 6 of Methods in Computational Physics, Volume 11,: Seismology, edited by Alder, B,, Fernbach, S., and Bolt,A. Academic Press, New York and London

Lysmer, J., and Kuhlemeyer, R.L., "Finite Dynamic Model for Infinite Media", Jour. Engineering Mechanics Division, ASCE, Vol.95, No. EM4, pp. 859-877, August 1969

Lysmer, J. and Richart, F.E., "Dynamic Response of Footings to Vertical Loading", Jnl. Soil Mech. and Found, Div., ASCE, Vol.92 No. SM1, 1966

Lysmer, J., Seed, H.B., Udaka, T. and Hwang, R., "Efficient Finite Element Analysis of Seismic Structure-Soil-Structure Interaction" Second ASCE Speciality Conference on Structural Design of Nuclear Power Plant Facilities, New Orleans, 1975

Lysmer, J., and Waas, G., "Shear Waves in Plane Infinite Structures", Jour. Engng. Mech. Div., ASCE, Vol.98, No. EM1, pp.85-105, February 1972

Miller, G.F., and Pursey, H., "On the partition of energy between elastic waves in a semi-infinite solid", Proc. Royal Society, London, A, v.233, (1955)

Newton, R. E., "Radiation Boundary Condition for Plane Strain", private communication, 1973

Newton, R.E., "Finite Element Analysis of Two-Dimensional Added Mass and Damping", Chapter 11, Finite Elements in Fluids Vol.1, Eds., Gallagher et al, Wiley 1975.

Orlanski, I., "A Simple Boundary Condition for Unbounded Hyperbolic Flows", J. of Computational Physics, 21, 251-269, 1976

Pedro,J.O., "Finite Element Stress Analysis of Arch Dams", (Dimensionamen das barrageus abóda pelo metodo dos elementos finitos), LNEC, Proc. 46/11/5272, Lisbon,1976

Rayleigh, Lord, "On waves propagated along the plane surface of an elastic solid", London Mathematical Society Proc., 17 (1885) pp 4-11

Rellich,F.,"Uber das asymptotische Verhalten der Losungen von $\Delta u+\lambda u=0$ in unendlichen Gebieten" Jahresbericht der Deutschen Mathematiker Vereingigung Vol.53, 1943, pages 57-65

Richart, F.E., Hall, J.R., and Woods, R.D., "Vibrations of Soils and Foundations, Prentice-Hall, 1970

Richart, F.E., and Whitman, R.V., "Comparison of Footing Vibration Tests with Theory", J. Soil Mech. and Found. Div., Proc. ASCE, Vol.93, No. SM6, pp 143-168 (1967)

Robinson, A.R., "The Transmitting Boundary -- Again", Proc. Structural and Geotechnical Mechanics Symposium, University of Illinois at Urbana - Champaign, October 2-3, 1975

Roesset, J.M., and Ettouney, M.M., (1977),"Transmitting Boundaries: A Comparison," Int. J. Num. and Anal, Meth. in Geomech. Vol.1 (1977)

Saini, S.S., Zienkiewicz, O.C. and Bettess, P., "Coupled hydro-dynamic response of concrete gravity dams using finite and infinite elements" University College of Wales, Swansea, Department of Civil Engineering Research Report C/R/270/76

Seed, H.B., Lysmer, J. and Hwang, R., "Soil-Structure Interaction Analysis for Seismic Response."Proc. ASCE, 101, GT5, 439-459, 1975

Seed, H.B., Duncan, J.M., and Idriss, I.M., "Criteria and Methods for Static and Dynamic Analysis of Earth Dams", Proc. Symp. on Criteria and Assumptions for Numerical Analysis of Dams, Swansea 1975.

Shaw, R.P. "Boundary Integral Equation Methods Applied to Water Waves" AMD Vol.11 Boundary Integral Equation Method: Computational Applications in Applied Mechanics, 1975. ASME

Shaw, R.P. "Transient Scattering by a Circular Cylinder" Journal of Sound and Vibration, (1975) Volume 42, part 3, pages 295-304

Smith, W.D., "A Nonreflecting Plane Boundary for Wave Propagation Problems", Journal of Computational Physics, Vol.15, pp. 492-503, 1974

Smith, W.D., "The Application of Finite Element Analysis to Body Wave Propagation Problems", Geophysical Journal of the Royal Astronomical Society, Vol.42,pp.747-768, 1975

Sommerfeld, A., Jahresber. Deutschen Math. Vereinigung, Vol.21, 312 (1912)

Sommerfeld, A., "Partial Differential Equations in Physics" Academic Press, 1949

Stoker, U.V. "Water Waves" Interscience, New York, 1957

Urlich, C.M., and Kuhlemeyer, R.L., "Coupled Rocking and Lateral Vibrations of Embedded Footings," Canadian Geotechnical Journal, Vol.10, No. 2 pp.145-160, 1973

Vaish, A.K., and Chopra, A.K., "Earthquake Finite Element Analysis of Structure - Foundation Systems", Proc. ASCE, 100 EM6, 1101-1116,1974

Waas, G., "Linear Two-Dimensional Analysis of Soil Dynamics Problems in Semi-Infinite Layered Media", Ph.D. Thesis presented to the University of California at Berkeley, California, 1972

Waas, G., "Earth Vibration Effects and Abatement for Military Facilities- Analysis Method for Footing Vibrations through Layered Media", Technical Report S-71-14, U.S. Army Engineer Waterways Experiment Station, Vicksburg, Mississippi, September, 1972

Waas, G., and Lysmer, J., "Vibrations of Footings Embedded in Layered Media", Proc. WES Symposium on Applications of the Finite Element Method in Geotechnical Engineering, U.S. Army Engineer Waterways Experiment Station, Vicksburg, Mississippi, May 1972.

Whitman, R.V., and Richart, F.E., "Design Procedures for Dynamically Loaded Foundations,Proc. ASCE, 93 SM6, 169-193, 1967

Woods, R.D., "Screening of Surface Waves in Soils", J. Soil Mech.and Found. Div., Proc. ASCE, Vol.44, No. SM4, (July 1968)

Zienkiewicz, O.C. "The Finite Element Method and Boundary Solution Procedures as General Approximation Methods for Field Problems" World Congress on Finite Element Methods in Structural Mechanics, Bournemouth, October 12-17, 1975

Zienkiewicz, O.C., and Bettess, P. "Infinite elements in the study of fluid-structure interaction problems" Second International Symposium on Computing Methods in Applied Science and Engineering, Versailles, France, 15-17th December, 1975

Zienkiewicz, O.C., Bettess, P., and Kelly, D.W. (1978),"The Finite Element Method for Determining Fluid Loadings," Ch.4., Numerical Methods in Offshore Engineering, J. Wiley

Zienkiewicz, O.C., Kelly, D.W., and Bettess, P. "The Coupling of the Finite Element Method and Boundary Solution Procedures", Int. J. Num. Meth. Engng., Vol.11, 355-375, (1977)

Zienkiewicz, O.C. and Newton, R.E. "Coupled Vibrations in a Structure Submerged in a Compressible Fluid", Int. Symp. on Finite Element Techniques, Stuttgart, 1969

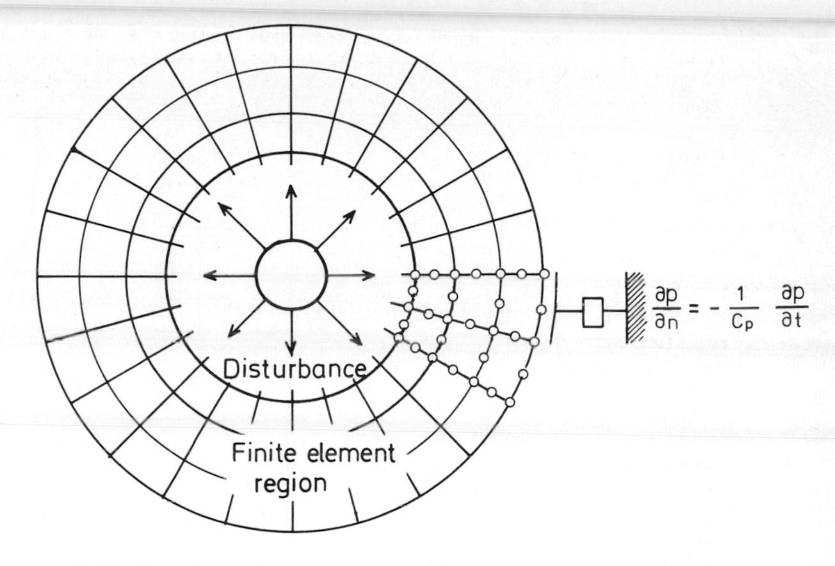

(a) Compression waves only

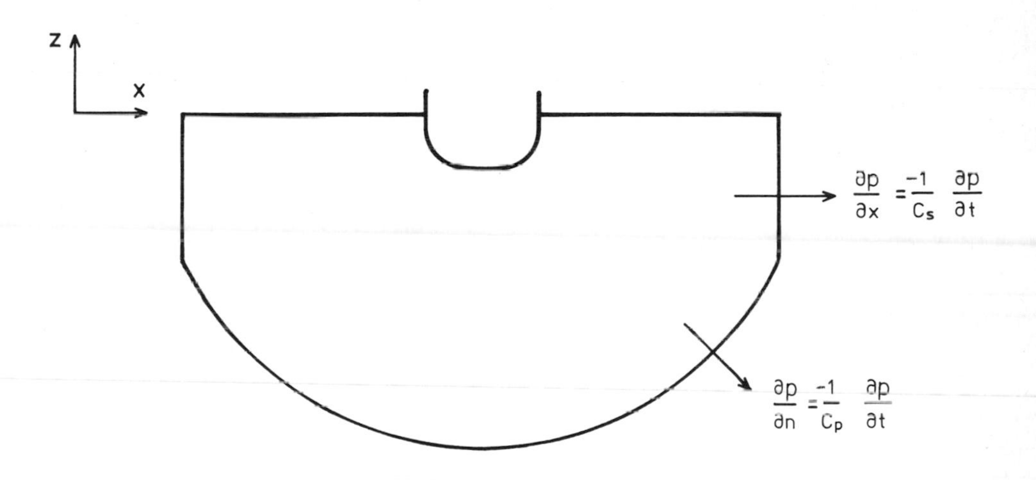

(b) Compression waves and surface waves

FIGURE 1 RADIATION CONDITIONS APPLIED AT FINITE DISTANCES

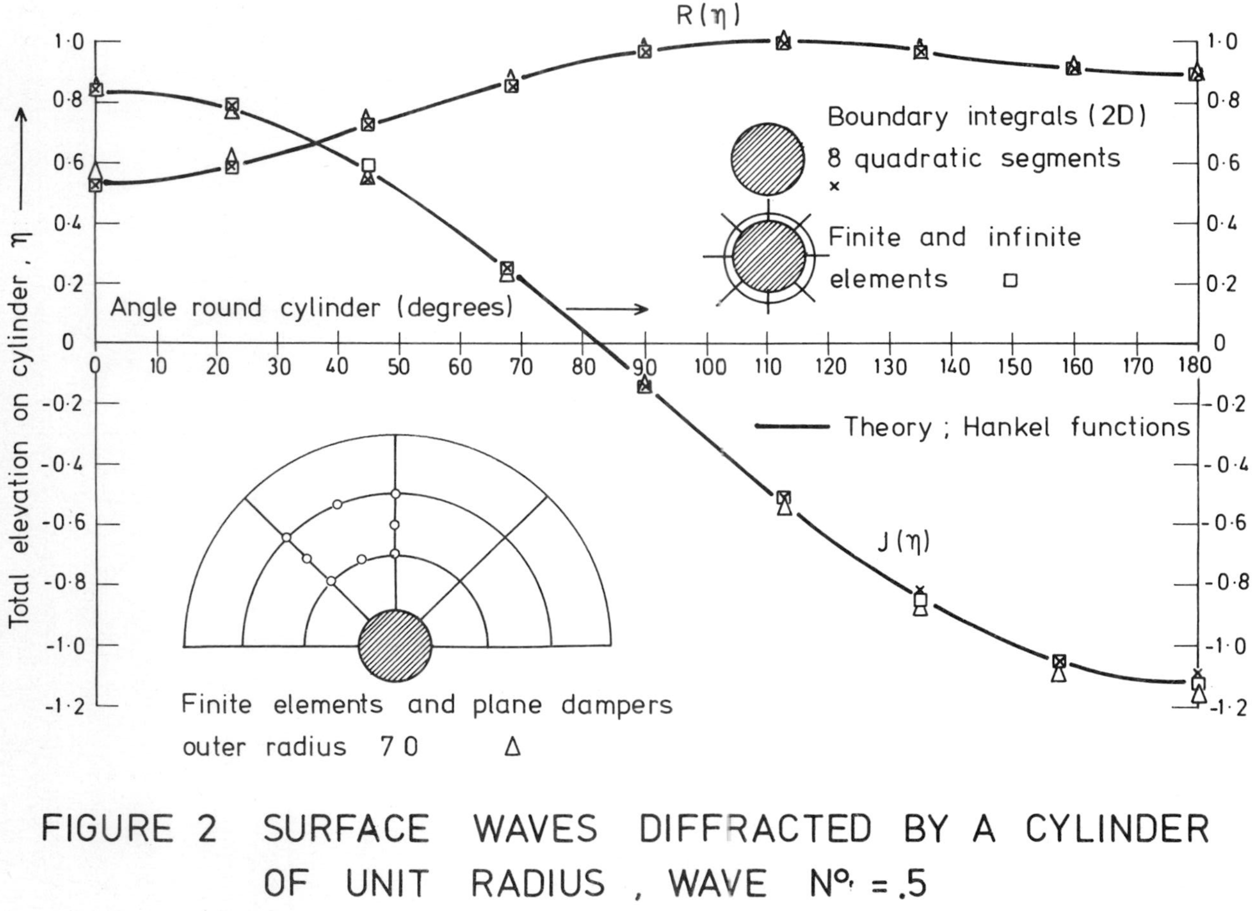

FIGURE 2 SURFACE WAVES DIFFRACTED BY A CYLINDER OF UNIT RADIUS, WAVE No. = .5

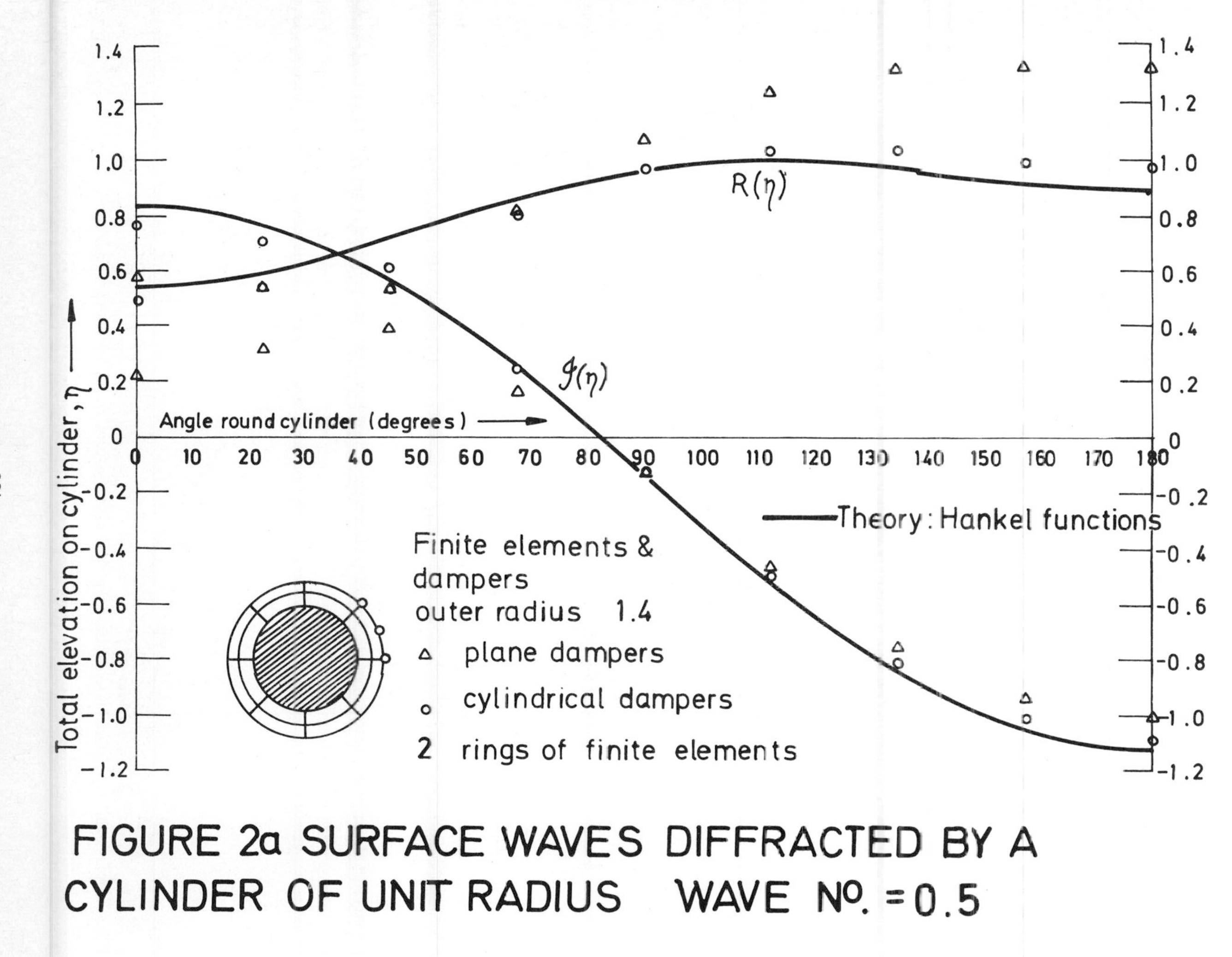

FIGURE 2a SURFACE WAVES DIFFRACTED BY A CYLINDER OF UNIT RADIUS WAVE No. = 0.5

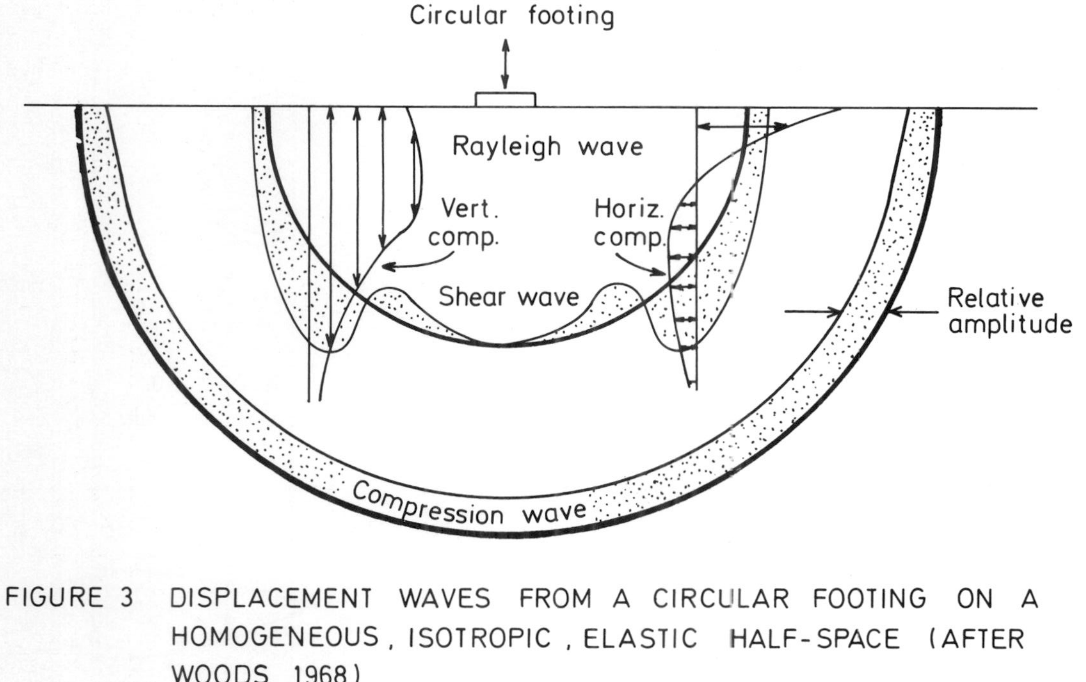

FIGURE 3 DISPLACEMENT WAVES FROM A CIRCULAR FOOTING ON A HOMOGENEOUS, ISOTROPIC, ELASTIC HALF-SPACE (AFTER WOODS 1968)

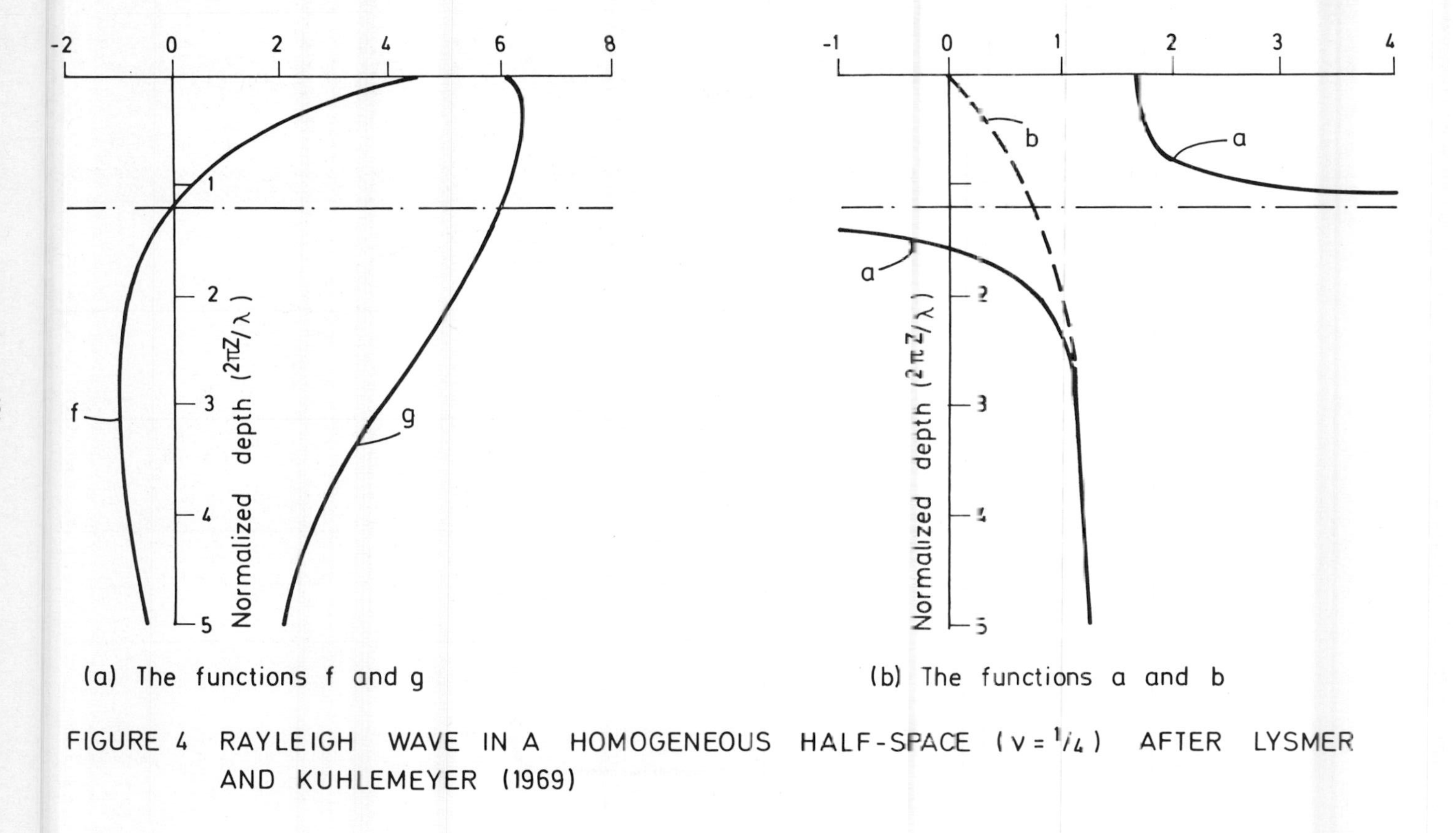

(a) The functions f and g

(b) The functions a and b

FIGURE 4 RAYLEIGH WAVE IN A HOMOGENEOUS HALF-SPACE ($\nu = 1/4$) AFTER LYSMER AND KUHLEMEYER (1969)

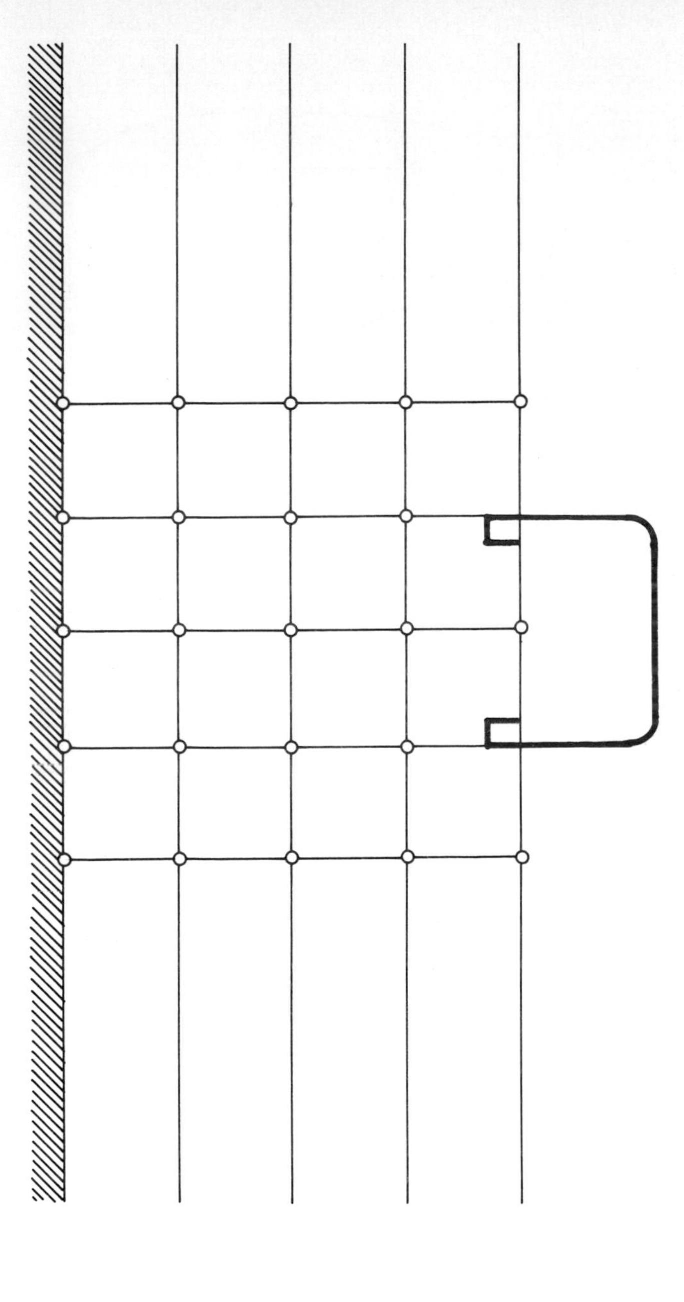

FIGURE 5 MODEL FOR LAYERED MEDIA IN AN INFINITE HALF-SPACE

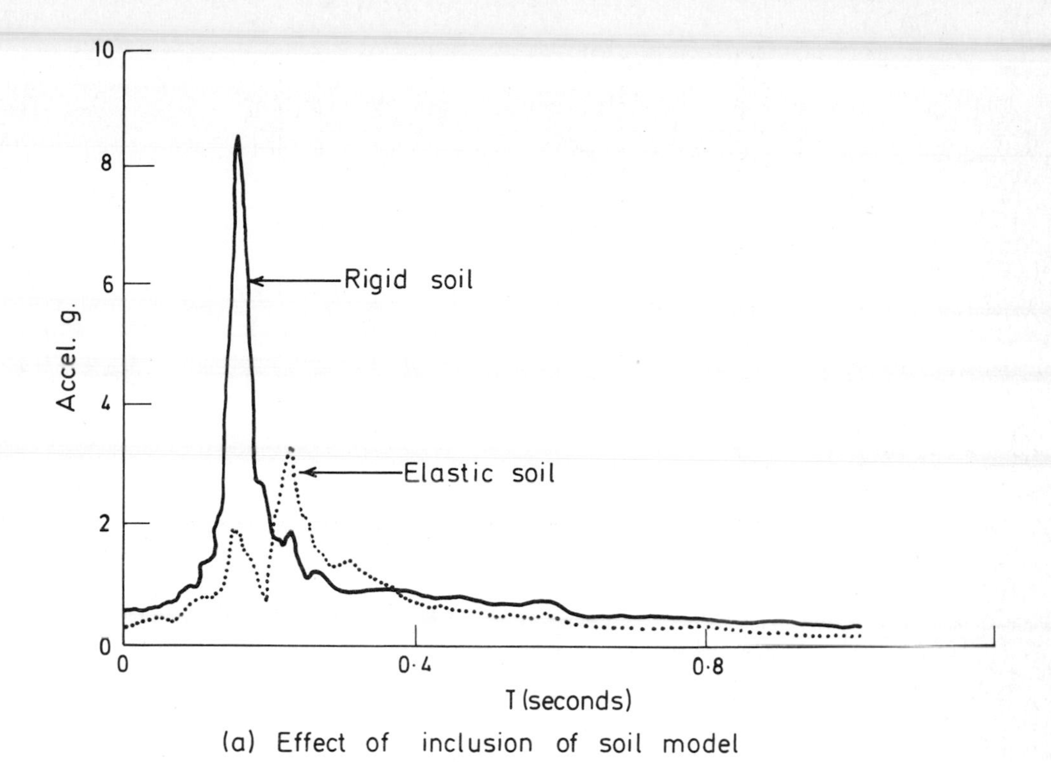

(a) Effect of inclusion of soil model

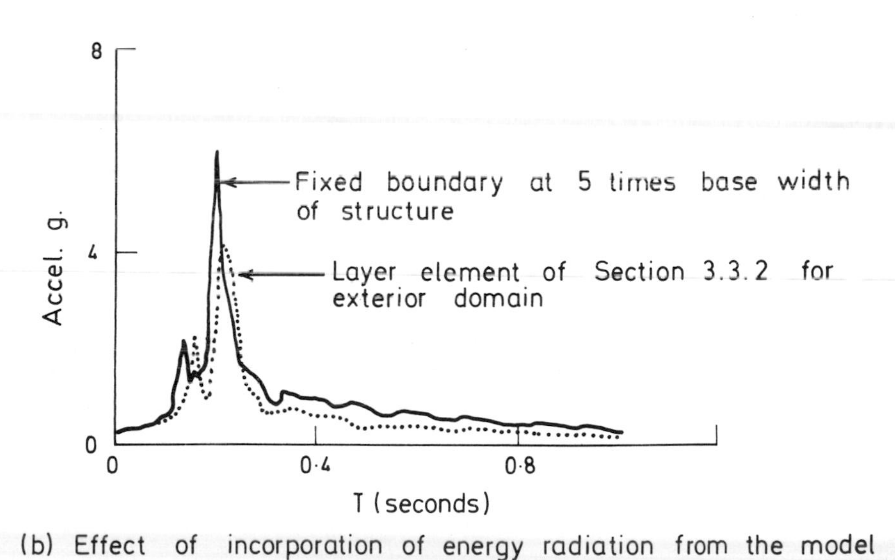

(b) Effect of incorporation of energy radiation from the model

FIGURE 6 EFFECT OF MODELLING ON RESPONSE SPECTRUM (AFTER ROESSET AND ETTOUNEY 1977)

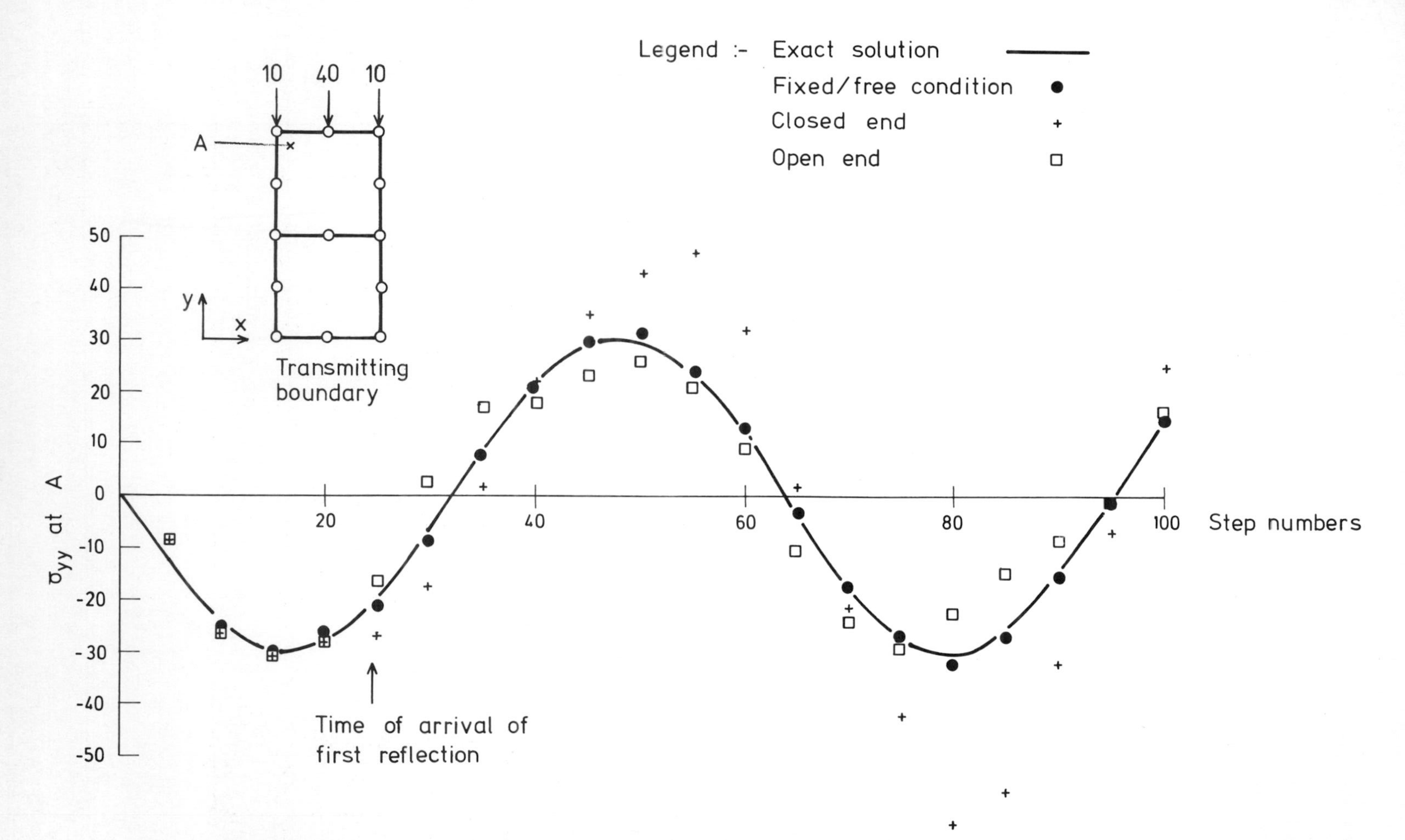

FIGURE 7 ONE-DIMENSIONAL ELASTIC WAVE PROPAGATION

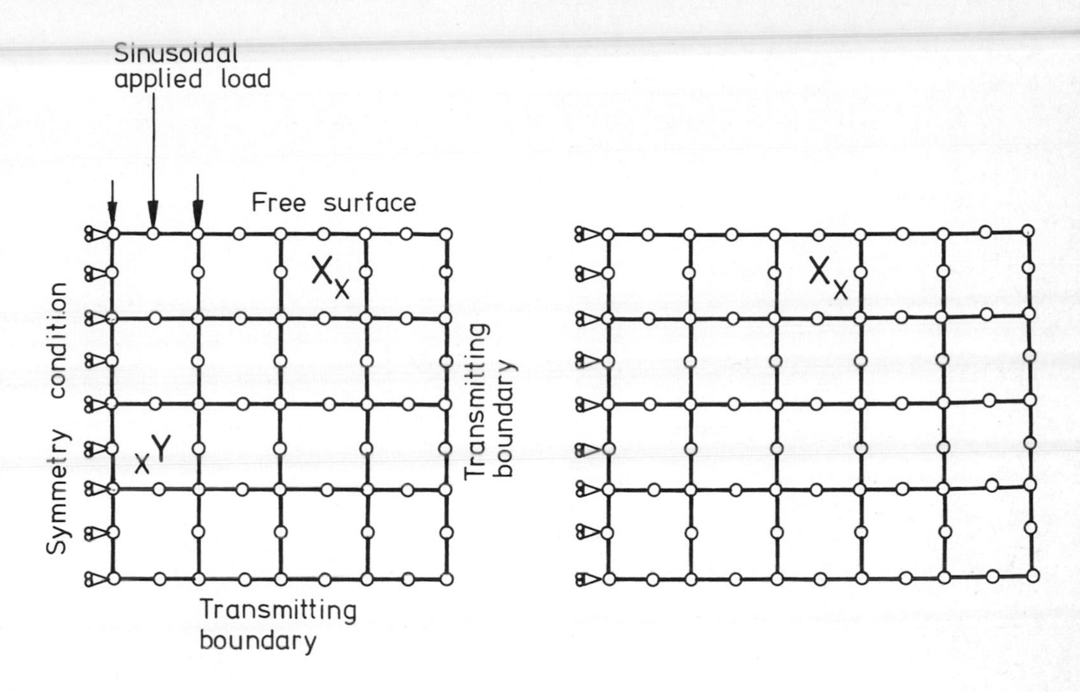

Mesh A

Mesh B

Mesh C

FIGURE 8(a) TWO DIMENSIONAL ELASTIC WAVE PROPAGATION FINITE ELEMENT MESHES

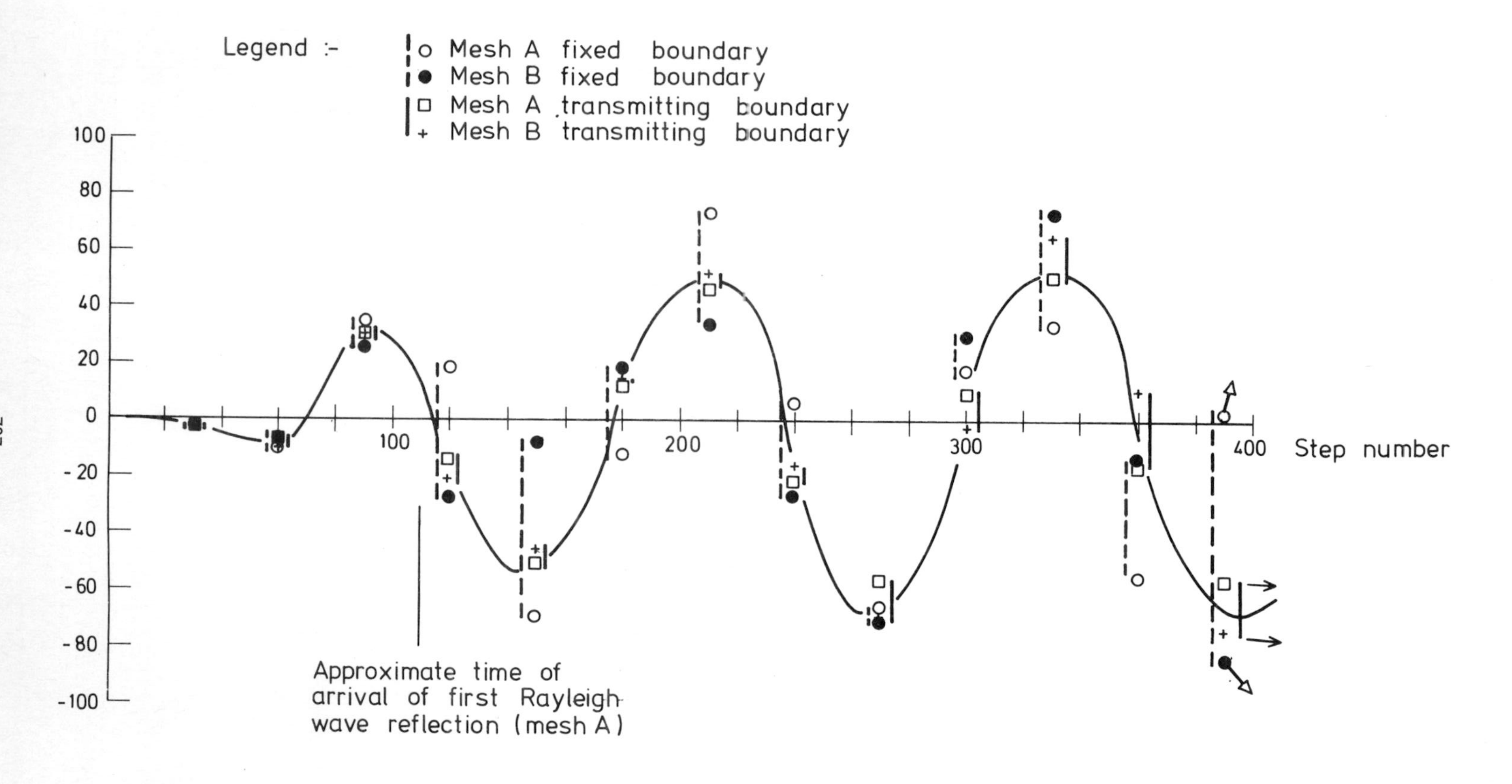

FIGURE 8(b) HORIZONTAL STRESS AT X FOR MESHES A & B

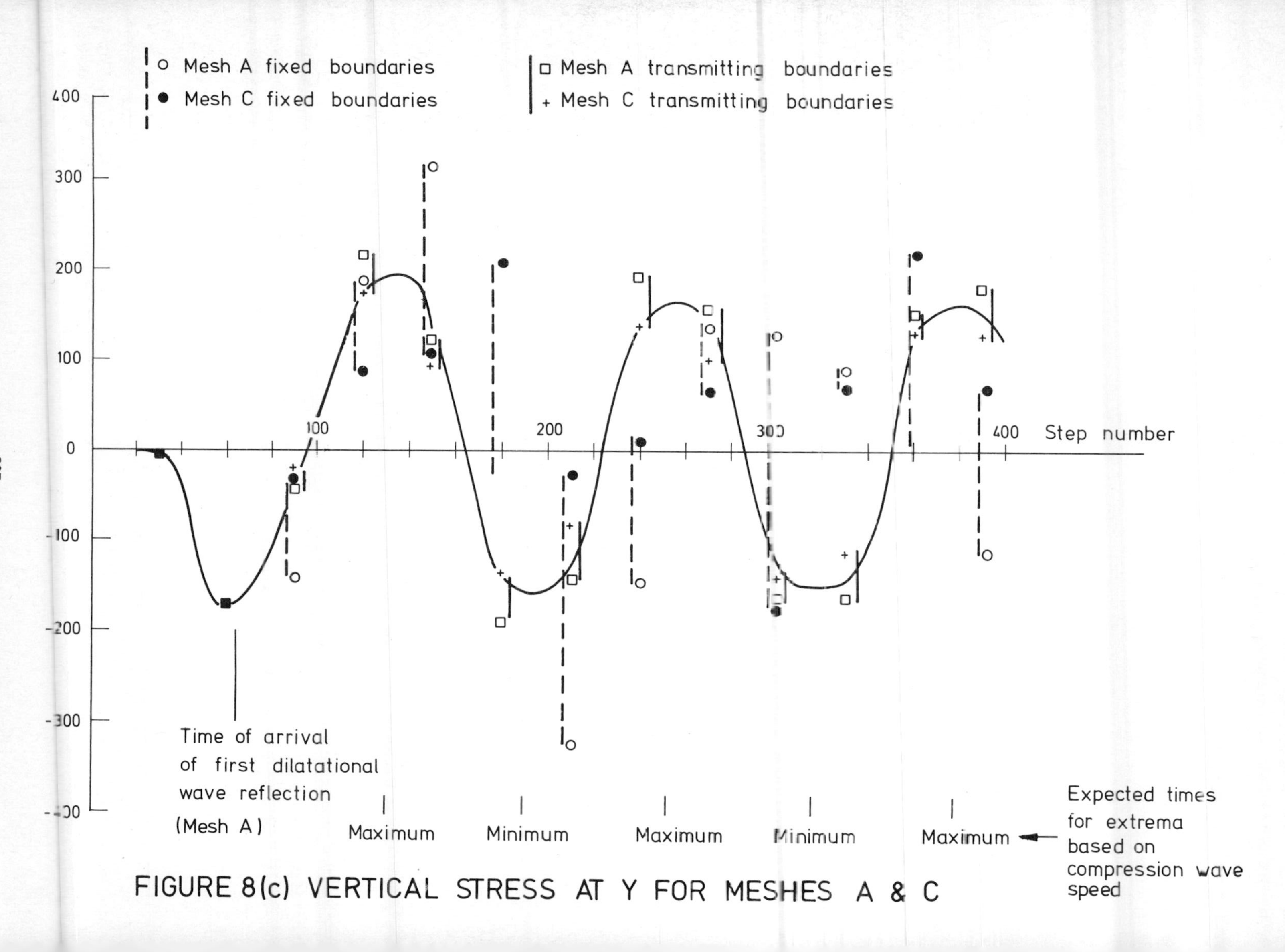

FIGURE 8(c) VERTICAL STRESS AT Y FOR MESHES A & C

TIME DEPENDANT PROBLEMS

PROBLEMES D'EVOLUTION

ON THE FINITE ELEMENT APPROXIMATION FOR

EVOLUTION EQUATIONS OF PARABOLIC TYPE

Hiroshi FUJITA and Takashi SUZUKI
Department of Mathematics
University of Tokyo
Hongo, Tokyo, Japan

1. Introduction

As a continuation of our previous works (Fujita [2 , 3], Fujita-Mizutani [4 , 5]), the present paper is again concerned with the operator theoretical study of the finite element approximation for partial differential equations of parabolic type. Actually we shall deal with initial value problems which can be reduced to an abstract initial value problem in the Hilbert space $X = L_2(\Omega)$ of the following form

$$\frac{du}{dt} + A(t)u = 0 \qquad (0 < t \leqq T), \tag{1.1}$$

$$u(0) = u_0. \tag{1.2}$$

Here t is the time variable which ranges over the interval $[0, T]$, T being a positive number. The initial value u_0 is a given function in $L_2(\Omega)$. For the sake of simplicity we consider the case of two space variables. Namely, we assume that Ω is a bounded domain in the plane R^2 with its piecewise smooth boundary $\partial\Omega$. For each t in $[0, T]$, $A(t)$ is an m-sectorial operator in X (e.g., see Kato [9]) defined through an elliptic differential operator L_t of the form

$$L_t[u] = -\sum_{i,j=1}^{2} \frac{\partial}{\partial x_i} a_{ij}(t,x)\frac{\partial u}{\partial x_j} + \sum_{j=1}^{2} b_j(t,x)\frac{\partial u}{\partial x_j} + c(t,x)u \tag{1.3}$$

and a suitable boundary condition to be imposed on functions belonging to $D(A(t))$, the domain of $A(t)$. For the most part in this paper, we take the Dirichlet boundary condition

$$u|_{\partial\Omega} = 0 , \tag{1.4}$$

while a few results will be mentioned near the end of this paper concerning the case of the Neumann boundary condition

$$(1.5) \qquad \frac{\partial u}{\partial \nu}\Big|_{\partial\Omega} = 0 ,$$

where

$$\frac{\partial}{\partial \nu} = \sum n_i a_{ij}(t,x) \frac{\partial}{\partial x_j} ,$$

that is, $\partial/\partial\nu$ is the differentiation along the conormal to $\partial\Omega$.

In this paper only the space variable x is discretized in the manner of the finite element method. In other words, we shall consider the semi-discrete approximation. As will be described in §2 more specifically, we triangulate Ω into small elements and by h we mean the size parameter of this triangulation. We use simplest trial functions, namely, "piecewise linear trial functions". The totality of these trial functions which satisfy the boundary condition is denoted by V_h. The approximate solution obtained by our semi-discrete approximation is denoted by u_h. Then our main result is the following; under certain regularity assumption on the coefficients and the triangulation, the rate of convergence of the approximation is optimal and we have the error estimate

$$(1.6) \qquad \|u(t) - u_h(t)\| \leqq C\frac{h^2}{t}\|u_0\| \qquad (t > 0),$$

provided that the boundary condition is the Dirichlet boundary condition. Here and hereafter C stands for various constants which may differ in one context. The norm $\|\ \ \|$ in (1.6) is the $L_2(\Omega)$-norm. We remark that in deriving (1.6) no smoothness assumption on the initial value u_0 is made and the self-adjointness of $A(t)$ is not assumed either. Thus (1.6) is just a generalization of our previous result, which was obtained for the case of $A(t) \equiv A$ Fujita [2] and Fujita-Mizutani [4, 5], to the case of the time dependent $A(t)$. As a matter of fact, we also derived a weaker estimate

$$(1.7) \qquad \|u(t) - u_h(t)\| \leqq C_\varepsilon \frac{h^{2-\varepsilon}}{t} \|u_0\| \qquad (t > 0)$$

previously (Fujita [3]). (1.6) may be regarded as a sharpening of (1.7). Here we should refer to the works by H. P. Helfrich in the same direction. He proved (1.6) first in 1974 for the case of $A(t) \equiv A$ and in 1975 for the case of time dependent $A(t)$, assuming the self-adjointness of $A(t)$ and that

$$V_h \subset \mathcal{D}(A(t)). \tag{1.8}$$

Hence our result relaxes his assumptions. Moreover, we note that some crucial tricks employed in the proof of our result are due to him.

§2. Preliminaries.

Concerning the coefficients of the differential operator L_t we assume that the followings.

(i) a_{ij}, b_j, c and the first derivatives in x of a_{ij} and b_j are continuous in $Q = [0, T] \times \bar{\Omega}$ and, moreover, are Hölder continuous in t with exponent θ uniformly with respect to x in Q, where θ is a parameter subject to $1/2 < \theta \leqq 1$.

(ii) L_t is uniformly elliptic. Namely, $a_{ij}(t,x) = \overline{a_{ji}(t,x)}$ and

$$\sum_{i,j}^{2} a_{ij}(t,x)\zeta_j\bar{\zeta}_i \geqq \delta_0|\zeta|^2 \qquad (\forall\, \zeta \in \mathbb{C}^2)$$

for some positive constant δ_0.

We put

$$V = H_0^1(\Omega) = \{u \in H^1(\Omega);\ \ u|_{\partial\Omega} = 0\}. \tag{2.1}$$

$H^j(\Omega)$ means the Sobolev space $W_2^j(\Omega)$ of complex valued functions $(j = 0, 1, \ldots)$, while the standard norm in $H^j(\Omega)$ will be written as $\|\ \ \|_j$. However, to represent $\|\ \ \|_{L^2(\Omega)}$ we write simply $\|\ \ \|$ instead of $\|\ \ \|_0$. We introduce a sesquilinear form $a_t(\ \ ,\ \ ): V \times V \longrightarrow \mathbb{C}$ by setting

$$a_t(v,w) = \int_\Omega \Big\{ \sum_{i,j} a_{ij}(t,x)\frac{\partial u}{\partial x_j}\,\overline{\frac{\partial v}{\partial x_i}} + \sum_j b_j(t,x)\frac{\partial u}{\partial x_j}\bar{v} + c(t,x)u\bar{v}\Big\}dx \tag{2.2}$$

for $v, w \in V$. Then we have

$$|a_t(v, w)| \leq c_1\|v\|_1\|w\|_1 \tag{2.3}$$

with some positive constant c_1 for all $t \in [0, T]$ and all $v, w \in V$. After replacing $c(t, x)$ by $c(t, x) + \lambda$ where λ is a sufficiently large constant, if necessary, we have

(2.4) $$\mathrm{Re}\; a_t(u, u) \geqq c_2 \|u\|_1^2 .$$

$A(t)$ is associated with $a_t(\;,\;)$ in the standard manner:

$$a_t(v, w) = (A(t)v, w) \quad (v \in D(A(t)), \quad w \in V).$$

According to the theory of elliptic differential equations,

(2.5) $$D(A(t)) = H^2(\Omega) \cap V = \mathcal{D} .$$

Hence the domain $D(A(t))$ is constant. Furthermore, the graph norm of $A(t)$ is equivalent to $\| \; \|_2$; namely, there exists a positive constant c_3 independent of t such that

(2.6) $$c_3^{-1} \|v\|_2 \leqq \|A(t)v\| \leqq c_3\|v\|_2.$$

The smoothness of the coefficients of L_t implies the following t-smoothness of $a_t(\;,\;)$;

(2.7) $$|a_t(v,w) - a_s(v,w)| \leqq c_4|t-s|^\theta \|v\|_1 \cdot \|w\|_1 ,$$

(2.8) $$|a_t(v,w) - a_s(v,w)| \leqq c_5|t-s|^\theta \|v\|_2 \cdot \|w\| ,$$

(2.9) $$|a_t(v,w) - a_s(v,w)| \leqq c_6|t-s|^\theta \|v\| \cdot \|w\|_2 .$$

From (2.6) and (2.8) it follows that

(2.10) $$\|A(t)A(s)^{-1} - I\| \leqq C|t-s|^\theta .$$

Therefore the simplest version of the generation theorem for the evolution equation of parablic type is applicable to (1.1) and (1.2), and the solution $u = u(t)$ is given by

$$u(t) = U(t, 0)u_0 .$$

Here $\{U(t,s)\}_{0 \leqq s \leqq t \leqq T}$ is the evolution operator generated by $\{A(t)\}_{0 \leqq t \leqq T}$. It is well-known that $U(t,s)$ satisfies the following inequalities (for instance, see Tanabe [13], Kato [8], Sobolevskii [11]);

$$(2.11)\qquad \|U(t,s)\| \leqq C \qquad (0 \leqq s \leqq t \leqq T)$$

$$(2.12)\qquad \|A(t)U(t,s)A(s)^{-1}\| \leqq C \qquad (0 \leqq s \leqq t \leqq T)$$

$$(2.13)\qquad \|A(t)U(t,s)\| \leqq C(t-s)^{-1} \qquad (0 \leqq s < t \leqq T)$$

$$(2.14)\qquad \|A(t)(U(t,s)-U(r,s))A(s)^{-1}\| \leqq C_\gamma (t-r)^\gamma (r-s)^{-\gamma}, \quad (0 \leqq s < r < t \leqq T)$$

for the parameter γ in $0 \leqq \gamma < \theta$.

The domain of the adjoint operator $A(t)^*$ is again equal to $\mathcal{D} = H^2(\Omega) \cap V$. $A(t)^*$ may be regarded as the operator associated with the adjoint form $a^*(v,w) = \overline{a(w,v)}$. Under our assumptions (2.6) as well as (2.10) holds true with $A(\cdot)$ replaced by $A(\cdot)^*$. We note also that $V(t,s) = U(T-s,T-t)^*$ is the evolution operator generated by $A(T-t)^*$.

We proceed to the approximation scheme. When Ω is a polygon we triangulate Ω regularly and adopt piecewise linear trial functions vanishing on the boundary. If $\partial\Omega$ is smooth and curved, we must use modified piecewise linear trial functions as in Zlámal [14]. In any case we have $V_h \subset V$. By $a_t^{(h)}(\ ,\ )$ we mean the restriction of $a_t(\ ,\ )$ on $V_h \times V_h$. Then $A_h(t): V_h \to V_h$ is an operator defined by the condition that

$$(2.15)\qquad a_t^{(h)}(\varphi_h, \psi_h) = (A_h(t)\varphi_h, \psi_h) \qquad (\varphi_h, \psi_h \in V_h)$$

Although an inequality like (2.6) is impossible for $A_h(t)$, we still have, for instance,

$$(2.16)\qquad \|U_h(t,s)\| \leqq C \qquad (0 \leqq s \leqq t \leqq T),$$

$$(2.17)\qquad \|A_h(t)U_h(t,s)\| \leqq C(t-s)^{-1} \qquad (0 \leqq s < t \leqq T),$$

$$(2.18)\qquad \|U_h(t,s)A_h(s)\| \leqq C(t-s)^{-1} \qquad (0 \leqq s < t \leqq T)$$

with a constant C independent of h, where $U_h(t,s)$ is the evolution operator in V_h generated by $A_h(t)$. This can be seen from the generation theorem by Fujie-Tanabe [1] in virtue of $1/2 < \theta \leqq 1$.

Our approximate solution $u_h : [0, T] \to V_h$ is determined by the conditions

$$\frac{d}{dt}(u_h, \varphi_h) + a_t(u_h, \varphi_h) = 0 \quad (\varphi_h \in V_h), \tag{2.19}$$

$$(u_h, \varphi_h) = (u_0, \varphi_h) \quad (\varphi_h \in V_h). \tag{2.20}$$

These conditions are equivalent to

$$\frac{du_h}{dt} + A_h(t)u_h = 0 \ , \tag{2.21}$$

$$u_h(0) = P_h u_0. \tag{2.22}$$

Here P_h is the orthogonal projection from X onto V_h. In terms of $U_h(t,s)$, u_h is given by

$$u_h(t) = U_h(t,0)u_0 \ .$$

Finally we remark that in our particular case where $A(t)$ and $A_h(t)$ are accretive, the constant C in (2.11) and (2.16) can be taken as $C = 1$.

§3. Lemmas concerning approximation for the boundary value problem.

We put

$$K_h(t) = A(t)^{-1} - J_h A_h(t)^{-1} P_h \qquad (t \in [0,T]) \tag{3.1}$$

where J_h is the canonical injection from V_h into X. However, if there is no fear of confusion, we shall omit J_h and write simply as

$$K_h(t) = A(t)^{-1} - A_h(t)^{-1} P_h. \tag{3.2}$$

The purpose of this section is to prepare some lemmas concerning $K_h(t)$. First of all, let us take an arbitrary $f \in X$ and put

$$v = A(t)^{-1} f \quad \text{and} \quad v_h = A_h(t)^{-1} P_h f \ .$$

Then we see that v is the solution of $A(t)v = f$ which is equivalent to the boundary problem consisting of $L_t[v] = f$ and $v|_{\partial\Omega} = 0$. v satisfies

$$(3.3) \qquad a_t(v, \varphi) = (f, \varphi) \qquad (\varphi \in V).$$

On the other hand, v_h satisfies

$$(3.4) \qquad a_t(v_h, \varphi_h) = (f, \varphi_h) \qquad (\varphi_h \in V_h).$$

Hence, v_h is the approximate solution for v by the finite element method. In other words, $K_h(t)$ is an operator which characterizes the error committed by the approximation for the boundary value problem.

From (3.3) and (3.4) follows that

$$a_t(v-v_h, \varphi_h) = 0 \qquad (\varphi \in V_h) ,$$

namely,

$$(3.5) \qquad a_t(K_h(t)f, \varphi_h) = 0 \qquad (\varphi \in V_h) .$$

The following lemma which gives the error estimate for the boundary value problem is essentially well-known. In fact, it can be proved in a standard manner by making use of (3.5), (2.3), (2.4), (2.6), Nitsche's trick and of the fact that

$$(3.6) \qquad \inf_{\varphi_h \in V_h} \|v - \varphi_h\|_1 \leqq c_7 h \|v\|_2 \qquad (v \in \mathcal{D})$$

for some constant $c_7 > 0$.

Lemma 3.1. As $h \longrightarrow 0$,

$$(3.7) \qquad \|K_h(t)f\|_1 \leqq Ch\|f\| \qquad (f \in X),$$

and

$$(3.8) \qquad \| K_h(t)f \| \leqq Ch^2\|f\| \qquad (f \in X),$$

where C is a constant independent of t. Similarly, by considering the adjoint operators we have

Lemma 3.2 As $h \longrightarrow 0$

$$(3.9) \qquad \|K_h(t)^*f\|_1 \leqq Ch\|f\| \qquad (f \in X),$$

(3.10) $$\|K_h(t)^*f\| \le Ch^2\|f\| \qquad (f \in X).$$

Incidentally, we note

(3.11) $$\|(I-P_h)v\| \le Ch^2\|v\|_2 \qquad (v \in \mathcal{D}),$$

which follows immediately from

(3.12) $$\inf_{\varphi_h \in V_h} \|v-\varphi_h\| \leqq c_8 h^2\|v\|_2 \qquad (v \in \mathcal{D}).$$

Lemma 3.3. We have for $t, s \in [0, T]$

(3.13) $$\|K_h(t)-K_h(s)\| \leqq Ch^2|t-s|^\theta.$$

Proof. Taking an arbitrary $f \in X$, we put

(3.14) $$w = K_h(t)f - K_h(s)f .$$

We are going to show

(3.15) $$\|w\|^2 \le Ch^2|t-s|^\theta\|w\|\cdot\|f\|.$$

We start with the following equality;

$$\begin{aligned}
\|w\|^2 &= a_t(w, (A(t)^*)^{-1}w)\\
&= a_t(w, (K_h(t)^* - (A_h(t)^*)^{-1}P_h)w)\\
&= a_t(K_h(t)f, K_h(t)^*w) - a_t(K_h(s)f, K_h(t)^*w)\\
&\quad - a_t(K_h(t)f, \psi_h(t)) + a_t(K_h(s)f, \psi_h(t))\\
&\equiv I_1 - I_2 - I_3 + I_4,
\end{aligned}$$

where $\psi_h(t) = (A_h(t)^*)^{-1}P_h w$. In view of (3.5) we notice that $I_3 = 0$. Also, by the property of $K_h^*(t)$ corresponding to (3.5), we have with $\varphi_h(t) = A_h(t)^{-1}P_h f$

$$\begin{aligned}
I_1 - I_2 &= a_t(A(t)^{-1}f - \varphi_h(t), K_h(t)^*w)\\
&\quad - a_t(A(s)^{-1}f - \varphi_h(s), K_h(t)^*w)
\end{aligned}$$

$$= a_t(A(t)^{-1}f - A(s)^{-1}f,\ K_h(t)^*w),$$

whence follows with the aid of Lemma 3.2 that

$$\begin{aligned}||I_1-I_2|| &\leqq c_5|t-s|^\theta||A(t)^{-1}f-A(s)^{-1}f||_2\cdot||K_h(t)^*w|| \\ &\leqq c_5|t-s|^\theta\cdot C||f||\cdot Ch^2||w|| \\ &\leqq Ch^2|t-s|^\theta||f||\cdot||w||\ .\end{aligned}$$

Nextly, by (3.5) we can write

$$I_4 = a_t(K_h(s)f,\ \psi_h(t)) - a_s(K_h(s)f,\ \psi_h(t)).$$

Then we have

$$\begin{aligned}I_4 = &\ a_t(K_h(s)f,\ (A(t)^*)^{-1}w)-a_t(K_h(s)f,K_h(t)^*w) \\ &- a_s(K_h(s)f,\ (A(t)^*)^{-1}w) + a_s(K_h(s)f,K_h(t)^*w)\end{aligned}$$

for $\psi_h(t) = (A(t)^*)^{-1}w - K_h(t)^*w$. Hence by means of (2.7) and (2.9) we have

$$\begin{aligned}||I_4|| &\leqq c_6|t-s|^\theta||K_h(s)f||\cdot||(A(t)^*)^{-1}w||_2 \\ &\quad + c_4|t-s|^\theta||K_h(s)f||_1\cdot||K_h(t)^*w||_1 \\ &\leqq C|t-s|^\theta(Ch^2||f||\cdot C||w||+Ch||f||\cdot Ch||w||) \\ &\leqq Ch^2|t-s|^\theta||f||\cdot||w||\end{aligned}$$

with the aid of Lemmas 3.1 and 3.2. Summing up the estimates thus obtained, we have (3.15) and establish the lemma. Q.E.D.

§4. Error estimates for the initial value problem.

Under the assumptions stated in §2 we claim

Theorem 4.1. Let φ be any function in $\mathcal{D}$. Then we have

$$(4.1)\qquad \| U(t,s)\varphi - U_h(t,s)P_h\varphi\| \leqq Ch^2\|\varphi\|_2 ,$$

for $0 \leqq s \leqq t \leqq T$.

Proof. Without loss of generality we assume $s < t$ and put

$$(4.2)\qquad \begin{cases} e_h^{(1)} = U(t,s)\varphi - P_hU(t,s)\varphi , \\ e_h^{(2)} = P_hU(t,s)\varphi - U_h(t,s)P_h\varphi . \end{cases}$$

By means of (3.11) we have

$$(4.3)\qquad \begin{aligned} h^{-2}\|e_h^{(1)}\| &\leqq C\|U(t,s)\varphi\|_2 \leqq C\|A(t)U(t,s)\varphi\| \\ &\leqq C\|A(t)U(t,s)A(s)^{-1}\|\cdot\|A(s)\varphi\| \\ &\leqq C\|\varphi\|_2 \end{aligned}$$

with the aid of (2.12) and (2.6). On the other hand we have

$$\begin{aligned} e_h^{(2)} &= \int_s^t \frac{\partial}{\partial r} U_h(t,r)P_hU(r,s)\varphi\, dr \\ &= \int_s^t U_h(t,r)(A_h(r)P_h - P_hA(r))U(r,s)\varphi\, dr \\ &= \int_s^t J_h(t,r)K_h(r)H(r,s)\varphi\, dr , \end{aligned}$$

where

$$J_h(t,r) = U_h(t,r)A_h(r)P_h ,$$

$$K_h(r) = A(r)^{-1} - P_hA_h(r)^{-1} \qquad \text{(as in §3)},$$

$$H(r,s) = A(r)U(r,s).$$

Rewriting the integrand of the integral above by telescoping, we have

$$(4.4) \qquad e_h^{(2)} = I_1 + I_2 + I_3$$

with

$$I_1 = \int_s^t J_h(t,r)K_h(r)(H(r,s)-H(t,s))\varphi\, dr,$$

$$I_2 = \int_s^t J_h(t,s)(K_h(r)-K_h(t))H(t,s)\varphi\, dr$$

and

$$I_3 = \int_s^t J_h(t,r)K_h(t)H(t,s)\varphi\, dr.$$

We are going to estimate $\|I_j\|$ $(j = 1,2,3)$. Firstly,

$$(4.5) \qquad \|I_1\| \leq \int_s^t C(t-r)^{-1}\cdot Ch^2\cdot\|(H(r,s)-H(t,s))\varphi\|dr$$

in virtue of (2.18) and Lemma 3.1. Furthermore we have according to § 2

$$\begin{aligned}\|(H(r,s)-H(t,s))\varphi\| &= \|(A(r)U(r,s)-A(t)U(t,s))\varphi\|\\ &\leq \|A(t)(U(t,s)-U(r,s))\varphi\|+\|(A(t)-A(r))U(r,s)\varphi\|\\ &\leq \|A(t)(U(t,s)-U(r,s))A(s)^{-1}\|\cdot\|A(s)\varphi\|\\ &\quad + \|(A(t)A(r)^{-1}-I)\|\cdot\|A(r)U(r,s)A(s)^{-1}\|\cdot\|A(s)\varphi\|\\ &\leqq C(t-r)^{1/2}(r-s)^{-1/2}\cdot C\|\varphi\|_2 + C(t-r)^{\theta}\cdot C\|\varphi\|_2\\ &\leqq C(t-r)^{1/2}(r-s)^{-1/2}\|\varphi\|_2 .\end{aligned}$$

Substituting this estimate into (4.5), we get

$$\begin{aligned}(4.6) \qquad \|I_1\| &\leqq Ch^2 \int_s^t (t-r)^{-1/2}(r-s)^{-1/2}\|\varphi\|_2\, dr\\ &= Ch^2 B(\tfrac{1}{2}, \tfrac{1}{2})\|\varphi\|_2 = Ch^2\|\varphi\|_2.\end{aligned}$$

Nextly, according to Lemma 3.3 we can estimate as

$$\begin{aligned}(4.7) \qquad \|I_2\| &\leqq \int_s^t C(t-r)^{-1}\cdot Ch^2(t-r)^{\theta}\|H(t,s)\varphi\|dr\\ &\leqq Ch^2(t-s)^{\theta}\|A(t)U(t,s)A(s)^{-1}\|\cdot\|A(s)\varphi\|\end{aligned}$$

$$\leqq Ch^2T^{\theta}\cdot C\cdot C\|\varphi\|_2 = Ch^2\|\varphi\|_2.$$

Finally, we have

$$I_3 = \int_s^t \frac{\partial}{\partial r} U_h(t,r)P_hK_h(t)H(t,s)\varphi\, dr$$

$$= [U_h(t,r)P_hK_h(t)H(t,s)\varphi]_{r=s}^{r=t}$$

$$= P_hK_h(t)H(t,s)\varphi - U_h(t,s)P_hK_h(t)H(t,s)\varphi .$$

From this follows, in view of $\|P_h\| = 1$ and $\|U_h(t,s)\| \leqq 1$, that

$$\|I_3\| \leqq Ch^2\|H(t,s)\varphi\| \leqq Ch^2\|\varphi\|_2. \tag{4.8}$$

Consequently, we end up with

$$\|e_h^{(2)}\| \leqq Ch^2\|\varphi\|_2$$

which yields the theorem when combined with (4.3). Q.E.D.

From Theorem 4.1 we get our main result Theorem 4.2 below, again under the assumptions stated in § 2.

Theorem 4.2. If $E_h(t,s) = U(t,s) - U_h(t,s)P_h$, then we have

$$\|E_h(t,s)\| \leqq Ch^2(t-s)^{-1} \qquad (0 \leqq s < t \leqq T). \tag{4.9}$$

In particular, the error between the exact solution $u(t) = U(t,0)u_0$ and the approximate solution $u_h(t) = U_h(t,0)P_hu_0$ admits of the following estimate;

$$\|u(t) - u_h(t)\| \leqq Ch^2t^{-1}\|u_0\| \qquad (t > 0). \tag{4.10}$$

Proof. Following Helfrich [6] we estimate $\|E_h(t,s)\|$ as below. Firstly we note that

$$E_h(t,s) = E_h(t,r)U(r,s) + U_h(t,r)P_hE_h(r,s)$$

for $0 \leqq s < r < t \leqq T$. Taking $r = (t+s)/2$, we have

$$\|E_h(t,r)U(r,s)\| \leqq \|E_h(t,r)A(r)^{-1}\|\cdot\|A(r)U(r,s)\| \tag{4.11}$$

$$\leqq Ch^2 \cdot C(r-s)^{-1} \leqq Ch^2(t-s)^{-1} ,$$

since (4.1) implies $\|E_h(t,s)A(s)^{-1}\| \leqq C$ in virtue of (2.6). On the other hand, from the equalities

$$U_h(t,r)P_hE_h(r,s) = U_h(t,r)A_h(r)A_h(r)^{-1}P_hE_h(r,s)$$

$$= U_h(t,r)A_h(r)P_h(A(r)^{-1}-K_h(r))E_h(r,s)$$

$$= U_h(t,r)A_h(r)P_hA(r)^{-1}E_h(r,s)$$

$$- U_h(t,r)A_h(r)P_hK_h(r)E_h(r,s)$$

we have

$$(4.12) \quad \|U_h(t,r)P_hE_h(r,s)\| \leqq C(t-r)^{-1}\|A(r)^{-1}E_h(r,s)\| + C(t-r)^{-1}\cdot 1\cdot Ch^2\cdot 2$$

in view of (2.18), (3.8) and $\|E_h(r,s)\| \leqq 2$. By considering the adjoint operators and noting that Theorem 4.1 is valid for these, we see that

$$\|A(r)^{-1}E_h(r,s)\| = \|E_h(r,s)^*(A(r)^*)^{-1}\| \leqq Ch^2.$$

Using this estimate in (4.12) and recalling that $r = (t+s)/2$, we get from (4.12)

$$\|U_h(t,r)P_hE_h(r,s)\| \leqq Ch^2(t-s)^{-1} ,$$

which yields (4.9) when combined with (4.11). Q.E.D.

§5. Remarks on the case of the Neumann boundary condition.

Since the conormal changes as $a_{ij}(t,x)$ changes, the domain $D(A(t))$ of $A(t)$ is not constant if the imposed boundary condition is the Nuumann boundary condition (1.5). This lack of the constancy of $D(A(t))$ causes quite a little difficulty in estimating the error. For instance, (2.14) is no longer true under these circumstances. Actually, $A(t)U(r,s)$ does not even make sense unless $t = r$.

Nevertheless, we can prove the following two theorems which we present here without proof. The proof will be given in a forthcoming paper by the second author (Suzuki [12]).

Theorem 5.1. Under the assumption stated in § 2, we have for the case of the Neumann boundary condition

(5.1) $$\|E_h(t,s)\| \leqq C_\delta(h^2/(t-s))^{1-\delta} \qquad (0 \leqq s < t \leqq T)$$

for any δ in $0 < \delta \leqq 1/2$ with a constant C_δ depending on δ.

Theorem 5.2. In addition to the assumptions of the preceding theorem, suppose that $A(t)$ is self-adjoint. Then we have

(5.2) $$\|E_h(t,s)v\|_1 \leqq Ch(t-s)^{-1}\|v\| \qquad (0 \leqq s < t \leqq T)$$

for any $v \in X$.

References

[1] Y. Fujie and H. Tanabe, On some parabolic equations of evolution in Hilbert space, Osaka J. Math., 10 (1973), 115-130.

[2] H. Fujita, On the finite element approximation for parabolic equations: an operator theoretical approach, Proceedings of the 2nd IRIA symposium on computing methods in applied sciences and engineering, 1975, December, Springer LN in Econ. and Math. Syst. 134, (1976), 171-192.

[3] H. Fujita, On the semi-discrete finite element approximation for the evolution equation $u_t + A(t)u = 0$ of parabolic type, Topics in Numerical Analysis, Vol.3 (Proceedings of Conference on Numerical Analysis, Dublin, 1976, August), Academic Press, London, to appear, 425-437.

[4] H. Fujita and A. Mizutani, On the finite element method for parabolic equations, I: Approximation of holomorphic semi-groups, J. Math. Soc. Japan, 28 (1976), 749-771.

[5] H. Fujita and A. Mizutani, Remarks on the finite element method for parabolic equations with higher accuracy, Proceedings of Japan-France seminar on functional analysis and numerical analysis, 1976, September, to appear.

[6] H. P. Helfrich, Fehlerabschätzungen für das Galerkinverfahren zür Lösung von Evolutionsgleichungen, Manus. Math., 13 (1974), 219-235.

[7] H. P. Helfrich, Lokale Konvergenz des Galerkinverfahrens bei Gleichungen vom parabolischen Typ in Hilberträumen, Thesis, 1975.

[8] T. Kato, Abstract evolution equations of parabolic type in Banach and Hilbert spaces, Nagoya Math. J., 5 (1961), 93-125.

[9] T. Kato, Perturbation Theory for Linear Operators, Springer, Berlin-Heidelberg-New York, 1966.

[10] P. A. Raviart, Multistep methods and parabolic equations, Proceedings of Japan-France seminar on functional analysis and numerical analysis, 1976, September, to appear.

[11] P. E. Sobolevskii, Parabolic type equations in Banach spaces, Trudy Moscow Math., 10 (1961), 297-350.

[12] T. Suzuki, An abstract study of Galerkin's method for the evolution equation $u_t + A(t)u = 0$ of parabolic type with the Neumann boundary condition, to apperar in J. Fac. Sci. Univ. Tokyo.

[13] H. Tanabe, On the equation of evolution in a Banach space, Osaka Math. J., 12 (1960), 363-376.

[14] M. Zlámal, Curved elements in the finite element method I, SIAM J. Numer. Anal., 10 (1973), 229-240.

SPECIAL APPLICATIONS OF HAMILTONS'S PRINCIPLE TO STRUCTURAL DYNAMICS

M. GERADIN
Chercheur Qualifié du F.N.R.S.
L.T.A.S.
University of Liège
Belgium

1. Introduction

Eigenvalue analysis is probably the less sensitive problem of structural analysis to discretization errors, due to the remarkable property of stationarity of Rayleigh's quotient with respect to errors on eigenmodes. Mainly for this reason there is, in structural dynamics, a still stronger tendency than in elastodynamics of by-passing variational theorems for constructing discretized models. Mass lumping in particular, although not justified from a variational point of view, has become a faily common practice.

It seems thus worthwile showing the richness contained in the variational theorems that can be deduced from Hamilton's principle, even for their application to geometrically non-linear dynamic problems. The time dimension makes their number still larger than elastostatics, but the only useful ones, as it is shown in section 2, operate on the following fields : displacements, velocities and stresses. As in elastostatics, all of them are specialized forms of a canonical principle operating on the three fields simultaneously. For a full demonstration of their filiation, the reader is referred to GERADIN and SANDER [15] .

Emphasis is made on their generalization to geometrically non-linear problems : section 3 is an attempt to underline the advantages that can be gained, and the limitations that are encountered in their implementation.

Section 4 and 5 are devoted to numerical experiments in plate vibration and non-linear dynamic response problems. Comparison is given between the solutions provided by Hamilton's principle, a mixed form of it operating on displacements and velocities and, for eigenvalue problems, Toupin's principle.

2. Variational principles for small and large displacements in elastodynamics

This section is an attempt to summarize the variational formulations that exist for structural dynamics. For geometrically non-linear analysis, the formulation used is in terms of Green strains and Kirchhoff-Trefftz tensors referenced to the initial configuration.The case of non-conservative and deformation-dependent loads has been discarded for simplicity, but could have been considered by adding to the functionals , when possible, the appropriate virtual work expressions.

2.1. HAMILTON's principle

Hamilton's principle, or displacement variational principle, states that for fixed-end values of the displacement field, the Lagrangian action of a conservative system

$$\mathcal{L}[u] = \int_{\tau_1}^{\tau_2} (T-U)d\tau \tag{1}$$

takes a stationary value on the trajectory of the motion.

T is the kinetic energy of the system:

$$T(u) = \frac{1}{2} \int_{V_o} \rho_o \dot{u}_i \dot{u}_i \, dV_o \tag{2}$$

and U, its potential energy, can be split into distinct parts:
the strain energy

$$U_1 = \int_{V_o} W(\varepsilon_{ij}) \, dV_o = \int_{V_o} W(D_i u_j) dV_o \tag{3}$$

results from the integration of the strain energy $W(\varepsilon_{ij})$ per unit of volume of reference over the domain V_o. The notation $W(D_i u_j)$ indicates that the strain energy density is a function of the displacement derivatives only. The region V_o is that occupied by the body in its initial state. If large displacements are considered, the strain energy density of the hyperelastic material can be expressed as a function of the Green strain tensor

$$\varepsilon_{ij} = \frac{1}{2} (D_i u_j + D_j u_i + D_m u_i D_m u_j) \tag{4}$$

to which the Kirchhoff-Trefftz stress tensor is associated by the constitutive relations

$$\sigma_{ij} = \frac{\partial W}{\partial \varepsilon_{ij}} \tag{5}$$

The second contribution to the potential energy, U_2, results from conservative body loads and surface tractions on the part S_σ of the boundary

$$U_2 = - \int_{S_\sigma} u_j \, \bar{t}_j(\tau) \, dS - \int_{V_o} u_j \, \bar{X}_j \, (\tau) \, dV \tag{6}$$

On the remaining part S_u of the boundary, the displacements are prescribed functions of time :

$$u_j = \bar{u}_j(\tau) \quad \text{on} \quad S_u \quad \text{at any time } \tau. \tag{7}$$

The variational derivatives of Hamilton's principle express equilibrium in the body and on the surface S_σ in a dynamical sense.

2.2. The canonical principle

One notices that Hamilton's principle contains two a priori requirements :

(a) compatibility in space : the strain field deduced from u_i by (4) has to be integrable, and on the part of the boundary where displacements are imposed,

$$u_i = \bar{u}_i \qquad \text{on } S_u \tag{8}$$

(b) compatibility in time : the time dependence between displacements and velocities

$$v_i = \dot{u}_i \qquad \text{in } V \tag{9}$$

is automatically assumed.

The canonical principle is one in which the conditions (4), (3) and (9) are incorporated. To construct it from Hamilton's principle by the method of FRIEDRICHS transformations, one proceeds in three steps :

(i) the continuity requirements (4), (3) and (9) are relaxed by adding to (1) the appropriate dislocation potentials with a lagrangian multiplier.

(ii) the meaning of the lagrangian multipliers utilized in (i) is restored by using the appropriate variational derivatives of the functional.

(iii) to eliminate the strain variables in profit of stresses, the complementary energy density $\phi(\sigma_{ij})$ is introduced by the contact transformation

(iii) By the variation δv_j one restores the dependence (9) between displacements and velocities.

The functional (13) is particularized to geometrically linear problems simply by noticing that, for infinitesimal displacements,

$$\frac{1}{2} \sigma_{ij} \ (D_i u_j + D_j u_i + D_i w_m \ D_j u_m) \simeq \sigma_{ij} \ D_i u_j \qquad \text{and}$$

$$t_j = n_i \ (\sigma_{ij} + \sigma_{im} \ D_m \ u_j) \simeq n_i \ \sigma_{ij} \ .$$

The only a priori conditions that remain in the canonical principle concern :

(i) the constitutive equations (11)

(ii) the symmetry of the Kirchhoff-Trefftz tensor $\sigma_{ij} = \sigma_{ji}$ which implies that rotational equilibrium is satisfied everywhere. A still more general principle could be derived by relaxing this condition, as it will be briefly mentioned in the next section.

The continuity requirements on the three fields are the following :

displacements u_i	C_1 continuity
velocities v_i	no continuity
stresses σ_{ij}	C_o continuity

2.2. Two-field variational principles

Two field variational principles are obtained by specializing the canonical principle to the cases where one set of Euler's equations are taken as essential conditions. The remaining two fields are left independent.

2.2.a. Principle operating on displacements and stresses

The principle operating on displacements and stresses is obtained by restoring the dependence (9) between displacements and velocities. It can be written

$$\delta \mathcal{F} \left[u,\sigma\right] = \delta \int_{\tau_1}^{\tau_2} F \left[u,\sigma\right] \, d\tau = 0$$

with the two-field functional

$$\phi(\sigma_{ij}) = \sigma_{ij}\,\varepsilon_{ij} - W(\varepsilon_{ij}) \tag{10}$$

The derivation of (10) with respect to stresses yields the constitutive relations in inverse form

$$\varepsilon_{ij} = \frac{\partial \phi}{\partial \sigma_{ij}} \tag{11}$$

The resulting variational principle operates on displacements, stresses and velocities

$$\delta \mathcal{J}[u,v,\sigma] = \delta \int_{t_1}^{t_2} F\,[u,v,\sigma]\; d\tau = 0 \tag{12}$$

with the functional

$$F[u,v,\sigma] = \int_{V_o} \Big[\phi(\sigma_{ij}) - \frac{1}{2}\,\sigma_{ij}\,(D_i u_j + D_j\, u_i + D_i u_m\, D_j u_m) + \bar{X}_i u_i - \frac{1}{2}\,\rho\; v_i v_i + \rho\; v_i \dot{u}_i\Big]\, dV_o + \int_{S_\sigma} \bar{t}_i\; u_i\; d\,S + \int_{S_u} t_j\;(u_j - \bar{u}_j)\; dS \tag{13}$$

where the t_j are the lagrangian surface tractions on the undeformed boundary

$$t_j = n_i\;(\sigma_{ij} + \sigma_{im}\, D_m\, u_j) \tag{14}$$

Its Euler equations provide the full set of equations of structural dynamics :

(i) the variation δu_i yields the dynamic equilibrium equations in the form given by Signorini :

$$\begin{aligned} &D_i\;(\sigma_{ij} + \sigma_{im}\, D_m\, u_j) + \bar{X}_j - \rho \dot{v}_j = 0 \qquad \text{in } V_o \\ &n_i\;(\sigma_{ij} + \sigma_{im}\, D_m\, u_j) + \bar{t}_j = 0 \qquad \text{on } S_\sigma \end{aligned} \tag{15}$$

(ii) Varying the stresses σ_{ij} and the surface tractions t_j restores the compatibility in space :

$$\begin{aligned} &\frac{\partial \phi}{\partial \sigma_{ij}} - \frac{1}{2}\,(D_i u_j + D_j u_i + D_i u_m\, D_j u_m) = 0 \text{ in } V_o \\ &u_j = \bar{u}_j \qquad \text{on } S_u \end{aligned} \tag{16}$$

$$F[u,\sigma] = \int_{V_o} \left[\phi(\sigma_{ij}) - \frac{1}{2}\sigma_{ij}(D_iu_j + D_ju_i + D_iu_m D_ju_m) + \bar{X}_i u_i + \frac{1}{2}\rho \dot{u}_i \dot{u}_i\right] dV + \int_{S_\sigma} \bar{t}_i u_i \, dS + \int_{S_u} t_i (u_i - \bar{u}_i) \, dS \tag{17}$$

It was formulated independently by FRAEIJS de VEUBEKE [6] and REISSNER [5] for static analysis of geometrically linear problems.

2.2.b. Principle operating on displacements and velocities (HUGHES)

It corresponds to the case where compatibility in space is rendered essential. It is obtained the most rapidly from Hamilton's principle by relaxing the dependence between displacements and velocities. The functional of this second two-field variational theorem of elastodynamics [14]

$$\delta \mathcal{J}[u,v] = \delta \int_{\tau_1}^{\tau_2} F[u,v] \, d\tau = 0$$

takes the form

$$F[u,v] = \int_{V_o} \left[- W(D_iu_j) + \bar{X}_i u_i - \frac{1}{2}\rho v_i v_i + \rho v_i \dot{u}_i\right] dV + \int_{S_\sigma} \bar{t}_i u_i \, dS \tag{18}$$

As it allows for independent assumptions on displacements and velocities, it can be used for reducing in a consistent way the number of parameters in the representation of the kinetic energy.

2.2.c. Principle operating on stresses and velocities (REISSNER)

The third two-field variational principle corresponds to the case where dynamic equilibrium (15) is an essential condition. As shown by FRAEIJS de VEUBEKE [11] , the existence of an explicit form of the functional $F[v,\sigma]$ is however limited to the geometrically linear case.

The functional of REISSNER's principle [1,2] , valid only for geometrically linear structures, is

$$F[v,\sigma] = \int_V \left[\phi(\sigma_{ij}) - \frac{1}{2}\rho v_i v_i\right] dV - \int_{S_u} t_i \bar{u}_i \, dS \; . \tag{19}$$

The formal independence of the stress and velocity fields in the associated variational principle is however seriously restricted in practice by the requirement of satisfying a priori dynamic equilibrium. As one has to consider infinitesimal displacements, the coupling between the two fields is reintroduced by

$$D_i \sigma_{ij} + \bar{X}_j - \rho \dot{v}_j = 0 \qquad \text{in V.} \tag{20}$$

One way of generalizing the functional (19) to finite elastic displacements, based on the polar decomposition of the Jacobian matrix of the transformation, was suggested by FRAEIJS de VEUBEKE [11] . The corresponding variational principle operates then one three fields : the Piola stress tensor, the velocities v_j and the matrix of finite rotations associated to the deformation. The resulting functional will not be explicited here.

2.3. Complementary energy principle (TOUPIN)

Just as Reissner's principle, the complementary energy principle is limited to linear elastodynamics.

The requirement to satisfy the dynamic equilibrium equations (20) suggests the introduction of an impulse field θ_{ij} [5] such that

$$v_j = \frac{1}{\rho} D_i \theta_{ij} \tag{21}$$

Hence, in the absence of body forces

$$\theta_{ij} = [\theta_{ij}]_{\tau_1} + \int_{\tau_1}^{\tau} \sigma_{ij} \, d\tau \qquad \text{and} \qquad \dot{\theta}_{ij} = \sigma_{ij} \, . \tag{22}$$

The resulting functional is

$$F(\theta) = \int_V \left[\phi(\dot{\theta}_{ij}) - \frac{1}{2\rho} D_i \theta_{ij} D_m \theta_{mj}\right] dV - \int_{S_u} t_j \bar{u}_j \, dS \tag{23}$$

with the modified expression of the surface tractions.

Euler's equations of this principle are obtained by variation of the impulses. They appear to be a disguised form of the time derivatives of the compatibility equations of linear elasticity :

$$\begin{aligned} &-\frac{d}{d\tau}\left(\frac{\partial \phi}{\partial \dot{\theta}_{ij}}\right) + \frac{1}{2\rho}\left(D_i D_m \theta_{mj} + D_j D_m \theta_{mi}\right) = 0 && \text{in V} \\ &-\frac{1}{\rho} D_i \theta_{ij} + \frac{d\bar{u}_j}{d\tau} = 0 && \text{on } S_u \end{aligned} \tag{24}$$

2.4. Special purpose two-field variational principles

Special forms of the two-field variational principles are useful in the finite element context to derive the so-called hybrid elements. The motivation for using these principles is always the desire to make independent assumptions inside the domain constituted by a finite element and along its boundary for the choice of connection modes.

Such principles have been introduced by PIAN [7] and PIAN and TONG [8], and play an important role in the justification of elements based on engineering intuition but violating the rules of the general principles.

The first of these principles allows for assumptions on the displacements inside the boundary and along the boundary S_σ that are independent of the assumptions on the lagrangian surface tractions along S_u, as recalled by the notation

$$\delta \mathcal{F}[u, t(S_u)] = \delta\int_{\tau_1}^{\tau_2} F\,[u, t(S_u)] = 0 .$$

The corresponding functional is

$$F\,[u, t(S_u)] = \int_{V_o} \left[\frac{1}{2}\rho\dot{u}_i\dot{u}_i - W(D_i u_j) + \bar{X}_i u_i\right] dV + \int_{S_\sigma} \bar{t}_i\, u_i\, dS + \int_{S_u} t_i\,(u_i - \bar{u}_i)\, dS . \tag{25}$$

The complementary functional obviously exists in the geometrically linear case. One then assumes that dynamic equilibrium is a priori satisfied only in the volume. The resulting principle operates on the impulses θ_{ij} in the volume and on the displacements on S_σ

$$\delta \mathcal{J}[\theta, u(S_\sigma)] = \delta\int_{\tau_1}^{\tau_2} F[\theta, u(S_\sigma)]\, d\tau = 0$$

with the functional

$$F\,[\theta, u(\sigma)] = \int_V \left[\phi(\dot{\theta}_{ij}) - \frac{1}{2\rho} D_i\, \theta_{ij}\, D_m\, \theta_{mj}\right] dV + \int_{S_\sigma} u_j\,(\bar{t}_j - n_i\,\dot{\theta}_{ij})\, dS - \int_{S_u} n_i\,\dot{\theta}_{ij}\,\bar{u}_j\, dS . \tag{26}$$

The filiation of the various principles is schematized on the flowchart next page.

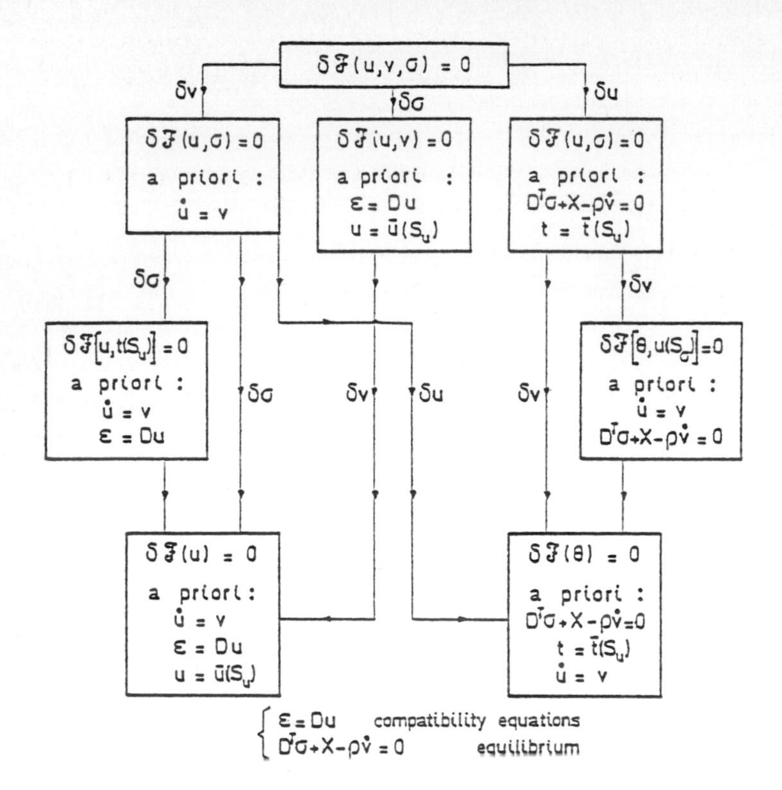

FILIATION OF THE VARIATIONAL PRINCIPLES

3. Practical use of the variational principles of elastodynamics

Before exhibiting in section 4 the numerical properties of some of the variational theorems described in section 2, it seems worthwile attempting to make a comparative study of their respective properties and limitations according to the use that can be made of them. A summary of these is given in the table below.

Canonical and F $[u,\sigma]$ principles

Up to now, they have always been considered of no practical interest from the point of view of finite element discretization. It seems, however, that this assertion should be reconsidered for non-linear analysis. The observation that the complementary energy remains a quadratic form of the stresses for geometrically non-linear problems indicates that a mixed discretization could perhaps provide a more adequate approach than a purely kinematical modelling.

F $[u,v]$ principle

The F $[u,v]$ functional has been regarded independently by HUGHES [14] and GERADIN, SANDER and NYSSEN [16] as a consistent way of lumping masses. The results obtained with it in eigenvalue analysis are quite promising, as it will be shown on numerical examples. However, the (u,v) approach does not give a systematic convergence to the eigenspectrum by either upper or lower bounds.

Its main drawback seems to be the practical impossibility of implementing it together with explicit time integration schemes in non-linear problems.

Complementary energy principles

A few global observations concern the variational principles based on the complementary energy (REISSNER, TOUPIN and PIAN) :

- they cannot be implemented for geometrically non-linear problems;
- they do not allow for an easy handling of other body loads than inertia forces;
- according to the dynamic equilibrium equation (20), the actual stress field contains two sets of stress modes |10] : an infinite set of self-stressing modes, solutions of the homogeneous equilibrium equation $D_i \sigma_{ij} = 0$, and particular solutions of the non-homogeneous equation (20).

The two first observations indicate that the practical application of complementary energy principles is limited to eigenvalue analysis. The last observation has two consequences :

First, for 2 - and 3 - dimensional elastic structures, the number of self-stressing modes included in a finite model is generally much larger than that of stress modes that equilibrate the body loads. This gives a significant reduction of the number of parameters in the representation of the kinetic energy, just as with the F $[u,v]$ functional.

Secondly, the convergence properties of equilibrium models are strongly influenced by the balance of particular stress modes and self-stressing modes. The smaller is the ratio of the number of particular stress modes to that of self-stressing modes, the larger is the chance of getting a convergence behavior to the eigenvalues by lower bounds |12| .

VARIATIONAL PRINCIPLE	FUNCTIONAL	EULER EQUATIONS	NON-LINEAR ELASTICITY	REDUCTION OF INERTIA FORCES	PRACTICAL INTEREST	OTHER LIMITATIONS
CANONICAL	$F[u,v,\sigma]$	• compatibility in space • compatibility in time • equilibrium	yes	yes	?	——
FRAEIJS DE VEUBEKE	$F[u,\sigma]$	• compatibility in space • equilibrium	yes	no	?	——
REISSNER	$F[v,\sigma]$	• compatibility in space	no	yes	?	body loads
HUGHES	$F[u,v]$	• compatibility in space • equilibrium	yes	yes	yes	non linear explicit schemes
HAMILTON	$F[u]$	• equilibrium	yes	no	yes	——
TOUPIN	$F[\sigma]$	• compatibility in space	no	yes	yes	body loads
"ANTI-PIAN"	$F[u, t\vert S_u]$	• equilibrium in V • compatibility on S_u	yes	no	?	——
PIAN	$F[\sigma, u\vert S_\sigma]$	• compatibility in V • equilibrium on S_σ	no	yes	yes	body loads

APPLICABILITY OF VARIATIONAL PRINCIPLES IN ELASTODYNAMICS

4. Plate vibration problems

The test example adopted is a cantilever plate. Its analysis has been performed for various skew angles : 0°, 15°, 30°, 45°, and with three different finite element models corresponding to distinct assumptions (fig. 1).

The plate bending conforming model (CQ)

The element used in the displacement approach is the purely conforming Fraeijs de Veubeke's quadrangle [9] . This well-known element is obtained from an assembling of four triangular regions with cubic deflection. The generalized displacements of the element are the deflection w and the two slopes $\frac{\partial w}{\partial x}$ and $\frac{\partial w}{\partial y}$ at each vertex, and the slopes $\frac{\partial w}{\partial n}$ normal to the external interfaces. The resulting model has 16 degrees of freedom. It is denoted CQ in the tables and diagrams.

The plate bending mixed (u,v) model (MXQ)

The mixed model, based on separate approximations of displacements and velocities, is derived from the purely conforming quadrangle CQ described above. It is noted MXQ. A linear variation of velocities over the whole element has been adopted to discretize the kinetic energy : account is thus taken of the only inertia forces associated to rigid body motions. The resulting element possesses 19 d.o.f. : 16 generalized displacements, and 3 generalized velocities.

As shown in reference [16] , all the generalized displacements can be eliminated before solving the eigenvalue problem, from what results a considerable economy in eigenvalue computation.

The plate bending equilibrium triangle (EQT)

This element results from the discretization of Toupin's principle. The field of self-stressing modes is obtained by assuming linearly varying moments. The resulting variables are 3 concentrated loads at vertices, 2 bending moments and 1 Kirchhoff shear load on each interface.The particular solution of the equation of dynamic equilibrium is introduced next by superimposing a cubic moment fields that provides a linear variation of velocities [10,13] . The resulting model has 15 degrees of freedom: 12 generalized connecting forces, and 3 internal parameters in terms of which the inertia forces of the element are represented. A considerable economy in eigenvalue computation can thus be expected with this EQT model just as with MXQ elements, since both are based on a similar approximation of the kinetic energy.

The numerical results

In the displacement analysis using CQ elements, the finest grid consists of 6x6 elements and 231 d.o.f., from which 37 are fixed. The results collected in table 1 confirm the upper boundness guaranteed by the kinematical approach. A significant deterioration of the finite element solution is observed when increasing the skewness of the plate.

The equilibrium approach (tables 2 and 3) underlines the same phenomenon. Distinct mesh patterns have been used in the square and skew cases. In the latter case, the finest mesh uses 50 elements and 420 d.o.f. . Only 150 of them contribute to the eigenvalue problem.

Note that in most cases the computed frequencies appear to be lower approximations, except when the discretization yields to a too rough representation of inertia forces.

Another important factor, when using triangular elements, is the skewness of the plate. The best subdivision is obtained, as expected, using 2 elements per quadrangle, with their common edge along the shorter diagonal.

Mixed models exhibit a not so systematic behavior as conforming and equilibrium elements. In most cases, they give a convergence by lower bounds, but such a behavior is not guaranteed.

The analysis with MXQ elements has been pushed up to a 8x8 mesh, with a total of 352 generalized displacements that are eliminated statically and 192 velocity parameters in terms of which the eigenvalue problem is solved. The corresponding results are given in table 4.

All the remarks that can be made about the comparison of the respective models strike up from the figure 2 which represents, for mode 1, the convergence as a function of three factors :

- the number of d.o.f. in terms of which the eigenvalue problem is solved;
- the skewness of the plate: the curves associated to different skew angles have been shifted horizontally.
- the type of model used.

The equilibrium models would normally give three distinct convergence curves according to the mesh subdivision adopted. Only the most favorable case (subdivision ⧄) has been represented on figure 1.

The results reported in table 5 allow a comparison with analyzes achieved by other authors. Among others, note that the experimental eigenfrequencies given by Barton [4] , even after the correction compensating the virtual inertia due to aerodynamic forces, are not bracketed by the bounds obtained from the dual analysis. The advantage offered by several simultaneous analyses appears clearly in this case, as it allows for a better understanding of the discrepancies between experimental and numerical results. The differences are very likely due to the difficulty of achieving a perfectly clamped support.

The comparison of the various numerical results confirm the impression that the "consistent way of lumping masses" provided by mixed models can yield to a

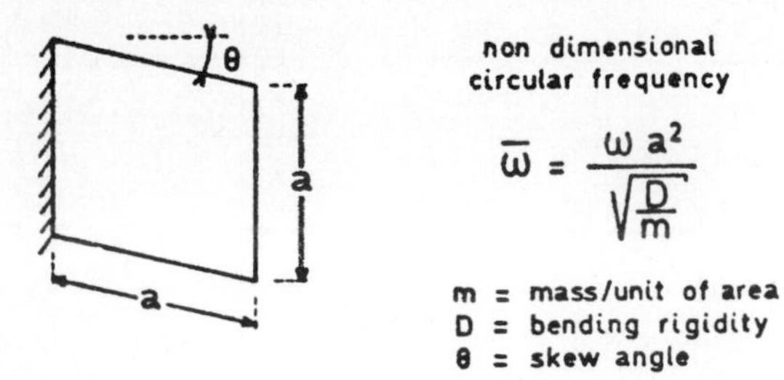

FIG. 1 SKEW CANTILEVER PLATE

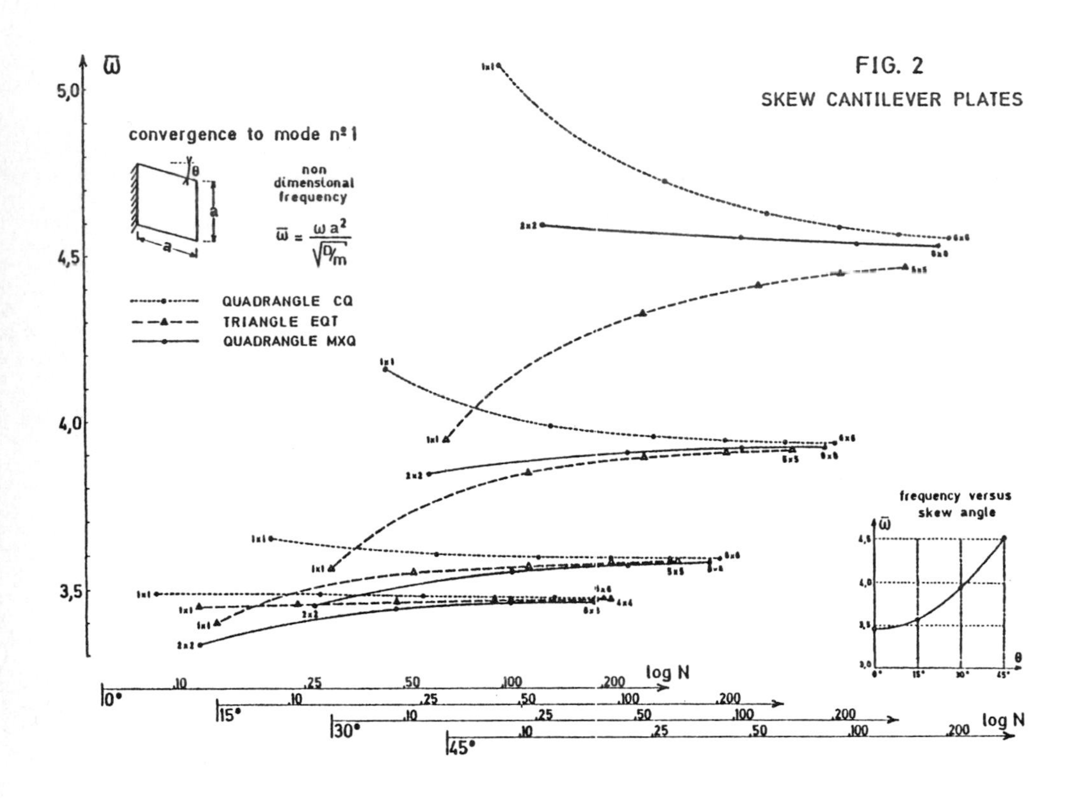

FIG. 2
SKEW CANTILEVER PLATES

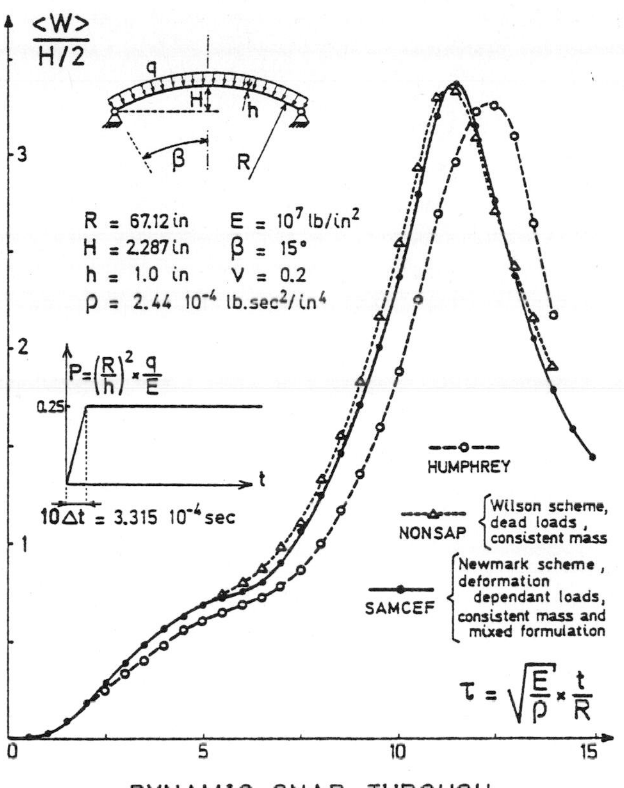

DYNAMIC SNAP-THROUGH
OF A SHALLOW CIRCULAR ARCH

FIG.3

significant saving of computational effort in free vibration analysis. The experimental confirmation of the same efficiency should still be made in linear and non-linear structural transient response.

5. Dynamic snap buckling of a shallow arch

The problem considered is that of a circular arch submitted in a very short time to pressure loading, as indicated on figure 3 . The material is assumed to be isotropic linear elastic. The analysis was performed using purely conforming elements with consistent mass matrix first, and then with elements obtained from the discretization of the F [u,v] functional. The purpose was to evaluate the effect of a crude representation of inertia in non-linear dynamics.
The same basic element has been utilized in both cases: the isoparametric quadrangle with 8 nodes. In the mixed approach, a linear variation of velocities is assumed, giving thus 3 velocity parameters to represent the kinetic energy.

The solutions obtained in the present analysis are denoted SAMCEF on figure 3
The plotted deflection ratio is defined as

$$\Delta = \frac{\langle w \rangle}{\frac{H}{2}} = \frac{\text{average normal deflection}}{\text{average rise of arch.}}$$

In both cases the pressure load was considered as deformation dependent, and the time step used is indicated on the figure. The purely conforming solution is based on Newmark's iteration scheme with equilibrium iterations within a time step.
In the mixed approach, using the F [u,v] functional, the problem becomes of first order in time : the trapezoïdal rule was adopted and implemented as described in [16] . The two curves corresponding to the [u] and [u,v] models could not be dissociated on the figure. The NONSAP solution [17] , which was obtained using the same number of 8 node elements with consistent mass matrix, is given for comparison. Wilson's scheme with equilibrium iterations was used with a time step of same size, but the pressure was applied as a dead load. The small discrepancy between the NONSAP and SAMCEF solutions is probably due to this difference in assumptions. The comparison with Humphrey's solution [18] deserves some explanation: Humphrey's solution is based on the assumption of shallowness, and is the result of a series expansion with a limited number of terms. Several factors can thus justify the apparently large discrepancy observed. Our main concern, in performing this last numerical experiment, was to compare the mixed F [u,v] approach with that based on the F [u] functional in geometrically non-linear problems. The results obtained are self-explanatory: the cruder representation of inertia forces contained in the F[u,v] models does not alter the response of geometrically non-linear problems. The F [u,v] method can thus be regarded as a step further on the way of getting consistently lumped mass matrices.

2 subdivisions

MODE nº	1 x 1 ⊠	1 x 1 ⊛	2 x 2 ⊠	2 x 2 ⊛	3 x 3 ⊠	3 x 3 ⊛	4 x 4 ⊠
1	3,447	3,453	3,462	3,466	3,467	3,469	3,468
2	7,850	8,246	8,410	8,470	8,475	8,494	8,492
3	20,74	20,76	21,16	21,21	21,24	21,26	21,26
4	25,07	26,73	26,89	27,07	27,10	27,16	27,16
5	28,86	28,92	30,14	30,72	30,75	30,89	30,87

TABLE 2 EQUILIBRIUM APPROACH (EQT) : θ=0°

	1x1	2x2	3x3	4x4	5x5	6x6
0°	3,489 9,222 26,96 35,29 43,75	3,486 8,606 21,50 27,63 31,96	3,479 8,535 21,38 27,39 31,27	3,476 8,518 21,33 27,28 31,08	3,474 8,513 21,31 27,24 31,02	3,473 8,511 21,30 27,22 30,99
15°	3,652 9,376 27,88 35,21 47,42	3,602 8,872 22,66 26,78 35,14	3,591 8,765 22,42 26,52 34,38	3,587 8,731 22,32 26,41 34,10	3,586 8,717 22,28 26,37 33,99	3,585 8,710 22,26 26,36 33,94
30°	4,159 10,12 28,72 40,09 57,33	3,988 9,683 26,22 27,06 43,03	3,954 9,532 25,75 26,36 42,06	3,943 9,480 25,56 26,12 41,67	3,938 9,456 25,47 26,03 41,52	3,935 9,443 25,41 25,99 41,45
45°	5,070 12,19 31,42 48,87 72,04	4,722 11,47 28,43 34,15 56,91	4,624 11,37 27,65 32,62 52,35	4,582 11,32 27,37 32,18 51,38	4,560 11,30 27,25 31,98 51,04	4,547 11,29 27,18 31,86 50,90

TABLE 1 CONFORMING APPROACH (CQ)

3 types of subdivision: 2 x 2 ⧅ ; 2 x 2 ⧄ ; 2 x 2 ⊠

	1 x 1 ⧅	1 x 1 ⧄	1 x 1 ⊠	2 x 2 ⧅	2 x 2 ⧄	2 x 2 ⊠	3 x 3 ⧅	3 x 3 ⧄	3 x 3 ⊠	4 x 4 ⧄	4 x 4 ⊠	5x5 ⧄
15°	3,378 7,374 11,82 22,35 39,37	3,400 7,733 16,02 20,18 47,97	3,502 8,096 21,46 24,03 32,46	3,537 8,377 21,23 25,60 31,17	3,549 8,478 21,64 25,58 32,54	3,567 8,563 21,97 25,98 33,03	3,567 8,571 21,84 25,98 32,89	3,565 8,607 22,00 26,12 33,32	3,570 8,651 22,13 26,24 33,59	3,576 8,632 22,05 25,18 33,41	3,574 8,649 22,11 26,25 33,60	3,579 8,667 22,16 26,29 33,71
30°	3,374 7,917 10,95 23,11 37,92	3,562 8,681 19,17 21,08 45,15	3,709 8,884 22,89 25,58 36,79	3,758 8,929 22,07 25,10 34,88	3,348 9,169 24,05 25,33 40,02	3,830 9,257 24,56 25,53 40,33	3,857 9,166 23,91 25,32 39,14	3,892 9,297 24,76 25,68 40,82	3,906 9,338 24,97 25,98 41,03	3,907 9,266 24,57 25,60 40,34	3,907 9,338 24,97 25,78 41,03	3,914 9,365 25,09 25,85 41,18
45°	3,344 8,302 11,66 22,80 40,01	3,942 10,26 23,35 26,88 43,01	4,084 10,43 26,27 31,56 40,71	4,002 10,34 22,69 25,83 35,32	4,321 11,00 25,57 30,37 48,09	4,370 11,06 26,01 30,40 48,79	4,235 10,82 24,67 28,54 46,37	4,406 11,14 26,35 30,84 49,97	4,432 11,15 26,53 30,98 50,27	4,441 11,00 26,58 29,87 48,16	4,441 11,18 26,60 31,06 50,41	4,460 11,20 26,72 31,17 50,56

TABLE 3 EQUILIBRIUM APPROACH (EQT) : θ = 15°, 30°, 45°

	2 x 2	4 x 4	6 x 6	8 x 8
0°	3,332 8,031 19,33 24,98 30,59	3,439 8,367 20,91 26,71 30,63	3,457 [illegible] 21,14 26,99 30,84	3,462 [illegible] 21,21 27,08 30,90
15°	3,455 8,304 20,20 25,15 32,57	3,552 8,579 21,82 26,06 33,42	3,569 8,641 22,07 26,22 33,68	3,575 8,665 22,14 26,27 33,76
30°	3,845 9,153 22,97 26,78 38,90	3,908 9,331 24,74 26,01 40,72	3,920 9,375 25,07 25,94 41,05	3,923 9,391 25,17 25,93 41,17
45°	4,594 10,96 28,40 31,41 53,82	4,552 11,16 27,35 30,89 51,32	4,533 11,25 27,14 31,27 50,79	4,524 11,24 27,08 31,40 50,71

TABLE 4 MIXED APPROACH (MXQ)

(1) THEORY (RAYLEIGH - RITZ) AND EXPERIMENT - BARTON, JNL APP. MECH., 1951
(2) PARALLELOGRAM ELEMENT (4x4, 75 D.O.F.) : DAWE, JNL OF STRAIN ANALYSIS, 1966
(3) TRIANGULAR ELEMENT (8 ELEMENTS, 18 D.O.F.) : ZIENKIEWICZ, INT. JNL OF SOLIDS AND STRUCTURES, 1968
(4) DISPLACEMENT AND EQUILIBRIUM APPROACHES WITH CQ AND EQT ELEMENTS
(5) MIXED APPROACH WITH MXQ ELEMENTS

	theoretical (1)	test (1)	corrected test (1)	Dawe (2)	Zienkie--wicz (3)	CQ 6x6 (4)	EQT 5x5 (4)	MXQ 6x6 (5)
15°	3,60 8,87 – – –	3,38 8,63 21,49 26,04 33,01	3,44 8,68 – – –	3,59 8,71 21,59 – –	3,57 8,60 21,75 – –	3,58 8,71 22,26 26,36 33,94	3,58 8,67 22,16 26,29 33,71	3,58 8,67 22,14 26,27 33,76
30°	3,96 10,19 – – –	3,82 9,23 24,51 25,54 40,64	3,88 9,33 – – –	3,95 9,42 25,56 – –	3,98 9,19 24,56 – –	3,93 9,44 25,41 25,99 41,45	3,91 9,37 25,09 25,85 41,18	3,92 9,39 25,17 25,93 41,17
45°	4,82 13,75 – – –	4,26 11,07 26,52 30,13 50,19	4,33 11,21 – – –	4,59 11,14 27,48 – –	4,67 11,01 27,56 – –	4,55 11,29 27,18 31,86 50,90	4,46 11,20 26,72 31,17 50,56	4,52 11,24 27,08 31,40 50,71

TABLE 5 SKEW CANTILEVER PLATES COMPARISON OF RESULTS

References

1. E. REISSNER, "Note on the method of complementary energy", Jnl Math. Phys., vol 27, 2, pp 159-160, 1948.
2. E. REISSNER, "Complementary energy procedure for flutter calculations", Jnl Aero. Sci., vol 16, 2, pp 316-317, 1949.
3. E. REISSNER, "On a variational theorem in elasticity", Jnl Math. Phys., vol 29, 1950.
4. M.V. BARTON, "Vibration of rectangular and skew cantilever plates" Jnl Appl. Mech., 18, 129-134, 1951.
5. R.A. TOUPIN, "A variational principle for the mechanical type analysis of mechanical systems", Jnl Appl. Mech., June 1952, pp 151-153.
6. B.M. FRAEIJS de VEUBEKE, "Displacements and equilibrium models in the finite element method", chap. 9 in Stress Analysis, 1965.
7. T.H.H. PIAN, "Derivation of element stiffness matrices by assumed stress distributions", AIAA Jnl, vol 2, 1333-1336, 1964.
8. T.H.H. PIAN and P.TONG, "Rationalization in deriving element stiffness matrices by assumed stress approach", Proc. Sec. Conf. on Matrix methods in Structural Mech., AFB, OHIO, USAF report AFFDL-TR-68-150, 441-469, 1968.
9. B.M. FRAEIJS de VEUBEKE, " A conforming element for plate bending", Int. Jnl. Sol. Struct., 4.
10. M. GERADIN, "Computational efficiency of equilibrium models in eigenvalue analysis", IUTAM Symposium on High Speed Computing of Elastic Structures, vol. 2, pp 589-623, Congrès et Colloques de l'Université de Liège, 1971.
11. B.M. FRAEIJS de VEUBEKE, "A new variational principle for finite elastic displacements", Int. Jnl. Sci., vol. 10, pp 745-763, 1972.
12. S. IDELSOHN, "Influence du nombre des modes d'autotension dans l'analyse dynamique des plaques au moyen d'une famille d'éléments statiquement admissibles" LTAS, Université de Liège, rapport VF-21, 1972.
13. M. GERADIN, "Analyse dynamique duale des structures par la méthode des éléments finis", Coll. Pub. Fac. Sc. Appl., n°36, 1973.
14. T.J.R. HUGHES, "A reduction scheme for problems of structural dynamics", Int. Jnl. sol. Struct, vol 12, pp 749-767, 1976.
15. M. GERADIN and G. SANDER, "Variational principles in elastodynamics" LTAS, Université de Liège, Report VA-18, 1977.
16. M. GERADIN, G. SANDER and C. NYSSEN, "Mixed variational formulations in linear and non-linear structural dynamics", Symp. on Appl. of Computer Methods in Engineering, Los Angeles, 23-26 August, 1977.
17. K.J. BATHE, E. RAMM and E.L. WILSON, "Finite element formulations for large deformation dynamic analysis", Int. Jnl. Num. meth. engg., vol. 9, 353-386, 1975.
18. J.S. HUMPHREY, "On dynamic snap buckling of shallow arches" AIAA Jnl, 4, 878-886, 1966.

NON-LINEAR PROBLEMS, BIFURCATION

PROBLEMES NON LINEAIRES, BIFURCATION

Constructive Methods for Bifurcation and Nonlinear Eigenvalue Problems*

Herbert B. Keller
Applied Mathematics
101-50
California Institute of Technology
Pasadena, California
91125

1. Introduction. In [4] new methods are presented for solving nonlinear eigenvalue problems. Here we briefly outline some of these methods and indicate improvements on them. In particular a new proof of the existence of bifurcations results from the scheme we advocate for computing bifurcation branches. The inclusion of linearized stability analysis is also indicated. These techniques have been applied to the Navier-Stokes equations for flow between rotating disks and other problems and the interesting results will be reported elsewhere.

The basic equilibrium problem is assumed given in the form:

$$G(u, \lambda) = 0 , \quad (1.1)$$

where $G : \mathbb{B} \times \mathbb{R} \to \mathbb{B}$ for some Banach space $\mathbb{B}$, and we seek arcs or branches of solutions, say $u = u(\lambda)$, as depicted in Figure 1.

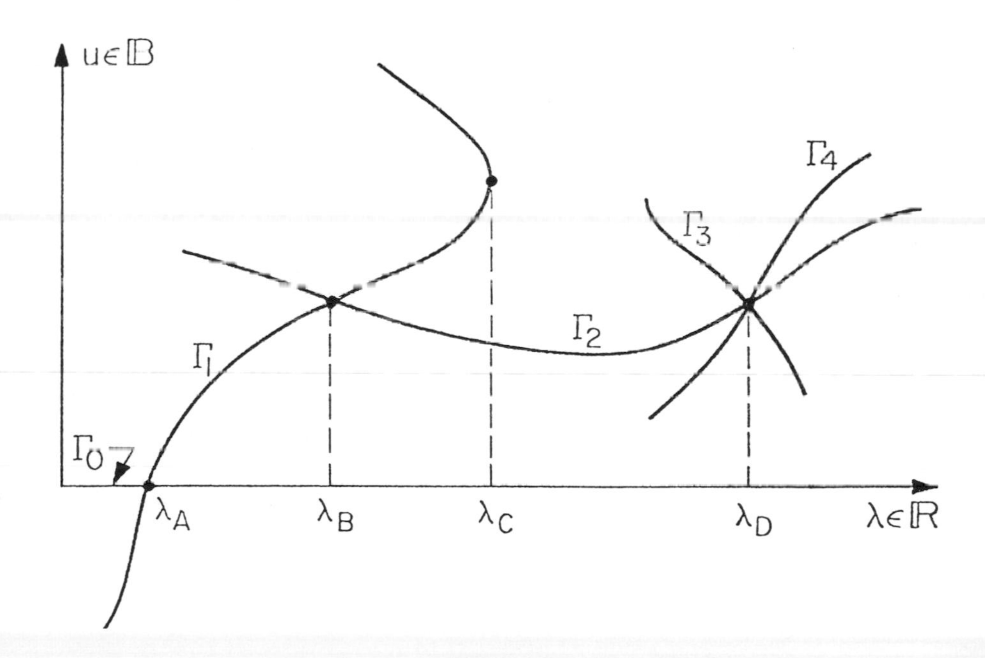

Figure 1.
Sketch of solution branches, limit points and bifurcating branches.

* This work was supported by the USERDA under Contract AT(04-3-767) and by the USARO under contract DAAG 29-75-C-0009.

There are inherent difficulties in trying to solve directly for branches parametrized in this way. Specifically limit points, as at $\lambda = \lambda_c$ in Figure 1 and bifurcation points, as at $\lambda = \lambda_a$, λ_b and λ_d, cause most constructive methods (i.e. Euler-Newton continuation) to fail. In particular if (u_o, λ_o) represents any such point then the Frechet derivative of G with respect to u,

(1.2) $$G_u^o \equiv G_u(u_o, \lambda_o)$$

is singular there. This causes difficulties and frequently failure of Newton-like methods.

Since the parametrization of solution arcs is arbitrary we seek solutions in the form: $u = u(s)$, $\lambda = \lambda(s)$ with parameter $s \in \mathbb{R}$. In addition to (1.1) we impose some normalization

(1.3) $$N(u, \lambda, s) = 0 \; .$$

Here $N : \mathbb{B} \times \mathbb{R}^2 \to \mathbb{R}$ is at our disposal and we choose it, as shown in Section 2, to make s an approximation to "arclength" on the solution branches. Under appropriate conditions this procedure eliminates limit points, allows us to determine bifurcation points accurately (by bisection for example) and enables us to switch branches at bifurcation (by changing the normalization, N).

The Frechet derivative of (G, N) with respect to (u, λ) has the form

(1.4) $$\frac{\partial(G, N)}{\partial(u, \lambda)} = \begin{pmatrix} G_u & G_\lambda \\ N_u & N_\lambda \end{pmatrix} .$$

Thus our procedure leads to a linearization which operates on an enlarged linear space: $\mathbb{B} \times \mathbb{R}$. As a result we refer to this procedure as "inflation". Crucial properties of the inflated linear operator are contained in the following simple lemma whose proof is contained in [5].

Lemma 1.5. Let $\mathbb{B}$ be a Banach space and let the linear operator $\mathcal{A} : \mathbb{B} \times \mathbb{R}^\nu \to \mathbb{B} \times \mathbb{R}^\nu$ have the form:

$$\mathcal{A} \equiv \begin{pmatrix} A & B \\ C^* & D \end{pmatrix} \quad \text{where} \quad \begin{cases} A : \mathbb{B} \to \mathbb{B}; & B : \mathbb{R}^\nu \to \mathbb{B}; \\ C^* : \mathbb{B} \to \mathbb{R}^\nu; & D : \mathbb{R}^\nu \to \mathbb{R}^\nu . \end{cases}$$

i) If A is nonsingular then $\mathcal{A}$ is nonsingular iff

(1.5) a) $$D - C^* A^{-1} B$$

is nonsingular.

ii) If $\mathcal{A}$ is singular with

(1.5) b) $$\dim \mathcal{N}(A) = \operatorname{codim} R(A) = \nu$$

then $\mathcal{A}$ is nonsingular iff

(1.5) $$c_0)\dim R(B) = \nu\ ,\quad c_1)R(B)\cap R(A) = 0\ ,$$

$$c_2)\dim R(C^*) = \nu\ ,\quad c_3)\mathcal{N}(A)\cap\mathcal{N}(C^*) = 0\ .$$

iii) If A is singular with $\dim\mathcal{N}(A) > \nu$ then $\mathcal{A}$ is singular.

This lemma has many applications in bifurcation theory, see [5] and Section 5, but here we only use the case with $\nu = 1$. Then conditions (1.5c) simply reduce to

(1.6) $$B \notin R(A) \quad\text{and}\quad C^* \notin R(A^*)\ .$$

2. Arclength Continuation. A natural parametrization for a solution branch $[u(s), \lambda(s)]$ is to use some form of arclength. Thus for any fixed $\Theta \varepsilon(0, 1)$ we can take for (1.3) the form:

(2.1)a) $$N_\Theta(u, \lambda, s) \equiv \Theta\ \|\dot{u}(s)\|^2 + (1-\Theta)|\dot{\lambda}(s)|^2 - 1 = 0$$

This form is not practical for computations so we use approximations to it. If $[u_0, \lambda_0]$ is a known solution to (1.1) then we set $[u(s_0), \lambda(s_0)] \equiv [u_0, \lambda_0]$ and use for $|s-s_0| < \delta$, say,

(2.1)b) $$N_2(u, \lambda, s) \equiv \Theta\ \|u(s) - u(s_0)\|^2 + (1-\Theta)|\lambda(s)-\lambda(s_0)|^2 - (s-s_0)^2 = 0$$

Alternatively if in addition to $[u(s_0), \lambda(s_0)]$ we know $[\dot{u}(s_0), \dot{\lambda}(s_0)]$ satisfying (2.1a) at $s = s_0$ then we can use:

(2.1)c) $$N_3(u, \lambda, s) = \Theta\dot{u}^*(s_0)[u(s)-u(s_0)] + (1-\Theta)\ \dot{\lambda}(s_0)[\lambda(s)-\lambda(s_0)] - (s-s_0) = 0.$$

Here $\dot{u}^*(s_0)$ is the element in $\mathbb{B}^*$ such that $\dot{u}^*(s_0)\ \dot{u}(s_0) = \|\dot{u}(s_0)\|^2$. We call N_2 and N_3 pseudo-arclength normalizations and proceed to show some of their properties.

A solution $[u_0, \lambda_0]$ of (1.1) is a regular point if G_u^0 of (1.2) is nonsingular and in addition there exist $[\dot{u}_0, \dot{\lambda}_0]$ such that

(2.2) $$a)\ G_u^0\ \dot{u}_0 + G_\lambda^0\dot{\lambda}_0 = 0\ ,\qquad b)\ \|\dot{u}_0\|^2 + |\dot{\lambda}_0|^2 > 0\ .$$

The solution is a normal limit point if (2.2) holds, G_u^0 is singular and satisfies:

(2.3) $$a)\ \dim\mathcal{N}(G_u^0) = \operatorname{codim} R(G_u^0) = 1,\qquad b)\ G_\lambda^0 \notin R(G_u^0)\ .$$

Note that (2.2a) and (2.3b) imply $\dot{\lambda}_0 = 0$. We have now

Theorem 2.3. Let $[u_0\lambda_0]$ be a regular point or a normal limit point. Let $G(u, \lambda)$ have two continuous derivatives in some sphere about $[u_0, \lambda_0]$. Then using $N \equiv N_3$ of (2.1c) there exists a smooth arc of solutions $[u(s), \lambda(s)]$ of

$$(2.4) \qquad G(u, \lambda) = 0, \qquad N(u, \lambda, s) = 0$$

on $|s-s_o| \leq \rho$ for some $\rho > 0$. On this solution arc the Frechet derivative (1.4), which is

$$(2.5) \qquad \mathcal{A}(s) \equiv \begin{pmatrix} G_u(s) & G_\lambda(s) \\ \Theta \dot{u}_o^* & (1-\Theta)\dot{\lambda}_o \end{pmatrix}$$

with $G_u(s) \equiv G_u(u(s), \lambda(s))$, etc., is nonsingular.

Proof. The proof uses Lemma 1.5 and is given in [4].
The corresponding result using $N \equiv N_2$ easily follows from the assumed smoothness. Since $\mathcal{A}(s)$ is nonsingular a number of constructive procedures can be used to prove the existence of the solution $x(s)$. Thus any smooth solution branch that is composed entirely of regular points and normal limit points can be determined by our pseudo-arclength normalization procedure. The numerical implementation of these procedures is indicated in [4] and [6] where Euler-Newton continuation is employed. That is given the solution $[u(s), \lambda(s)]$ at some value of s, we compute the tangent vector $[\dot{u}(s), \dot{\lambda}(s)]$ from

$$(2.6) \qquad \begin{aligned} &G_u(u(s), \lambda(s))\,\dot{u}(s) + G_\lambda(u(s), \lambda(s))\,\dot{\lambda}(s) = 0 \\ &N_u(u(s), \lambda(s), s)\dot{u}(s) + N_\lambda(u(s), \lambda(s), s)\,\dot{\lambda}(s) = -N_s(u(s), \lambda(s), s) \end{aligned}$$

Then we approximate the solution at $s + \Delta s$ by

$$(2.7) \qquad \begin{aligned} u^o(s + \Delta s) &= u(s) + \Delta s \dot{u}(s) \\ \lambda^o(s + \Delta s) &= \lambda(s) + \Delta s\, \dot{\lambda}(s) \end{aligned}$$

This is essentially one step of Euler's method to integrate (2.6).

With the initial iterate $[u^o(s + \Delta s), \lambda^o(s + \Delta s)]$ from (2.7) we employ Newton's method

$$(2.8) \qquad \begin{aligned} &G_u(u^\nu, \lambda^\nu)[u^{\nu+1}-u^\nu] + G_\lambda(u^\nu, \lambda^\nu)[\lambda^{\nu+1}-\lambda^\nu] = -G(u^\nu, \lambda^\nu) \\ &N_u(u^\nu, \lambda^\nu, s + \Delta s)[u^{\nu+1}-u^\nu] + N_\lambda(u^\nu, \lambda^\nu, s+ \Delta s)[\lambda^{\nu+1}-\lambda^\nu] = -N(u^\nu, \lambda^\nu, s+\Delta s) \end{aligned}$$

to solve (1.1), (1.3) at $s + \Delta s$.

3. Continuation Past and Location of Isolated Singular Points. A solution $x(s) \equiv \begin{pmatrix} u(s) \\ \lambda(s) \end{pmatrix}$ of (2.4) is a singular point if $\mathcal{A}(s)$ in (2.5) is singular. It is an isolated singular point if $\mathcal{A}(s^1)$ is nonsingular for all s^1 in a deleted neighborhood of s. Bifurcation points are, as we shall see, isolated singular points. Note that according to our definition, normal limit points are not singular points. Under mild smoothness conditions we show how continuation procedures can "jump" over isolated singular points and thus be used to locate them by a bisection process.

With the notation

$$(3.0)\qquad a)\ x(s) \equiv \begin{pmatrix} u(s) \\ \lambda(s) \end{pmatrix}, \qquad b)\ P(x,s) \equiv \begin{pmatrix} G(u,\lambda) \\ N(u,\lambda,s) \end{pmatrix},$$

we have recalling (1.4), that

$$(3.0)\qquad c)\ \mathcal{A}(s) = P_x(x(s), s)\,.$$

Let $x(s)$ be a smooth arc of solutions of (2.4) or equivalently of

$$(3.1)\qquad P(x, s) = 0\,,$$

for $s_a \leq s \leq s_b$ on which only $x(s_o)$ for some $s_o \in (s_a, s_b)$ is an isolated singular point. Then $\mathcal{A}(s)$ is nonsingular for each $s \in [s_a, s_b] - \{s_o\}$. In particular then the tangent $\dot{x}(s_a)$ is uniquely defined as in (2.6) with $s = s_a$ or now as the solution of

$$(3.2)\qquad a)\ \mathcal{A}(s_a)\,\dot{x}(s_a) = -P_s(x(s_a), s_a)\,.$$

An approximation to $x(s)$ is given as in (2.7) by the tangent line:

$$(3.2)\qquad b)\ x^o(s) \equiv x(s_a) + [s - s_a]\,\dot{x}(s_a)\,,$$

and with this initial estimate we try to construct a solution using the iteration scheme (chord or special Newton method):

$$(3.3)\qquad \begin{aligned} &a)\ \mathcal{A}^o(s) \equiv P_x(x^o(s), s)\,; \\ &b)\ \mathcal{A}^o(s)[x^{\nu+1}(s) - x^\nu(s)] = -P(s^\nu(s), s)\,,\quad \nu = 0, 1, \ldots \end{aligned}$$

To assure convergence we need only show that $x^o(s)$ is in an appropriate domain of attraction about $x(s)$. This has been done in [4] (see Theorem 4.4) under the assumptions that:

$$(3.4)\qquad \begin{aligned} &a)\ \|P_x(x(s), s) - P_x(y, s)\| \leq K(s)\|x(s) - y\|\,,\quad \text{if}\quad \|y - x(s)\| \leq \rho(s)\,, \\ &b)\ \max_{s_a \leq t \leq s} \|\ddot{x}(t)\| \leq \kappa(s)\,, \\ &c)\ |s - s_a|^2 \leq 2r(s)/\kappa(s)\,, \end{aligned}$$

where for some $\Theta(s) < 1/3$:

$$(3.4)\qquad d)\ r(s) \equiv \min\left[\rho(s), \frac{\Theta(s)}{M(s)K(s)}\right],$$

and

$$(3.4)\qquad e)\ M(s) = \|\mathcal{A}^{-1}(s)\| \quad \text{for}\quad s \neq s_o.$$

The convergence is geometric with ratio $2\,\Theta(s)/[1-\Theta(s)]$. But of course $M(s) \to \infty$ as $s \to s_o$.

A sharper result than the above can be obtained when $x(s_o)$ is a simple bifurcation point. Then it can be shown, [3], that for some $M_o > 0$

$$\| \mathcal{A}^{-1}(s) \| \leq M_o / |s - s_o| \quad \text{for} \quad s \neq s_o. \tag{3.5}$$

This implies that there is a full conical neighborhood, with positive semi-angle at the vertex, about the solution arc through $x(s_o)$, in which $P_x(y, s)$ is non-singular, see Figure 2. This is considerable more than is implied by (3. 4) in which the cone semi-angle could vanish and reduce the region of attraction to a cusped shape. Newton's method can also be shown to converge in this case.

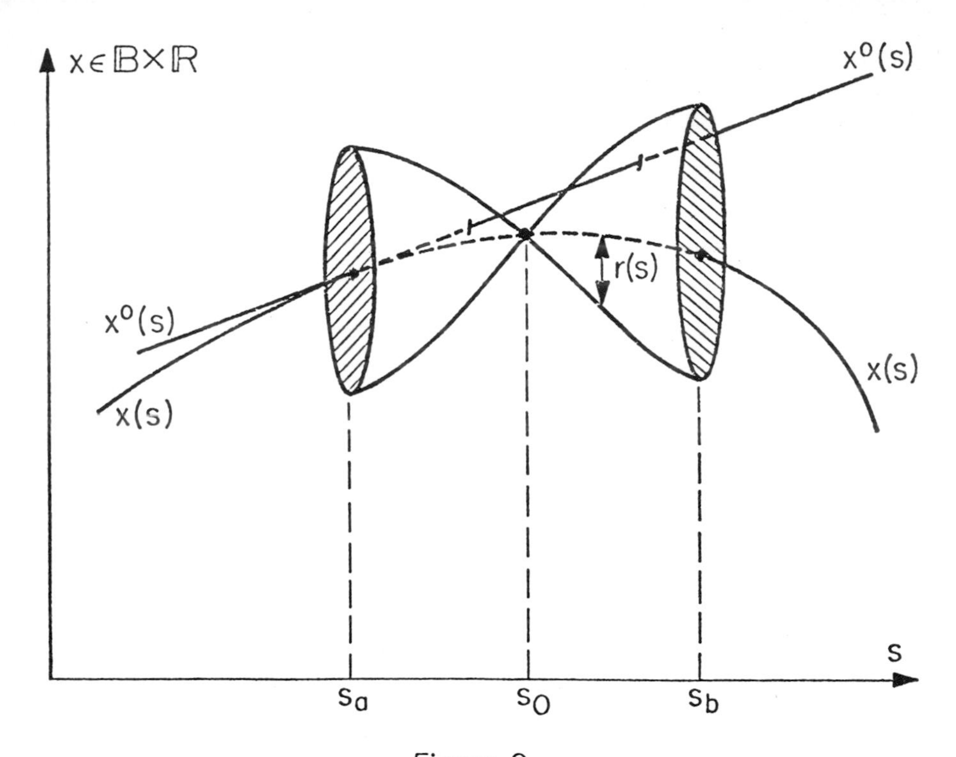

Figure 2

Technique for ''jumping'' over singular point, $x(s_o)$, on the solution branch $x(s)$.

It is a simple matter now to show that $x(s_o)$ can be determined by a bisection procedure [3] with $s_1 = s_a$ and $s_\nu < s_o < s_{\nu+1}$ for $\nu = 1, 3, 5, \ldots$ Each new tangent line, through the new solution $x(s_\nu)$ will have a smaller chord lying outside the cone. In the limit the tanget through $x(s_o)$, the bifurcation point, is entirely contained within the cone (locally of course). Indeed it is this final configuration, or rather a close approximation to it, that furnishes one of the best techniques for computing the bifurcating branch of solutions, see Figure 3.

4. **Switching Branches at Bifurcation Points.** Bifurcation points are singular points at which two or more smooth branches of solutions of (1.1) have nontangential intersections. In particular at a simple bifurcation point $[u_o, \lambda_o]$ we must have:

$$\text{(4.0)} \quad \text{a)} \ \dim \mathcal{N}(G_u^o) = \operatorname{codim} R(G_u^o) = 1 , \qquad \text{b)} \ G_\lambda^o \in R(G_u^o) .$$

We let

$$\text{(4.1)} \qquad \mathcal{N}(G_u^o) \equiv \operatorname{span}\{\phi_1\} , \qquad (G_u^{o*}) \equiv \operatorname{span}\{\psi_1^*\} , \quad \psi_1^* \phi_1 = 1 ,$$

Then (4.0)b) implies that there is a unique element $\phi_o \in \mathbb{B}$ such that:

$$\text{(4.2)} \qquad G_u^o \phi_o + G_\lambda^o = 0 , \ \psi_1^* \phi_o = 0 .$$

Let $[u(s), \lambda(s)]$ by any smooth branch of solutions of (1.1) through the bifurcation point, say with $u(s_o) = u_o, \lambda(s_o) = \lambda_o$.

Then since

$$\text{(4.3)} \quad \text{a)} \ G_u^o \dot u(s_o) + G_\lambda^o \dot\lambda(s_o) = 0 ,$$

it follows from (4.1)-(4.2) that

$$\text{(4.3)} \quad \text{b)} \ \dot u(s_o) = \alpha_1 \phi_1 + \alpha_o \phi_o ,$$

with

$$\text{(4.3)} \quad \text{c)} \ \alpha_o = \dot\lambda(s_o) , \ \alpha_1 = \psi_1^* \dot u(s_o) .$$

Differentiating the identity $G(u(s), \lambda(s)) = 0$ twice it follows that

$$G_u^o \ddot u_o = -[G_{uu}^o \dot u_o \dot u_o + 2G_{u\lambda}^o \dot u_o \dot\lambda_o + G_{\lambda\lambda}^o \dot\lambda_o \dot\lambda_o] - G_\lambda^o \ddot\lambda_o .$$

Since both $G_u^o \ddot u$ and $G_\lambda^o \ddot\lambda$ are in $R(G_u^o)$ the same must be true of the bracketed term. Thus $\psi_1^*[---] = 0$ must hold for this term and so using (4.3) it follows that the scalars $[\alpha_o, \alpha_1]$ must satisfy:

$$\text{(4.4)} \quad \text{a)} \ a\alpha_1^2 + 2b\alpha_1\alpha_o + c\alpha_o^2 = 0 ,$$

where

$$\text{(4.4)} \quad \text{b)} \ a \equiv \psi_1^* G_{uu}^o \phi_1 \phi_1 , \quad b \equiv \psi_1^*[G_{uu}^o \phi_o + G_{u\lambda}^o] \phi_1 ,$$

$$c \equiv \psi_1^* [G_{uu}^o \phi_o \phi_o + 2G_{u\lambda}^o \phi_o + G_{\lambda\lambda}^o] .$$

Since the tangent $[\dot u(s_o) , \dot\lambda(s_o)]$ to every smooth branch through the bifurcation point must have the form (4.3) it follows that (4.4a) must have a pair of distinct roots $[\alpha_o, \alpha_1]$ (since (4.4a) is homogeneous, roots are distinct if they are not scalar multiples of each other). If $[\alpha_o, \alpha_1]$ is one root of (4.4a) then the other root is distinct if and only if

$$(4.5)\qquad \begin{aligned} &\text{a)}\quad a\alpha_1 + b\alpha_0 \neq 0 \quad \text{when } \alpha_0 \neq 0\,, \\ &\text{b)}\quad b \neq 0 \quad \text{when } \alpha_0 = 0\,. \end{aligned}$$

If $[u_1(s), \lambda_1(s)]$ is the smooth solution branch through $[u_0, \lambda_0]$ on which the tangent $[\dot{u}_1(s_0), \dot{\lambda}_1(s_0)]$ is determined by the root $[\alpha_0, \alpha_1]$ then condition (4.5) can be written as

$$(4.6)\qquad \begin{aligned} &\text{a)}\quad \psi_1^*[G^0_{uu}\dot{u}_1(s_0) + G^0_{u\lambda}\dot{\lambda}_1(s_0)]\,\phi_1 \neq 0 \quad \text{when } \dot{\lambda}_1(s_0) \neq 0\,, \\ &\text{b)}\quad \psi_1^*[G^0_{uu}\phi_0 + G^0_{u\lambda}]\,\phi_1 \neq 0 \quad \text{when } \dot{\lambda}_1(s_0) = 0\,. \end{aligned}$$

This is essentially the form of the bifurcation condition given in [1]. With appropriate smoothness we can give a new constructive proof of the existence of a bifurcating branch under conditions (4.0), (4.1) and (4.6).

The details of this new bifurcation proof are contained in [3] and the basic geometric idea is sketched in Figure 3.

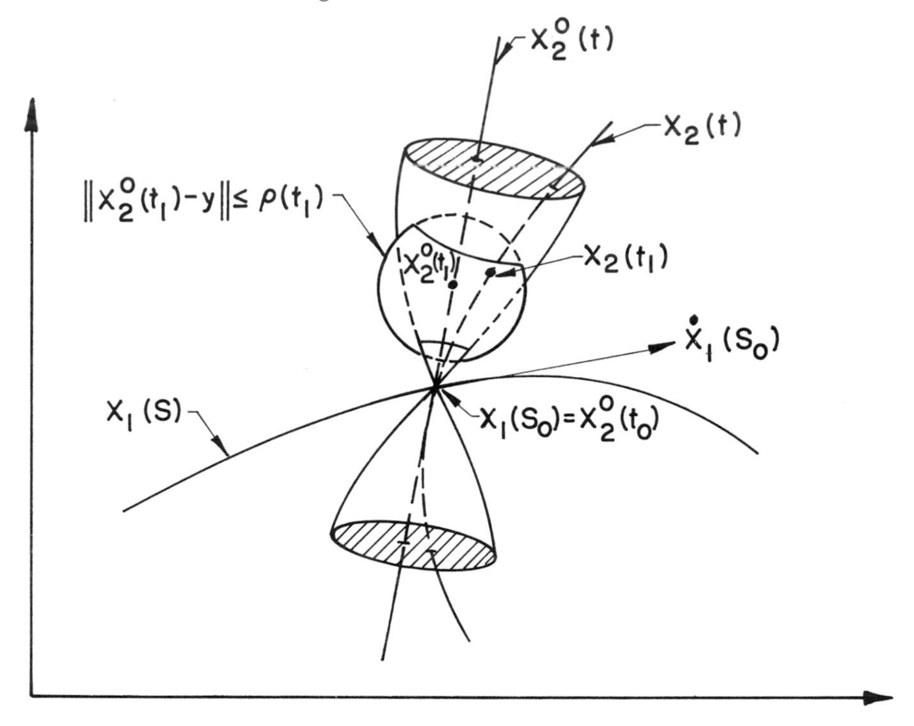

Figure 3.
Determination of $x_2(t_1)$, a point on the branch $x_2(t)$ bifurcating from $x_1(s)$ at $x_1(s_0)$.

We assume that we have located $x_1(s_o)$, the potential bifurcation point, on the given or computed solution arc $x_1(s)$ with tangent $\dot{x}_1(s_o) = [\alpha_1 \phi_1 + \alpha_o \phi_o, \alpha_o]$. Then the "other" root of (4.4a) is determined and with it we construct $x_2^o(t)$, the tangent "line" to the bifurcating branch $x_2(t)$ through $x_1(s_o)$. With $x_2^o(t_o) = x_1(s_o)$ we then can show, for $|t-t_o|$ sufficiently small that (3.1) has a root in the ball $\|x_2^o(t) - y\| \leq \rho(t)$. Furthermore $\rho(t)$ can be taken so small that $x_1(s)$ has no points in this ball. Thus a distinct branch of solutions has been determined. The proof thus sketched is closely patterned after the computational procedures used to actually construct bifurcating branches.
In [4] three additional procedures for determining bifurcating branches have been presented. The present one has been used extensively in numerous computations. It is the method most easily extended to treat multiple bifurcations. Techniques for the actual numerical implementation of this method, including the computation of the coefficients (a, b, c) in (4.4b) and of the elements ϕ_1 and ψ_1^* of (4.1) are discussed in [4] and [6].

5. Stability and Exchange of Stability. The class of problems indicated in (1.1) are frequently steady states of time dependent problems of the form

$$(5.0) \qquad B\frac{\partial U}{\partial t} = G(U, \lambda) .$$

Here B is a linear operator on $\mathbb{B}$ into $\mathbb{B}$; frequently it is a projection into some subspace of $\mathbb{B}$. Given an arc of solutions $[u(s), \lambda(s)]$ of (1.1) it is required to determine the temporal stability of each point as a solution of (5.0). This is usually done by linearized stability analysis in which a solution of (5.0) is sought in the form: $\lambda = \lambda(s)$,

$$(5.1) \qquad U(t,s) = u(s) + \epsilon e^{\sigma t} v(s) \quad , \quad \|v(s)\| = 1 .$$

Expanding about $\epsilon = 0$ and retaining only terms up to first order in ϵ yields the eigenvalue problem

$$(5.2) \qquad [G_u(u(s), \lambda(s)) - \sigma B] v(s) = 0 \quad , \quad \|v(s)\| = 1 .$$

If all eigenvalues $\sigma = \sigma(s)$ of (5.2) have $\mathrm{Re}\,\sigma < 0$ for a given s we say that $u(s)$ is linearly stable. If at least one eigenvalue has $\mathrm{Re}\,\sigma > 0$ then $u(s)$ is linearly unstable. If the least real part of all eigenvalues is zero for a given s then $u(s)$ has neutral stability.

At a limit point or a bifurcation point, of (1.1) the value $\sigma = 0$ is an eigenvalue of (5.2) with $v = \phi_1$. If this does indeed represent neutral stability then we can determine the local stability properties of the branches in a neighborhood of that limit point or bifurcation point.
If $[u(s), \lambda(s)]$ is a smooth branch of solutions of (1.1) and $\sigma(s_o)$ is a B-simple

eigenvalue, see [2], of $G_u(s_o) \equiv G_u(u(s_o), \lambda(s_o))$ at a fixed value of $s=s_o$ then there is a smooth family $\sigma(s)$ of B-simple eigenvalues of $G_u(s)$ for $|s-s_o| < \delta$, say. This result is proven in [2] for special parameter variations. However a trivial application of Lemma 1.5 gives us the more general result and the existence of the smooth family of eigenfunctions $v(s)$, see [3], [5]. Numerically it is a simple matter to compute the continuation of the eigenvalue $\sigma(s)$ and eigenfunction $v(s)$ for which $\sigma(s_o) = 0$ and $v(s_o) = \phi_1$. The basic technique simply employs inverse iteration as in computing ϕ_1, see [4] and [6]. We can also determine the stability properties analytically following the ideas in [2]. With $\sigma = \sigma(s)$ in (5.2) we obtain by differentiation with respect to s :

$$[G_u(s) - \sigma(s)B]\, \dot{v}(s) + [\dot{G}_u(s) - \dot{\sigma}(s)B]\, v(s) = 0 .$$

Evaluating this at $s=s_o$, the limit or bifurcation point, and operating on the left with ψ_1^* gives, on recalling that $\psi_1^* G_u^o = 0$, $\sigma(s_o) = 0$ and $v(s_o) = \phi_1$:

$$\dot{\sigma}(s_o) = \frac{\psi_1^*[G_{uu}^o \dot{u}(s_o) + G_{u\lambda}^o \dot{\lambda}(s_o)]\phi_1}{\psi_1^* B \phi_1} . \tag{5.3}$$

This quantity is easily computed and even simplifies considerably. More details can be found in [3], [5] and [6]. For example it easily follows that at a simple bifurcation branch on which $\dot{\lambda}(s_o) \neq 0$ there must be an exchange of stability, see [5].

REFERENCES

[1] Crandall, M. G. and Rabinowitz, P. H., Bifurcation from simple eigenvalues, J. Funct. Anal. 8 (1971) 321-340.

[2] Crandall, M. G. and Rabinowitz, P. H., Bifurcation, perturbation of simple eigenvalues and linearized stability, M. R. C. Tech. Report No. 1295, Madison, 1972.

[3] Decker, D. W., Some topics in bifurcation theory, PhD. thesis, Caltech, Pasadena, 1978.

[4] Keller, H. B., Numerical solution of bifurcation and nonlinear eigenvalue problems, in Applications of Bifurcation Theory (P. Rabinowitz, ed.) Academic Press, New York (1977) 359-384.

[5] Keller, H. B., Bifurcation theory and nonlinear eigenvalue problems, unpublished Lecture Notes, Caltech, Pasadena, 1976.

[6] Szeto, R., The flow between rotating coaxial disks, PhD. thesis, Caltech, Pasadena, 1978.

FINITE ELEMENT APPROXIMATIONS TO BIFURCATION PROBLEMS OF TURNING POINT TYPE

F. Kikuchi
Institute of Space and Aeronautical Science
University of Tokyo
Komaba, Meguro-ku, Tokyo, 153 Japan

1. Introduction

Various types of bifurcation phenomena may be observed in problems of stability limit or loss of stability of nonlinear physical systems. We will consider in this paper a special type of bifurcation called to be of *turning point type*, which is the most fundamental and popular in bifurcation problems.

As a model problem, let us consider the solution $u = u(x)$ of

$$- d^2u/dx^2 = \lambda e^u \quad \text{for } |x| < 1/2 , \quad u(-1/2) = u(1/2) = 0 , \qquad (1.1)$$

where λ is a real parameter. The solutions of this problem can be expressed explicitly : We can show that there is a positive number λ_0 such that the solutions exist for $\lambda \leqq \lambda_0$ but do not for $\lambda > \lambda_0$. Furthermore, there exist two solutions for $\lambda \in (0,\lambda_0)$. If we consider a graph of λ versus $u(0)$ (in general, values of u at an appropriate point), then λ attains maximum at $\lambda = \lambda_0$, which is a turning point as will be pictured in Fig. 2 of section 7. This type of point $\{\lambda_0,u_0\}$ (u_0 = solution of (1.1) for $\lambda = \lambda_0$) is a critical point in the sense defined later, and may be regarded as a kind of bifurcation point. Numerical methods such as the finite element method are usually employed for the analysis, since the problems are highly nonlinear in general. However, we often encounter difficulties in numerical solutions just due to the fact that we deal with critical points. We will consider numerical analysis of the finite element method applied to turning point type bifurcation, together with some numerical results.

As for numerical analysis of bifurcation problems, see, for example, Weiss[1], Kikuchi[2], Yamaguchi-Fujii[3], Bolley[4], Mizutani[5], and, especially for the problems of turning points, Simpson[6,7]. General analytical results of bifurcation problems can be found in Keller[8], Keller and Antman[9], Kransnosel'skii et al.[10], Sattinger[11], and Nirenberg[12]. As an example of bifurcation problem of turning point type in nonlinear elasticity, we may refer to the work made by Keller and Wolfe[13].

2. Preliminaries

We will consider real functions defined on a bounded domain Ω of R^n $(n = 1,2,3)$. For simplicity, Ω is assumed to be a polyhedral domain. We will use the real Sobolev spaces $H^m(\Omega)$ and $H_0^m(\Omega)$ equipped with the same norm $\|\ \|_m$ $(m = 0,1,2,\ldots)$. The space $H^0(\Omega)$ is identical to $L_2(\Omega)$, whose norm and inner product will be denoted by $\|\ \|$ and $(\ ,\ )$, respectively. As a fundamental space, we will employ

$$X = H_0^1(\Omega) \cap C(\overline{\Omega}) \quad , \tag{2.1}$$

where $C(\overline{\Omega})$ is the space of continuous functions over $\overline{\Omega}$ equipped with the maximum norm $|\ |$, and the norm of X is given by

$$\| u \|_X = \| u \|_1 + |u| \qquad \text{for} \quad u \in X. \tag{2.2}$$

Let us consider the problem : For a given $g \in L_2(\Omega)$, find $u \in H_0^1(\Omega)$ such that

$$\langle u,\bar{u}\rangle = (g,\bar{u}) \qquad \text{for all} \quad \bar{u} \in H_0^1(\Omega) \quad , \tag{2.3}$$

where $\langle\ ,\ \rangle$ is defined by

$$\langle u,\bar{u}\rangle = \sum_{i=1}^{n} (\partial u/\partial x_i, \partial \bar{u}/\partial x_i) \quad \text{with} \quad x = (x_1,\ldots,x_n) \ . \tag{2.4}$$

As is well known, u exists uniquely in $H_0^1(\Omega)$ for each $g \in L_2(\Omega)$ with the estimation $\| u \|_1 \leqq C \| g \|$ (C = universal positive constant). Furthermore, we will assume stronger regularity for u :

(A1) *The solution* u *defined by* (2.3) *belongs to* $H^2(\Omega) \cap H_0^1(\Omega) \subset X$ *with the estimation*

$$\| u \|_2 \leqq C \| g \| \ . \tag{2.5}$$

Since Ω is a polyhedral domain, the present assumption cannot be justified unconditionally except for $n = 1$. For $n = 2$, the regularity condition is met if Ω is a convex domain. As for general results, see [14].

3. Continuous problem

Our problem is to find a real number λ and a real function $u(x)$ such that

$$- \Delta u = \lambda f(u) \quad \text{in } \Omega \ , \qquad u = 0 \quad \text{on } \partial\Omega \ , \tag{3.1}$$

where Δ is Laplacian, $\partial\Omega$ boundary of Ω, f a real function on R^1, and $f(u)$ implies $f(u)(x) = f(u(x))$ for $x \in \Omega$. (In (3.1), the term $\lambda f(u)$ is homogeneously linear in λ. However, we can extend our theory to include nonlinearity of λ.) We assume the following on f.

(A2) *The function* f *is twice continuously differentiable on* R^1 *and its second order derivative* f'' *is Lipschitz continuous in the sense*

$$|f''(t_1) - f''(t_2)| \leqq C|t_1 - t_2| \quad \textit{for any} \quad t_1, t_2 \in [-t_0, t_0] , \tag{3.2}$$

where t_0 *is an arbitrary positive constant, and* C *may depend on* t_0 *but not on* t_1 *and* t_2.

Let us express (3.1) in a weak form :

(I) *Find a pair* $\{\lambda, u\} \in R^1 \times X$ *such that*

$$\langle u, \bar{u}\rangle = \lambda(f(u), \bar{u}) \quad \textit{for all } \bar{u} \in H_0^1(\Omega) \quad . \tag{3.3}$$

Notice that f(u) can be regarded as an operator from X into $L_2(\Omega)$ (actually, also into $C(\bar{\Omega})$), and f(u) is twice continuously differentiable for u in the Fréchet sense. We will write the derivatives as f_u and f_{uu}.

Let $\{\lambda_0, u_0\} \in R^1 \times X$ be a pair that satisfies (3.3). From (A1), u_0 belongs to $H^2(\Omega)$ since $f(u_0) \in L_2(\Omega)$. We will call $\{\lambda_0, u_0\}$ a *critical point* of (3.3) provided that there exists a non-zero function $\phi \in H_0^1(\Omega)$ which satisfies

$$\langle \phi, \bar{u}\rangle - \lambda_0(f_u(u_0)\phi, \bar{u}) = 0 \qquad \text{for all} \quad \bar{u} \in H_0^1(\Omega) \quad . \tag{3.4}$$

Notice here that $f_u(u_0)$ is extended to a bounded linear operator in $L_2(\Omega)$. The condition (3.4) implies that the linearized operator associated with (3.3) has a zero eigenvalue at $\{\lambda_0, u_0\}$. Assume the following.

(A3) *The considered zero eigenvalue is simple and the eigenfunction* ϕ *is normalized in such a way that*

$$\|\phi\| = 1 \quad . \tag{3.5}$$

Clearly, $\phi \in H^2(\Omega) \cap H_0^1(\Omega) \subset X$ from (A1) and (3.4). Define :

$$M = \{g \in L_2(\Omega) \text{ such that } (g, \phi) = 0\} \quad , \tag{3.6}$$

$$X_M = X \cap M = \{g \in X \text{ such that } (g, \phi) = 0\} \quad , \tag{3.7}$$

$$Pg = g - (g, \phi)\phi \in M \quad \text{for} \quad g \in L_2(\Omega). \tag{3.8}$$

Let us consider the following auxiliary problem.

(II) *For a given* $g \in L_2(\Omega)$, *find* $v \in H_0^1(\Omega) \cap M$ *such that*

$$\langle v,\bar{u}\rangle - \lambda_0(f_u(u_0)v,\bar{u}) = (Pg,\bar{u}) \quad \textit{for all} \quad \bar{u} \in H_0^1(\Omega) \ . \tag{3.9}$$

As may be easily shown, the above problem has no solution if the term Pg is replaced with g ($\notin M$). We can prove that there exists a unique solution $v \in H_0^1(\Omega) \cap M$ to the problem (II) for each $g \in L_2(\Omega)$. Furthermore, v belongs to $H^2(\Omega)$ (and hence to X_M) with the estimation

$$\| v \|_2 \leq C \| g \| \quad . \tag{3.10}$$

Thus we can use an operator $Q : L_2(\Omega) \to X_M$ defined by the relation $Qg = v$. Qg actually belongs to $H^2(\Omega)$ and satisfies

$$\| Qg \|_2 \leq C \| g \| \quad \text{for} \quad g \in L_2(\Omega) \ . \tag{3.10'}$$

Let us express the solution u of (3.3), if it exists, in the form

$$u = u_0 + \varepsilon\phi + v \quad \text{with } \varepsilon = (u - u_0, \phi) \text{ and } v \in X_M \ . \tag{3.11}$$

Substituting the above into (3.3), we have

$$\langle v,\bar{u}\rangle - \lambda_0(f_u(u_0)v,\bar{u}) = (\lambda f(u) - \lambda_0 f(u_0) - \lambda_0 f_u(u_0)(u - u_0),\bar{u}) \tag{3.12}$$

for all $\bar{u} \in H_0^1(\Omega)$. From the consideration on the problem (II), we have as the solvability condition of the above

$$\lambda = \lambda_0 + \lambda_0(f(u_0) + f_u(u_0)(u - u_0) - f(u),\phi)\big/(f(u),\phi) \tag{3.13}$$

if $(f(u),\phi) \neq 0$. The present relation can be used to determine λ from u. We will employ the following assumption.

(A4) *At the considered critical point* $\{\lambda_0, u_0\}$, *it holds that*

$$\alpha_1 \equiv (f(u_0),\phi) \neq 0 \quad , \tag{3.14}$$

$$\alpha_2 \equiv -(f_{uu}(u_0)\phi^2,\phi)\lambda_0/\alpha_1 \neq 0 \quad , \tag{3.15}$$

where $f_{uu}(u_0)\phi^2$ *implies* $(f_{uu}(u_0)\phi)\phi$.

Our aim is to construct a (local) path of solutions to (3.3) which goes through $\{\lambda_0, u_0\}$. Since $\{\lambda_0, u_0\}$ is a critical point, we cannot generally use λ as a path parameter. Instead, we will show that ε in (3.11) can be used as a path parameter to obtain $u = u(\varepsilon)$ in the form

$$u(\varepsilon) = u_0 + \varepsilon\phi + v(\varepsilon) \quad , \quad v(\varepsilon) \in X_M \quad , \tag{3.16}$$

while the relation (3.13) is available to decide $\lambda = \lambda(\varepsilon)$.

Let us give some definitions :

$$Z = \{\{\varepsilon, v\} \in R^1 \times X_M \text{ such that } (f(u_0 + \varepsilon\phi + v), \phi) \neq 0\} \quad , \tag{3.17}$$

$\Lambda : Z \to R^1$;

$$\Lambda(\varepsilon, v) = \lambda_0 + \frac{(f(u_0) + f_u(u_0)(u - u_0) - f(u), \phi)}{(f(u), \phi)} \lambda_0 \quad , \tag{3.18}$$

$S : Z \to M$;

$$S(\varepsilon, v) = \Lambda(\varepsilon, v) f(u) - \lambda_0 f(u_0) - \lambda_0 f_u(u_0)(u - u_0) \quad , \tag{3.19}$$

$T : Z \to X_M$;

$$T(\varepsilon, v) = Q\, S(\varepsilon, v) \quad , \tag{3.20}$$

where

$$u = u_0 + \varepsilon\phi + v \quad \text{for } \{\varepsilon, v\} \in Z \quad . \tag{3.21}$$

We can easily show that $\{\varepsilon, v\} \in R^1 \times X_M$ belongs to Z if $|\varepsilon|$ and $\| v \|_X$ are small enough due to the assumption (3.14).

Our method of constructing the path of solutions $\{\lambda(\varepsilon), u(\varepsilon)\}$ is to apply the *implicit function theorem* (essentially the principle of contraction mappings) to the equation

$$v = T(\varepsilon, v) \quad \text{with } \lambda = \Lambda(\varepsilon, v) \quad \text{for } \{\varepsilon, v\} \in R^1 \times X_M \; , \tag{3.22}$$

which is equivalent to (3.3) for $|\varepsilon|$ and $\| v \|_X$ small enough.

To this end, we should prepare some lemmas with respect to differentiability of the operators $\Lambda(\varepsilon, v)$ and $S(\varepsilon, v)$ in appropriately chosen metrics. We do not give the details here, since the description becomes lengthy. However, it is to be noted that Λ and S are twice continuously Fréchet differentiable with respect to ε and v and the second order derivatives are Lipschitz continuous, when $|\varepsilon|$ and $\| v \|_X$ are small enough. Furthermore, we have

$$\Lambda(0,0) = \lambda_0, \quad S(0,0) = 0, \quad \Lambda_\varepsilon(0,0) = 0, \quad \Lambda_v(0,0) = 0,$$

$$S_\varepsilon(0,0) = 0, \quad S_v(0,0) = 0, \quad \Lambda_{\varepsilon\varepsilon}(0,0) = \alpha_2 \; (\neq 0) \quad , \tag{3.23}$$

where the subscripts ε and v denote $\partial/\partial\varepsilon$ and $\partial/\partial v$, respectively. We can use the implicit function theorem with the relation $T = Q\,S$ and the boundedness of Q taken into account. Then we have :

Theorem 1 Let $\varepsilon_0 > 0$ be small enough. Then there exists for $|\varepsilon| \leqq \varepsilon_0$ a pair $\{\lambda(\varepsilon), u(\varepsilon)\} \in R^1 \times (H^2(\Omega) \cap H_0^1(\Omega)) \subset R^1 \times X$ which satisfies (3.3), where $u(\varepsilon)$ is a function of ε in the form (3.16) and $\lambda(\varepsilon)$ is a function of ε

subjected to (3.13). For each ε with $|\varepsilon| \leqq \varepsilon_0$, this pair is a unique solution of (3.3) in a certain small neighborhood of $\{\lambda_0, u_0\}$ (with respect to the metric of $R^1 \times X$). Furthermore, $\{\lambda(\varepsilon), u(\varepsilon)\}$ is twice continuously differentiable for ε in the metric of $R^1 \times H^2(\Omega)$ (also of $R^1 \times X$), and the second order derivative is Lipschitz continuous in the sense

$$|\lambda''(\varepsilon_1) - \lambda''(\varepsilon_2)| \leqq C|\varepsilon_1 - \varepsilon_2| \ , \ \| u''(\varepsilon_1) - u''(\varepsilon_2) \|_2 \leqq C|\varepsilon_1 - \varepsilon_2| \quad (3.24)$$

for $|\varepsilon_1|, |\varepsilon_2| \leqq \varepsilon_0$, where " implies $d^2/d\varepsilon^2$. It also holds that

$$\{\lambda(0), u(0)\} = \{\lambda_0, u_0\} \ , \quad \lambda'(0) = 0 \ ,$$
$$u'(0) = \phi \ , \quad \lambda''(0) = \alpha_2 \ (\neq 0) \ , \quad (3.25)$$

where ' is $d/d\varepsilon$. ////

The present theorem implicates that $\lambda(\varepsilon)$ is almost parabolic around $\{\varepsilon, \lambda\} = \{0, \lambda_0\}$ as pictured in Fig. 1. Thus, λ—ε curve has a turning point at $\lambda = \lambda_0$. This is why the considered bifurcation point is called to be of *turning point* type. We can also see that the term $\varepsilon\phi$ dominates in $u - u_0$, when the absolute value of ε is sufficiently small. As was mentioned in section 1, this type of bifurcation phenomena are frequently observed in nonlinear problems.

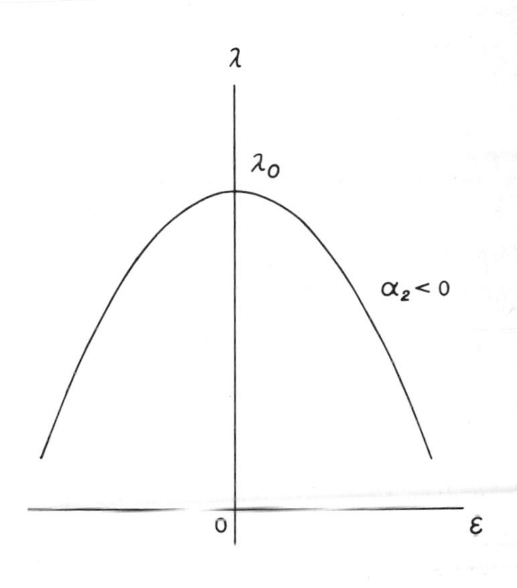

Fig. 1 $\lambda - \varepsilon$ *curve*

4. Discrete problem based on the finite element method

We will consider a family of triangulations $\{T^h\}$ defined over $\bar{\Omega}$, where each finite element is assumed to be an open n-simplex. Since Ω is a polyhedron, the whole domain can be triangulated without any overlappings and gaps. We assume $\{T^h\}$ to be *regular* and *quasi-uniform*. Ruoughly speaking, these conditions imply that the simplices are not too flat, and that the simplices in each T^h are of comparable sizes. The representative mesh size of each T^h will be denoted by h. The regularity of $\{T^h\}$ also implicates that we can find a sequnce from $\{T^h\}$ such that $h \downarrow 0$. In the sequel, we only consider the case where h is sufficiently small.

We will use the simplest finite element method called *piecewise linear approximation* [15] : Each trial function is continuous over $\bar{\Omega}$, and linear function of $x_1,\ldots,x_n$ in every simplex $T \in T^h$. To cope with the homogeneous Dirichlet conditions, we impose trial functions to vanish on $\partial\Omega$. Then the space X^h of trial functions can be regarded as a finite dimensional subspace of X.

Consider the discrete analog of (2.3) : For a given $g \in L_2(\Omega)$, find $u_h \in X^h$ such that

$$\langle u_h,\bar{u}_h\rangle = (g,\bar{u}_h) \quad ; \quad \forall\, \bar{u}_h \in X^h . \tag{4.1}$$

As is well known, u_h exists uniquely in X^h and satisfies the following error estimation for u defined by (2.3), cf. [15] :

$$\| u_h - u \|_\alpha \leqq C\, h^{2-\alpha} \| g \| \quad (\alpha = 0,1) \quad , \qquad |u_h - u| \leqq C\, h^{2-n/2} \| g \| . \tag{4.2}$$

The discrete problem corresponding to (3.3) is given by :

$(I)_h$ *Find a pair* $\{\lambda_h,u_h\} \in R^1 \times X^h$ *such that*

$$\langle u_h,\bar{u}_h\rangle = \lambda_h(f(u_h),\bar{u}_h) \quad ; \quad \forall\, \bar{u}_h \in X^h . \tag{4.3}$$

Our aim is to construct a path of solutions to (4.3) in the vicinity of $\{\lambda_0,u_0\}$ in a completely parallel manner to the continuous case. To this end, we first define $u_{h0} \in X^h$ by the relation

$$\langle u_{h0},\bar{u}_h\rangle = \lambda_0(f(u_0),\bar{u}_h) \quad ; \quad \forall\, \bar{u}_h \in X^h . \tag{4.4}$$

Then we have :

$$\| u_{h0} - u_0 \|_\alpha \leqq C\, h^{2-\alpha} \quad (\alpha = 0,1) \quad , \qquad | u_{h0} - u_0| \leqq C\, h^{2-n/2} . \tag{4.5}$$

We also use the discrete eigenvalue problem : Find a pair $\{\mu_h,w_h\} \in R^1 \times X^h$ such that $w_h \neq 0$ and

$$\langle w_h,\bar{u}_h\rangle - \lambda_0(f_u(u_0)w_h,\bar{u}_h) = \mu_h(w_h,\bar{u}_h) \quad ; \quad \forall\, \bar{u}_h \in X^h . \tag{4.6}$$

Then we can find a pair $\{\nu_h,\phi_h\} \in R^1 \times X^h$ which satisfies the above and

$$\| \phi_h \| = 1 \quad , \quad (\phi_h,\phi) \geqq 0 \quad , \quad |\nu_h| \leqq C\, h^2 \quad ,$$
$$\| \phi_h - \phi \| \leqq C\, h^{2-\alpha} \quad (\alpha = 0,1) \quad , \qquad |\phi_h - \phi| \leqq C\, h^{2-n/2} . \tag{4.7}$$

Furthermore, ν_h is simple for h small enough, and hence the eigenpair $\{\nu_h,\phi_h\}$ is a nice approximation to the original one $\{0,\phi\}$. Usually, the eigenvalue ν_h is the smallest one in absolute value.

Define :

$$M^h = \{g \in L_2(\Omega) \text{ such that } (g,\phi_h) = 0\} \quad , \tag{4.8}$$

$$X_M^h = X^h \cap M^h = \{g \in X^h \text{ such that } (g,\phi_h) = 0\} \quad , \tag{4.9}$$

$$P_h g = g - (g,\phi_h)\phi_h \in M^h \text{ for } g \in L_2(\Omega) \quad . \tag{4.10}$$

Let us consider the discrete version of (II) :

$(II)_h$ *For a given* $g \in L_2(\Omega)$, *find* $v_h \in X_M^h$ *such that*

$$\langle v_h,\bar{u}_h\rangle - ((\lambda_0 f_u(u_0) + \nu_h)v_h,\bar{u}_h) = (P_h g,\bar{u}_h) \quad ; \forall \bar{u}_h \in X^h. \tag{4.11}$$

As in the continuous case, the above has no solution if the term $P_h g$ is simply g ($\notin M^h$). We can prove (for h small enough, as already mentioned) that v_h exists uniquely in X_M^h and satisfies

$$\| v_h \|_X \leqq C\| g \| \quad , \tag{4.12}$$

$$\| v_h - Qg \| \leqq C h^{2-\alpha}\| g \| \quad (\alpha = 0,1) \quad , \quad |v_h - Qg| \leqq C h^{2-n/2}\| g \| .$$

Therefore, we can use an operator $Q_h : L_2(\Omega) \to X_M^h$ defined by $Q_h g = v_h$, which is an approximation of Q introduced in section 3.

As in the continuous case, let us express the solution $u_h \in X^h$ of (4.3) in the form

$$u_h = u_{h0} + \varepsilon\phi_h + v_h \quad \text{with } \varepsilon = (u_h - u_{h0},\phi_h) \text{ and } v_h \in X_M^h \quad , \tag{4.13}$$

and try to obtain v_h and λ_h as functions of ε.

Define :

$$Z^h = \{\{\varepsilon,v_h\} \in R^1 \times X_M^h \text{ such that } (f(u_{h0} + \varepsilon\phi_h + v_h),\phi_h) \neq 0 \quad , \tag{4.14}$$

$$\Lambda_h : Z^h \to R^1 \ ;$$

$$\Lambda_h(\varepsilon,v_h) = \lambda_0 + \frac{(\lambda_0 f(u_0) + (\lambda_0 f_u(u_0) + \nu_h)(u_h - u_{h0}) - \lambda_0 f(u_h),\phi_h)}{(f(u_h),\phi_h)} \quad , \tag{4.15}$$

$$S_h : Z^h \to M^h$$

$$S_h(\varepsilon,v_h) = \Lambda_h(\varepsilon,v_h)f(u_h) - \lambda_0 f(u_0) - (\lambda_0 f_u(u_0) + \nu_h)(u_h - u_{h0}) \quad , \tag{4.16}$$

$$T_h : Z^h \to X_M^h$$

$$T_h(\varepsilon,v_h) = Q_h S_h(\varepsilon,v_h) \quad , \tag{4.17}$$

where

$$u_h = u_{h0} + \varepsilon\phi_h + v_h \quad \text{for } \{\varepsilon,v_h\} \in Z^h \quad . \tag{4.18}$$

Now we can use the implicit function theorem to

$$v_h = T_h(\varepsilon, v_h) \quad \text{with} \quad \lambda_h = \Lambda_h(\varepsilon, v_h) \quad \text{for } \{\varepsilon, v_h\} \in Z^h \,. \tag{4.19}$$

As the continuous case, the above is equivalent to (4.3) when $|\varepsilon|$ and $\| v_h \|_X$ are small enough. We can show that Λ_h and S_h are twice continuously differentiable with respect to ε and v_h, with Lipschitz continuous second order derivatives. Furthermore, we have

$$|\Lambda_h(0,0) - \lambda_0| \leqq C h^2 \,, \quad \| S_h(0,0) \| \leqq C h^2 \,, \quad |\Lambda_{h\varepsilon}(0,0)| \leqq C h^2 \,,$$

$$|\Lambda_{hv}(0,0) w_h| \leqq C h^2 \| w_h \|_X \quad (\text{or } C h \| w_h \|) \,, \quad \| S_{h\varepsilon}(0,0) \| \leqq C h^2 \,,$$

$$\| S_{hv}(0,0) w_h \| \leqq C h^2 \| w_h \|_X \quad (\text{or } C h \| w_h \|) \,, \quad |\Lambda_{h\varepsilon\varepsilon}(0,0) - \alpha_2| \leqq C h^2 \,, \tag{4.20}$$

where $w_h \in X_M^h$, and the suffixes ε and v imply $\partial/\partial\varepsilon$ and $\partial/\partial v_h$, respectively.

We have :

<u>Theorem 2</u> Let ε_0 be small enough. Then there exists for $|\varepsilon| \leqq \varepsilon_0$ a pair $\{\lambda_h(\varepsilon), u_h(\varepsilon)\} \in R^1 \times X^h$ which satisfies (4.3), where $u_h(\varepsilon)$ is a function of ε in the form $u_h(\varepsilon) = u_{h0} + \varepsilon \phi_h + v_h(\varepsilon)$ with $v_h(\varepsilon) \in X_M^h$, and $\lambda_h(\varepsilon)$ is a function of ε determined from the relation $\lambda_h(\varepsilon) = \Lambda_h(\varepsilon, v_h(\varepsilon))$. For each ε with $|\varepsilon| \leqq \varepsilon_0$, this pair is a unique solution of (4.3) for a certain small neighborhood of $\{\lambda_{h0}, u_{h0}\}$. $\{\lambda_h(\varepsilon), u_h(\varepsilon)\}$ is twice continuously differentiable as a $R^1 \times X^h$ valued function of ε, and the second order derivative is Lipschitz continuous in the sense

$$\| u_h''(\varepsilon_1) - u_h''(\varepsilon_2) \|_X \leqq C|\varepsilon_1 - \varepsilon_2| \,, \quad |u_h''(\varepsilon_1) - u_h''(\varepsilon_2)| \leqq C|\varepsilon_1 - \varepsilon_2| \tag{4.21}$$

for $|\varepsilon_1|$, $|\varepsilon_2| \leqq \varepsilon_0$.

5. <u>Error analysis of the finite element solutions</u>

When h is small enough, we can consider the errors $\lambda_h(\varepsilon) - \lambda(\varepsilon)$ and $u_h(\varepsilon) - u(\varepsilon)$ for each ε. To this end, we must evaluate such differences as $\Lambda_h(\varepsilon, v_h) - \Lambda(\varepsilon, v)$, $S_h(\varepsilon, v_h) - S(\varepsilon, v)$ etc. for $v_h \in X_M^h$ and $v \in X_M$. For example, we have the following for $\varepsilon_0 > 0$ and $\delta_0 > 0$ small enough :

$$|\Lambda_h(\varepsilon, v_h) - \Lambda(\varepsilon, v)| \leqq C\{h^2 + (|\varepsilon| + \gamma_0) \| v_h - v \|\} \,,$$

$$\| S_h(\varepsilon, v_h) - S(\varepsilon, v) \| \leqq C\{h^2 + (|\varepsilon| + \gamma_0) \| v_h - v \|\} \,, \tag{5.1}$$

where $\| v \|_X \leqq \delta_0$, $\| v_h \|_X \leqq \delta_0$, $|\varepsilon| \leqq \varepsilon_0$, and $\gamma_0 = \max\{\| v_h \|_X, \| v \|_X\}$.
We only give the final results with the proofs omitted.

Theorem 3 Let ε_0 be small enough, and consider the paths of solutions $\{\lambda(\varepsilon),u(\varepsilon)\}$ to (3.3) and $\{\lambda_h(\varepsilon),u_h(\varepsilon)\}$ to (4.3) in a small neighborhood of $\{\lambda_0,u_0\}$ for ε with $|\varepsilon| \leqq \varepsilon_0$. Then

$$\| \frac{d^\beta u_h}{d\varepsilon^\beta}(\varepsilon) - \frac{d^\beta u}{d\varepsilon^\beta}(\varepsilon) \|_\alpha \leqq C h^{2-\alpha} \quad (\alpha = 0,1) \quad , \tag{5.2}$$

$$| \frac{d^\beta u_h}{d\varepsilon^\beta}(\varepsilon) - \frac{d^\beta u}{d\varepsilon^\beta}(\varepsilon) | \leqq C h^{2-n/2} \quad , \tag{5.3}$$

$$| \frac{d^\beta \lambda_h}{d\varepsilon^\beta}(\varepsilon) - \frac{d^\beta \lambda}{d\varepsilon^\beta}(\varepsilon) | \leqq C h^2 \quad , \tag{5.4}$$

where $\beta = 0,1,2$. Thus the path $\{\lambda_h(\varepsilon),u_h(\varepsilon)\}$ converges in $R^1 \times X$ to the exact one $\{\lambda(\varepsilon),u(\varepsilon)\}$ uniformly with respect to ε with $|\varepsilon| \leqq \varepsilon_0$. ////

From the estimation (5.4), we can see that the graph of $\lambda_h(\varepsilon)$ is almost parabolic near $\varepsilon = 0$ with the same convexity as $\lambda(\varepsilon)$. It is easy to see that $\lambda_h(\varepsilon)$ attains local maximum or minimum at a certain point near $\varepsilon = 0$, that is, $d\lambda_h/d\varepsilon = 0$ at $\varepsilon = \varepsilon_c$, where ε_c can be evaluated as

$$|\varepsilon_c| \leqq C h^2 \qquad (|\varepsilon_c| \leqq \varepsilon_0 \text{ for } h \text{ small enough}) \ . \tag{5.5}$$

Consider $\{\lambda_h^c,u_h^c\} \equiv \{\lambda_h(\varepsilon_c),u_h(\varepsilon_c)\}$. Then this is a *critical point* of (4.3) in the sense that there exists a non-zero function $w_h \in X^h$ such that

$$\langle w_h,\bar{u}_h\rangle - \lambda_h^c(f_u(u_h^c)w_h,\bar{u}_h) = 0 \quad , \quad \forall \bar{u}_h \in X^h \ . \tag{5.6}$$

The above implies that the linearized operator associated with (4.3) has a zero eigenvalue at $\{\lambda_h^c,u_h^c\}$. The zero eigenvalue is simple for h small enough. Define ϕ_h^c by

$$\phi_h^c = \frac{du_h}{d\varepsilon}(\varepsilon_c) \in X^h \ . \tag{5.7}$$

Then ϕ_h^c is an eigenfunction of (5.6) and satisfies

$$\| \phi_h^c - \phi_h \|_X \leqq C h^2 \ . \tag{5.8}$$

To see that $\{\lambda_h^c,u_h^c\}$ is a critical point, differentiate (4.3) with respect to ε after substituting $\{\lambda_h(\varepsilon),u_h(\varepsilon)\}$. Then we have

$$\langle du_h/d\varepsilon,\bar{u}_h\rangle - \lambda_h(f_u(u_h)du_h/d\varepsilon,\bar{u}_h) = d\lambda_h/d\varepsilon(f(u_h),\bar{u}_h) \ , \tag{5.9}$$

which coincides with (5.6) at $\varepsilon = \varepsilon_c$ with $w_h = du_h/d\varepsilon$, since $d\lambda_h/d\varepsilon$ vanishes there. Clearly, there is no other critical points around $\varepsilon = 0$.

6. Numerical method

The proof of the implicit function theorem that we have used is basically accomplished by the use of the principle of contraction mappings, which may be applied to numerical construction of the path for (4.3). However, we have employed $\{\lambda_0, u_0\}$, u_{h0} and $\{\nu_h, \phi_h\}$ in section 4, which we cannot know generally except in very rare cases. Nevertheless, we may often obtain $\{\lambda^*, u_h^*\} \in R^1 \times X^h$ as a rough approximation of $\{\lambda_0, u_0\}$, and we are able to consider the following eigenvalue problem instead of (4.6):

$$<w_h, \bar{u}_h> - \lambda^*(f_u(u_h^*)w_h, \bar{u}_h) = \mu_h(w_h, \bar{u}_h) \quad ; \quad \forall\ \bar{u}_h \in X^h . \qquad (6.1)$$

Then we can find an eigenpair $\{\nu_h^*, \phi_h^*\}$, which has essentially the same properties as $\{\nu_h, \phi_h\}$ except in the error estimations.

The solution of (4.3) is to be obtained in the form

$$u_h = u_h^* + \varepsilon\phi_h^* + v_h \ ; \quad v_h \in X^h \ , \quad (v_h, \phi_h^*) = 0 \ . \qquad (6.2)$$

Then the equations to deal with are

$$<v_h, \bar{u}_h> - (\lambda^* f_u(u_h^*) + \nu_h^*)v_h, \bar{u}_h)$$
$$= (\lambda_h f(u_h) - (\lambda^* f_u(u_h^*) + \nu_h^*)(u_h - u_h^*) - \lambda^* f(u_h^*), \bar{u}_h) \ ; \ \forall\ \bar{u}_h \in X^h, \qquad (6.3)$$

$$\lambda_h = \lambda^* + \frac{(\lambda^* f(u_h^*) + (\lambda^* f_u(u_h^*) + \nu_h^*)(u_h - u_h^*) - \lambda^* f(u_h), \phi_h^*)}{(f(u_h), \phi_h^*)} \ , \qquad (6.4)$$

The above equations can be usually solved by a simple iteration method for each (small) ε : Choose an initial approximation of v_h, say $v_h = 0$, and substitute it into the right-hand side of (6.3) with the solvability condition (6.4) taken into account. Then solving (6.3) with respect to v_h in the left-hand side, we may expect to have an improved approximation. If h and $|\varepsilon|$ are small enough and $\{\lambda_h^*, u_h^*\}$ is a nice approximation of $\{\lambda_0, u_0\}$, then the process has essentially the same structure as that considered in section 4, and hence gives a rapidly converging sequence of approximate solutions to (4.3). As for numerical techniques for solving (6.3) with (6.4), see [2].

7. Numerical experiment

Numerical experiments are performed to see the validity of our theory. The test problems are :

(A) $n = 1$, Ω = unit segment ($|x_1| < 1/2$) and $f(u) = e^u$,

(B) $n = 2$, Ω = unit squre ($|x_i| < 1/2$; $i = 1,2$) and $f(u) = e^u$,

(C) $n = 2$, Ω = unit square, and $f(u) = 1 + (u + u^2/2)/(1 + u^2/100)$,

where (B) and (C) are the same as treated by Simpson [7].

For the two-dimensional problems (B) and (C), we used rectangular element based on piecewise bilinear shape functions [15], for which we can obtain basically the same error estimates as those given in section 5.

For the problem (A), Ω is decomposed into N finite elements with equal length $h = 1/N$, while for (B) and (C), Ω is regarded as an assemblage of $N \times N$ square elements with equal side length $h = 1/N$. We only obtained solutions with symmetric properties, and hence only one half ($n = 1$) or a quarter ($n = 2$) of Ω was analyzed in actual computations. The five point Gauss quadrature formula (or its product form for two-dimensional problems) is employed for the integration of element matrices. We also tested the three point formula for comparison, but could not find significant differences, as far as the results of the present problems are concerned.

The numerical calculations were performed by the use of the scheme in section 6 with $\{\lambda^*, u_h^*\} = \{0,0\}$ in all cases. The method turned out to be stable for fairly wide range of ε in the present experiments. The results are summarized in the sequel.

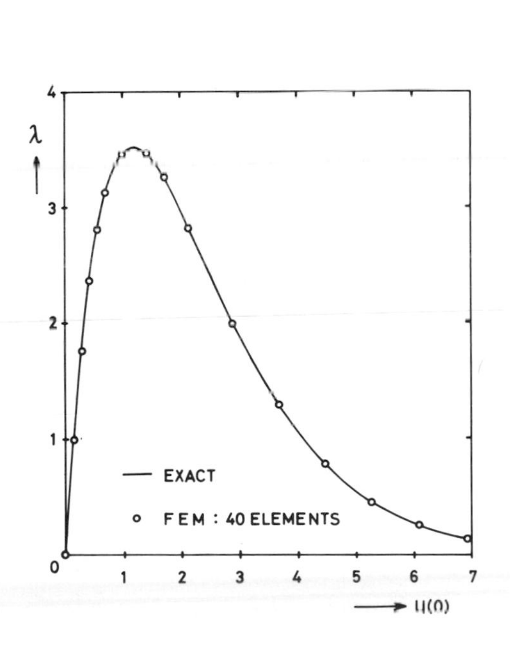

Fig. 2 λ – $u(0)$ curve for Problem (A)

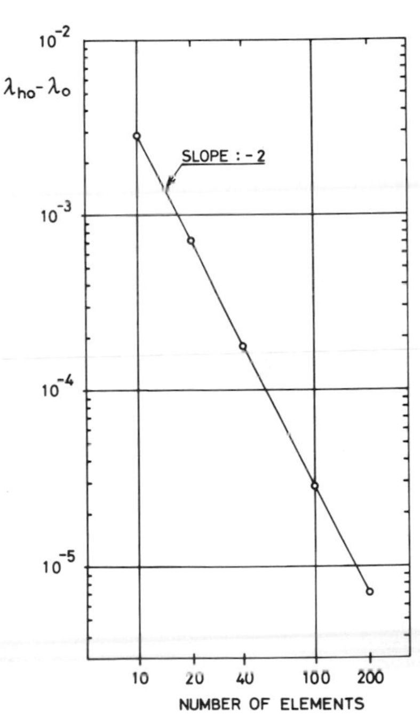

Fig. 3 Convergence character of $\lambda_{h0} - \lambda_0$ for Problem (A)

Problem (A) Figure 2 illustrates λ - u(0) curve based on the results for N = 40, which agrees well with the exact one. As the maximum value λ_0 of λ is known exactly (λ_0 = 3.51383072...), we can see convergence character of the approximate maximum value λ_{h0} of λ_h by increasing N. Plotted in Fig. 3 are values of $\lambda_{h0} - \lambda_0$ versus several values of N. Since the employed meshes are uniform, we can expect asymptotic behavior in errors. The results summarized in Fig. 3 give experimental evidence for this expectation, and $\lambda_{h0} - \lambda_0$ behaves like h^2 for smaller values of h.

Problem (B) Figure 4 depicts λ - u(0) curve for N = 40, where one turning point is observed, just the same as in Problem (A). We can get much numerical evidence for the existence of such a point in this problem, although the author is not aware of any analytical results. The maximum value λ_0 of λ is not available, but we can use the extrapolation technique to obtain improved results from the approximate values λ_{h0}, assuming that $\lambda_{h0} - \lambda_0$ is asymptotically proportional to h^2 for the present type of meshes (see [2]). We choose N = 10, 20, 40. The maximum values of λ_h are given in Table 1, together with the extrapolated values. We have two extrapolated values, which are in good agreement with each other as well as with the finite difference solution obtained by Simpson [7].

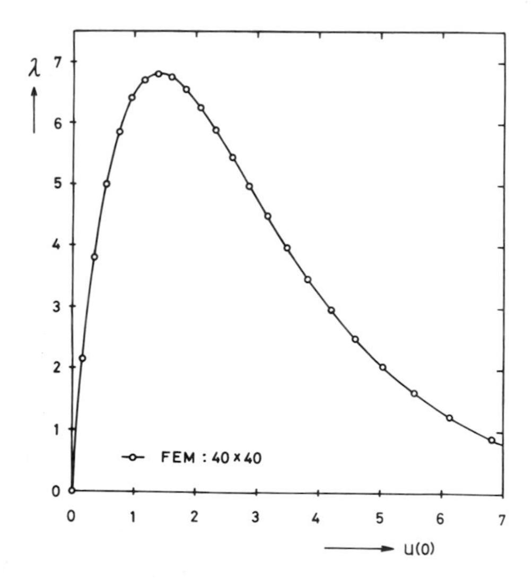

Fig. 4 λ - *u(0) curve for Problem (B)*

N	λ_{h0}	*extrapolated values*
10	*6.86186*	
		} *6.80810*
20	*6.82154*	
		} *6.80812*
40	*6.81148*	
Simpson :	*6.8082*	

Table 1 *Convergence character of* λ_{h0} *for Problem (B)*

<u>Problem (C)</u> Figure 5 gives λ - u(0) curve based on the results for N = 40, where two turning points are found : One is a local maximum point, and the other is a local minimum one. Again we tested the extrapolation technique by using the values for N = 10, 20 and 40, and the results are summarized in Table 2, where λ_{max} = local maximum of λ, and λ_{min} = local minimum of λ, respectively. The two series of extrapolated values are mutually in good agreement, but have slight difference from the results by Simpson.

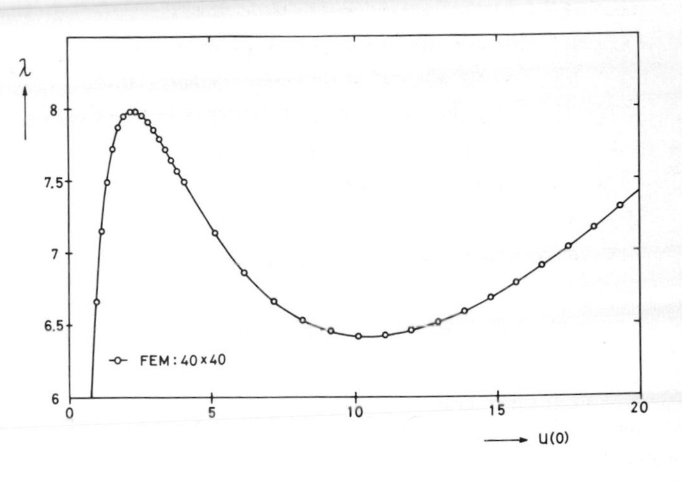

Fig. 5 λ - *u(0) curve for Problem (C)*

N	$\lambda_{h,max}$	*extrapolated values*	$\lambda_{h,min}$	*extrapolated values*
10	8.04628	} 7.98164	6.47054	} 6.41764
20	7.99780	} 7.98168	6.43086	} 6.41769
40	7.98571		6.42098	
Simpson :	7.957		*Simpson :*	6.423

Table 2 Convergence character of $\lambda_{h,max}$ and $\lambda_{h,min}$ for Problem (C)

8. Concluding remarks

We have made numerical analysis of finite element approximations to bifurcation problems of turning point type. One of the most important results that we have obtained is that turning points can be well approximated by numerical solutions, provided that the original problems have such points. Numerical solutions are also illustrated to see the effectiveness of our method. Our results can be extended in various ways. For example, we can deal with more general elliptic operators than $-\Delta$, and may discuss the effectiveness of higher order finite elements, which are commonly employed to improve convergence rates of numerical solutions. The complete proofs of our analysis will be published very soon [16].

References

[1] R. Weiss: Bifurcation in difference approximation to two-point boundary value problems, Math. Comp. 29 (1975) 746-760.

[2] F. Kikuchi: An iterative finite element scheme for bifurcation analysis of semi-linear elliptic equations, ISAS Report, University of Tokyo, No. 542 (1976) 203-231.

[3] M. Yamaguchi and H. Fujii: On numerical deformations of singularities in nonlinear elasticity, this Symposium (1977).

[4] C. Bolley: Etude numérique d'un problème de bifurcation, Thesis, Universite de Rennes (1977).

[5] A. Mizutani: On the finite element method for $\Delta u + \mu u - f(x,u) = 0$, to appear.

[6] R. B. Simpson: Existence and error estimates for solutions of a discrete analog of nonlinear eigenvalue problems, Math. Comp. 26 (1972) 190-211.

[7] R. B. Simpson: A method for the numerical determination of bifurcation states of nonlinear systems of equations, SIAM J. Numer. Anal. 12 (1975) 439-451.

[8] H. B. Keller: Nonlinear bifurcation, J. Diff. Eq. 7 (1970) 417-434.

[9] J. B. Keller and S. Antman (editors): Bifurcation Theory and Nonlinear Eigenvalue Problems, Benjamin (1969).

[10] M. A. Krasnosel'skii et al.: Approximate Solutions of Operator Equations, Wolters-Noordhoff (1972).

[11] D. H. Sattinger: Topics in Stability and Bifurcation Theory, Lecture Notes in Mathematics, #309, Springer (1973).

[12] L. Nirenberg: Topics in Nonlinear Functional Analysis, Courant Institute of Mathematical Sciences, New York University (1974).

[13] H. B. Keller and A. W. Wolfe: On the non-unique equilibrium states and buckling mechanism of spherical shells, SIAM J. 13 (1965) 674-705.

[14] P. Grisvard: Behavior of the solutions of an elliptic boundary value problem in a polygonal or polyhedral domain, in Numerical Solution of Partial Differential Equations-III, SYNSPADE 1975, edited by B. Hubbard, Academic Press (1976) 204-274.

[15] G. Strang and G. J. Fix: An Analysis of the Finite Element Method, Prentice-Hall (1973).

[16] F. Kikuchi: Numerical analysis of the finite element method applied to bifurcation problems of turning point type, to appear.

ON NUMERICAL DEFORMATION OF SINGULARITIES IN NONLINEAR ELASTICITY

Masaya Yamaguti
Department of Mathematics
Faculty of Science
Kyoto University

Hiroshi Fujii
Department of Computer Sciences
Faculty of Science
Kyoto Sangyo University

Kyoto, Japan

1. Introduction. Shallow Shell Theory and Numerical Analysis

Shells under external loadings often reveal some instability phenomena, such as the snap-through and the bifurcation bucklings. Stability of such arches and shells is one of the most important and challenging subjects in applied mechanics, and has been attracting attention of many researchers in the fields of both applied mechanics and mathematics. Since the early work of von Kármán and Tsien in 1939 [1], a variety of works has been published on the determination of critical loads for snap and bifurcation bucklings, together with their sensitivities to imperfections, and on initial post-bifurcation behaviors [2-8]. A remarkable fact is that numerical techniques play a crucial role in those works. Huang [4] was the first who made the use of the finite difference method to the analysis of bifurcation bucklings of thin shallow spherical shells. The finite element method has also been applied to the analysis of geometrically nonlinear problems, e.g., [7]. Among them, Endou, Hangai and Kawamata [8] persued a complete post-buckling analysis of shallow spherical shells. As is well known, shallow shells under external *normal* loadings have *nonlinear* fundamental paths. This fact explains the basic importance of numerical analysis in such problems. We remark that this shows a contrast to some bifurcation problems with *linear* fundamental paths, e.g., the buckling of a plate under lateral forces, possibly with small imperfections in normal forces, or in initial deformations. In such problems, modern operator theoretical approaches have been fruitful. See, [9-13].

The objective of the present paper is to provide a mathematical foundation to the numerical analysis of such stability problems in nonlinear elasticity. In other words, we discuss the numerical analysis of nonlinear systems with singularities such as snap and bifurcation bucklings. See, also [14] and [15].

Since the fundamental path itself is nonlinear, it must be computed numerically. This paper is, however, focussed on the question of *numerical* realization and deformation of singularities. We raise a question whether or not the structure of singularities can be realized *numerically*. We clarify how and in what schemes this realization is established.

We procced the study within the framework of a nonlinear operator equation in a Hilbert space V:

$$(Q): \qquad (I - L_\lambda)w + T(w) = \mu\cdot f + g_\lambda. \tag{1.1}$$

Here, w measures deflections of an arch or shell from its initial state; L_λ denotes a linear, compact operator, and T a nonlinear, continuous compact operator; μ is the loading parameter of a given external *normal* load f, while g_λ is a given function in V (possibly $\equiv 0$); λ corresponds to the lateral force applied to the edge, which is assumed to be a fixed constant.

The setting (Q) involves the shallow arch, and the shallow shell theory of von Kármán, Donnell and Marguerre. (See, e.g., [16], [11].) In fact, the von Kármán-Donnell-Marguerre equation is:

$$(P)_1: \quad \begin{aligned} \Delta^2\psi &= -\frac{1}{2}[w,w] - [W,w] \\ \Delta^2 w &= [W + w, \psi + \lambda\Psi] + \mu\cdot p \end{aligned} \qquad \text{in } \Omega \subset R^2, \tag{1.2}$$

where Δ^2 denotes the biharmonic operator, and

$$(P)_2: \qquad [w,\psi] = w_{xx}\psi_{yy} + w_{yy}\psi_{xx} - 2w_{xy}\psi_{xy}\ . \tag{1.3}$$

At the boundary $\partial\Omega$, the simply supported condition is imposed:

$$(P)_3: \qquad w = \Delta w = \psi = \Delta\psi = 0 \qquad \text{on } \partial\Omega. \tag{1.4}$$

Here, w represents the radial component of deflection, and W the known initial deflection; ψ is the Airy stress function, and Ψ is the known Airy function of the applied force to the edge; p is the external normal load on the shell with μ the loading parameter.

If we let $\quad V = H^2(\Omega)\cap H^1_o(\Omega)$, with inner product $< u,v >= \int\Delta u\cdot\Delta v$,

$B(u,v) = \Delta^{-2}[u,v]$, and $G = \Delta^{-2}$, $\qquad$ (1.5)

(P) is reduced to a pair of uncoupled nonlinear equations:

$$\psi = -\frac{1}{2}B(w,w) - B(W,w) \quad \text{and} \quad w = B(W+w,\psi+\lambda\Psi) + \mu\cdot Gp. \tag{1.6}$$

Eliminating ψ in these equations, we obtain a single operator equation of the form (Q). For details of formulations and properties of operators, see [11], [17].

Another example within the framework of (Q) is the shallow arch equation:

$$(P)': \qquad EI\frac{d^4w}{dx^4} - \frac{EA}{\ell}\{-\lambda + \frac{1}{2}\|\frac{dw}{dx}\|_o^2 + (\frac{dW}{dx}, \frac{dw}{dx})_o\}\frac{d^2}{dx^2}(W + w) = \mu\cdot p \tag{1.7}$$

$$\text{in } \Omega \equiv (-\frac{\ell}{2}, \frac{\ell}{2})$$

with appropriate boundary conditions, where $(\ ,\ )_o$ and $\|\cdot\|_o$ denote the inner product and the norm in $L^2(\Omega)$. It may not be difficult to reduce (P)' to an operator equation in $V = H^2(\Omega)\cap H^1_o(\Omega)$ or in $V = H^2_o(\Omega)$, and we leave it to the reader.

The plan of this paper is given in the following: We first discuss very quickly the classification of critical points, and then, we introduce the notion of symmetry invariances into (Q). Note that critical points treated here are only of the simple case, i.e., the case where the kernel of linearized operator is one dimensional. The numerical scheme considered is a class of finite element schemes, which primarily concerns with the compatible finite element spaces V^h. Our main results are described in § 7, § 8 and § 9, where we discuss numerical realization of critical points in V^h, i.e., of snap and bifurcation points. It is shown that a snap point is always realized numerically in V^h-space. With regards to bifurcations, this is generally not true. However, if the scheme preserves the axi-symmetry, which (Q) is assumed to have, then, the *axi-asymmetric* bifurcation is realized as itself in V^h. This implies that the nature of *"numerical imperfections"* is very non-generic. As will be seen in § 9, it is natural to expect that if the scheme destroys this symmetry, the bifurcation can no longer be realized, and it turns into a snap point. Also, axi-symmetric bifurcations (if exists) cannot generally be realized in V^h. In all cases, we obtain the orders of convergence of the numerical buckling load μ_c^h, of the buckling mode ϕ_c^h, and of the deflection w_c^h.

Because of limited space, proofs of the results and detailed examples are not present. For such details, please refer to [17], [18].

2. Snap and Bifurcation Bucklings. Classification

We begin by giving a brief note on classifications of singularities, which arise in many contexts in nonlinear elasticity. Detailed discussions and proofs are found in [17].

Let V be a real Hilbert space with inner product < , > and norm $\|\cdot\|_V$. Let $V \equiv R^1 \times V$. We consider the following operator equation:

$$\text{(Q)}: \qquad F(\mu,w) \equiv (I-L_\lambda)w + T(w) - \mu f - g_\lambda = 0, \tag{2.1}$$

where $F: V \to V$ is a continuous operator. We assume that T is a continuous, compact operator $V \to V$, such that $T(0) = 0$, and p-times continuously Fréchet differentiable with $p > 3$. L_λ is a linear compact operator $V \to V$. We assume also that L_λ and T'_w are self-adjoint. A solution of (Q) is a pair $(\mu,w) \in V$ such that $F(\mu,w) = 0$. Envisaging applications to numerical analysis, our object is, for a known solution $O \equiv (\mu_o,w_o) \in V$, to obtain all the paths in V which contain O. By a *path* we mean a connected subset of S, where S denotes the closure of the solutions of (Q) in V.

If we let $w = w_o + v$, and $\mu = \mu_o + \nu$ in (Q), we have

$$\begin{aligned}\text{(R)}: \qquad G(\nu,v) &\equiv (I-L_\lambda+T'_{w_o})v + \frac{1}{2}T''_{w_o}(v,v) + \frac{1}{6}T'''_{w_o}(v,v,v) + R(w_o,v) - \nu f \\ &\equiv L_o v + \frac{1}{2}Q_o(v,v) + \frac{1}{6}C_o(v,v,v) + R_o(v) - \nu f = 0\end{aligned} \tag{2.2}$$

where T'_{w_o}, T''_{w_o} and T'''_{w_o} denote the Fréchet derivatives of T at $w=w_o$, and R the remainder term. We say that $O \equiv (\mu_o,w_o)$ is an *ordinary point* of (Q) if L_o has a continuous inverse in V, and a *critical point* if not. By the implicit function theorem, we have

Lemma 2.1 ; If $O \equiv (\mu_o,w_o)$ is an ordinary point of (Q), then for $|\nu| \le \nu_o$ (ν_o: sufficiently small), there exists a unique, smooth C^p-function $v = v(\nu)$: $R^1 \to V$, such that

$G(\nu, v(\nu)) = 0$ and $v(0) = 0$.

Suppose now $C \equiv (\mu_c, w_c)$ is critical. Let L_c, Q_c, C_c and R_c denote the terms arising from C corresponding to L_o, Q_o, C_o and R_o, respectively. Since T and L_λ are compact, the kernel of L_c is of finite dimension. We consider only the *simple* critical case, i.e., the case of dim $\ker(L_c) = 1$.

Let $\Pi: V \to \ker(L_c)$ the orthogonal projection, and $\omega = I - \Pi: V \to [\ker(L_c)]^{\perp}$. Let $\phi_c \in \ker(L_c)$, such that $\|\phi_c\|_V = 1$. There is a bounded linear operator $L_c^{\dagger}$ on ωV such that $L_c \cdot L_c^{\dagger}\omega = \omega$.

We let

$$A_c \equiv \Pi Q_c(\phi_c, \phi_c)$$

$$B_c \equiv \Pi Q_c(\phi_c, L_c^{\dagger}\omega f)$$

$$C_c \equiv \Pi Q_c(L_c^{\dagger}\omega f, L_c^{\dagger}\omega f)$$

$$D_c \equiv \frac{1}{2}\{\Pi C_c(\phi_c, \phi_c, \phi_c) - 3\Pi Q_c(\phi_c, L_c^{\dagger}\omega Q_c(\phi_c, \phi_c))\}.$$

Definition 2.2 : A simple critical point $C \equiv (\mu_c, w_c)$ is called a *snap point* if $\Pi f \neq 0$. Moreover, if $A_c \neq 0$, C is called a *non-degenerate* snap point.

Definition 2.3 : A simple critical point $C \equiv (\mu_c, w_c)$ is called a *non-degenerate point of bifurcation*, if $\Pi f = 0$ and $B_c^2 - A_c C_c > 0$. Moreover, if $A_c \neq 0$, C is called a non-degenerate *asymmetric* point of bifurcation, and if $A_c = 0$, non-degenerate *symmetric* point of bifurcation. A non-degenerate symmetric point of bifurcation is called a *cusp* bifurcation if $D_c \neq 0$.

Note : A snap point may also be called a *limiting* point or a *turning* point.

Lemma 2.4 : Suppose $C \equiv (\mu_c, w_c)$ is a simple, non-degenerate snap point of (Q). Then, in a neighborhood of C, there is a unique smooth path, say (α), which meets C. In other words, there is a local parameter α, $|\alpha| \leq \alpha_o$ (α_o: sufficiently small), such that there exist uniquely smooth C^p-functions $\nu(\alpha)$ and $v(\alpha)$:

$$\nu(\alpha) = \frac{A_c}{2\Pi f}\alpha^2 + O(\alpha^3)$$

$$v(\alpha) = \alpha\phi_c + \nu(\alpha)L_c^{\dagger}\omega f + O(\alpha^3)$$

such that $G(\nu(\alpha), v(\alpha)) = 0$ for $|\alpha| \leq \alpha_o$. Furthermore, on this (α)-path, there is an eigenvalue $\zeta = \zeta(\alpha)$ of the linearized operator of G on the path (α), which is a smooth C^{p-1}-function of α, such that $\zeta(0) = 0$, and $\frac{\partial\zeta}{\partial\alpha}(0) \neq 0$.

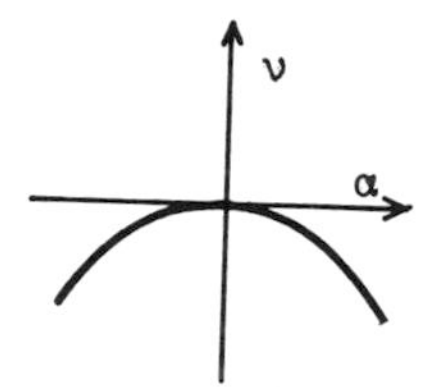

Fig. 1

Lemma 2.5 : Suppose $C \equiv (\mu_c, w_c)$ is a simple, non-degenerate, asymmetric point of bifurcation of (Q). Then, there exist two paths (μ_+) and (μ_-) in a neighborhood of C, which intersect at C. In other words, taking ν as the local parameter for $|\nu| \leq \nu_o$ (ν_o: sufficiently small), we have two smooth C^{p-2}-functions $\alpha_{\pm}(\nu)$, such that

$$\alpha_{\pm}(\nu) = \frac{-B_c \pm \sqrt{B_c^2 - A_c C_c}}{A_c}\nu + O(\nu^2).$$

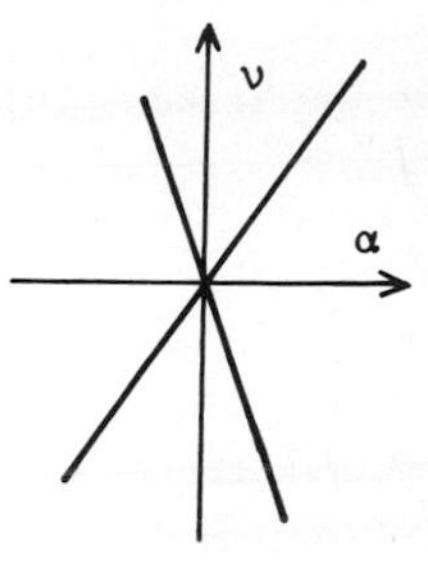

Fig. 2

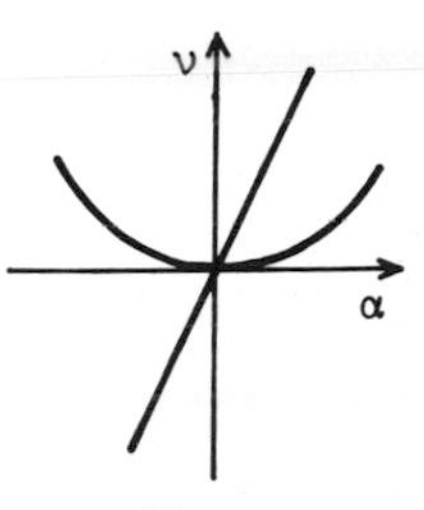

Fig. 3

The paths$(\mu_\pm)$are given by$(\mu_c+\nu,\ w_c+v_\pm(\nu))$, where

$$v_\pm(\nu) = \alpha_\pm(\nu)\phi_c + \nu L_c^\dagger \omega f + O(\nu^2).$$

Furthermore, there exist eigenvalues $\zeta^+(\nu)$ and $\zeta^-(\nu)$ of the linearized operators of G along (μ^+)- and (μ^-)-paths respectively, which are smooth C^{p-2}-functions of ν, such that

$$\zeta^+(0) = \zeta^-(0) = 0, \text{ and } \frac{\partial\zeta^+}{\partial\nu}(0)\cdot\frac{\partial\zeta^-}{\partial\nu}(0) < 0.$$

Lemma 2.6 : Suppose $C\equiv(\mu_c,w_c)$ is a simple, non-degenerate, symmetric point of bifurcation of (Q). Suppose that $D_c \neq 0$. Then, there exist two paths (μ) and (α) in a neighborhood of C, which intersect at C. In other words, for the (μ)-path, taking ν as the local parameter, we have a smooth C^{p-2}-function $\alpha=\alpha(\nu)$; $R^1 \to R^1$, such that

$$\alpha(\nu) = -\frac{C_c}{2B_c}\nu + O(\nu^2) \quad \text{for } |\nu| \leq \nu_o$$

(ν_o: sufficiently small), and the (μ)-path is given by $(\mu_c+\nu, w_c+v(\nu))$, where

$$v(\nu) = \alpha(\nu)\phi_c + \nu L_c^\dagger \omega f + O(\nu^2).$$

For $|\alpha| \leq \alpha_o$ (α_o: sufficiently small), the (α)-path is given by

$$\nu(\alpha) = -\frac{D_c}{3B_c}\alpha^2 + O(\alpha^3),$$

$$v(\alpha) = \alpha\phi_c + \nu(\alpha)L_c^\dagger \omega f + O(\alpha^3).$$

Furthermore, along the (μ)- and (α)-paths, there exist eigenvalues $\zeta^\mu(\nu)$ and $\zeta^\alpha(\alpha)$ of the linearized operator of G on the (μ)- and (α)-paths, respectively, each of which is a smooth C^{p-2}-function of ν and α, respectively, such that $\zeta^\mu(0) = 0$ and $\zeta^\alpha(0) = 0$. They satisfy the relations

$$\frac{\partial\zeta^\mu}{\partial\nu}(0) = B_c \neq 0, \ \frac{\partial\zeta^\alpha}{\partial\alpha}(0) = A_c = 0, \text{ and}$$

$$\frac{\partial^2\zeta^\alpha}{\partial\alpha^2}(0) = 2D_c = -3\frac{\partial\zeta^\mu}{\partial\nu}(0)\cdot\frac{\partial^2\nu}{\partial\alpha^2}(0) \neq 0.$$

The proof in this section is essentially due to the implicit function theorem, and to the Morse lemma, with the aid of the Lyapunov-Schmidt procedure. See, e.g., Nirenberg [19].

3. Symmetry Invariances

The notion of symmetry invariances of the problem (Q) plays a key role in some aspects of subsequent discussions. In fact, it may be easily seen that the shallow arch has an axis of symmetry at x=0, and that the corresponding operator equation is invariant under this symmetry. If the domain Ω has a certain axis of symmetry, which is represented by an operator S_k, the von Kármán-Donnell-Marguerre equation is also invariant under S_k, provided that Ψ, W and p have the same symmetry. We discuss here a particular case of symmetry, i.e., $S_k{}^2 = I$. The general theory will be given in a subsequent paper [18] with applications to shells of revolution. (See, also [13].)

Let us recall that our equation is :

$$\begin{aligned} F(\mu,w) &\equiv (I-L_\lambda)w + T(w) - \mu f - g_\lambda \\ &\equiv N(w) - \mu f - g_\lambda = 0. \end{aligned} \tag{3.1}$$

Definition 3.1 : (Q) is said to have S-*invariance* if there is a linear continuous operator S: V → V, such that

$$F(\mu,Sw) = S\cdot F(\mu,w) \quad \text{for any } (\mu,w) \in V. \tag{3.2}$$

Let us assume in the following that (Q) has S-invariance such that $S^2=I$. Then, if we define two operators $P_\pm$ by $P_\pm = (I \pm S)/2$, we have $P_+ + P_- = I$, $P_\pm{}^2 = P_\pm$ and $P_+P_- = P_-P_+ = 0$. Thus, V is represented as a direct sum $V=V_+ \oplus V_-$, where $V_\pm = P_\pm V$. Moreover, we have the following Lemma for (Q) with S-invariance such that $S^2=I$.

Lemma 3.2 : (*i*) For any $w_+ \in V_+$, $P_-N(w_+)P_+=0$ and $P_+N(w_+)P_-=0$. That is, the linearized operator of N(w) at $w=w_+$ is diagonal.

(*ii*) For any $w_+ \in V_+$, $P_-N(w_+) = 0$, and $P_-f = P_-g_\lambda = 0$. That is, $P_-F(\mu,(w_+,w_-)) = 0$ has a *trivial* solution $w_- \equiv 0$ for all $\mu \in R^1$ and $w_+ \in V_+$.

(*iii*)
$$P_-N''_{w_+}(u_-,v_-) = 0 \text{ for any } (u_-,v_-,w_+) \in V_-\times V_-\times V_+,$$
$$P_-N''_{w_+}(u_+,v_+) = 0 \text{ for any } (u_+,v_+,w_+) \in V_+\times V_+\times V_+.$$

A direct consequence of (i) and (ii) of the above Lemma is the existence of " *symmetric path*" of (Q).

Lemma 3.3 : Suppose $C_+\equiv(\mu_o,w_{o,+}) \in R^1\times V_+$ is an ordinary point of (Q). Then, the ordinary path which contains C_+ (*Cf*. Lemma 2.1) lies in $R^1\times V_+$.

Next, in a neighborhood of a critical point $C_+ \in R^1\times V_+$, we have the following

Lemma 3.4 : Let $C_+\equiv(\mu_c,w_{c,+}) \in R^1\times V_+$ be a simple critical point of (Q). Then, the corresponding eigenfunction ϕ_c lies either in V_+ or in V_-.

(*i*) (*Axi-symmetric snapping*) If $\phi_c \in V_+$, and $\langle\phi_c, f\rangle \neq 0$, C_+ is a snap point in $R^1\times V_+$, i.e., there exists a unique path in $R^1\times V_+$.

(*ii*) (*Axi-asymmetric bifurcation*) If $\phi_c \in V_-$, C_+ is a bifurcation point, since $\langle\phi_c, f\rangle = 0$. (Note that $f \in V_+$.) From (iii) of Lemma 3.2, $A_c=C_c=0$, which implies that C_+ is a symmetric point of bifurcation. Also, the (μ)-path lies in $R^1\times V_+$.

(*iii*) (*Axi-symmetric bifurcation*) If $\phi_c \in V_+$ and still $\langle\phi_c,f\rangle = 0$, we have an axi-symmetric bifurcation. Both of the two paths emerging from C_+ are in $R^1\times V_+$.

Firstly, the above Lemma shows that at least one smooth symmetric path emerges from any simple, non-degenerate critical point $C_+ \in R^1\times V_+$. Secondly, it shows that on this axi-symmetric path, every axi-asymmetric bifurcation point is *necessarily* a symmetric point of bifurcation, i.e., essentially a cusp bifurcation.

Finally, we may remark that the case (iii) scarcely occurs in actual problems.

4. Finite Element Scheme

We are now at the stage to discuss numerical approximation of the problem (Q). The scheme considered here is a class of finite element schemes (Q^h) described below. We note that our setting (Q^h) could include a mixed finite element scheme for appropriately chosen V and V^h as was proposed by Miyoshi in [20], our primal concern here is with the compatible finite element scheme. It may be also worthnoting that in many applications, e.g., shells of revolution, or shallow cylindrical shells, the approximate space V^h is often taken to be a *hybrid* space of finite element and Galerkin, or to be a pure Galerkin space, due to high geometric symmetry of the problem. See, e.g., [6], [8]. In such cases, our setting (Q^h) may have to be modified accordingly, but it appears that the essential framework of the theory is still valid. Such applications, as well as the detailed proof are found in [17], [18].

We first give some assumptions on (Q) and on the solution w, taking in mind applications to arch and shell problems.

Let U, V and W be real, separable Hilbert spaces such that

$$W \subsetneq V \subsetneq U \quad \text{(with the injections continuous and dense)}$$

with norm $\|\cdot\|_U$, $\|\cdot\|_V$ and $\|\cdot\|_W$, respectively. Recall that the inner product of V is given by < , >. We assume the following to (Q):

(*i*) (*Regularity of the solution*) $w \in W$ for any (μ,w) such that $F(\mu,w) = 0$.

(*ii*) L_λ, T'_w are linear continuous from U into V, and from V into W, i.e., for $\forall w \in W$,

$$\|T'_w\|_{V\to W} \le c_1, \quad \|T'_w\|_{U\to V} \le c_2, \quad \|L_\lambda\|_{V\to W} \le c_3 \quad \text{and} \quad \|L_\lambda\|_{U\to V} \le c_4.$$

Let V^h be finite element subspaces of V ($h \to o$), and let P^h be the orthogonal projection from V onto V^h, i.e., $\langle (I-P^h)u, v^h \rangle = 0$, for all $v^h \in V^h$. We assume that there exist two constants ℓ and m such that for any $u \in W$,

$$\|(I-P^h)u\|_V \le Ch^{\ell}\|u\|_W \quad \text{and} \quad \|(I-P^h)u\|_U \le Ch^{\ell+m}\|u\|_W. \tag{4.1}$$

Let $\qquad k = \min(\ell, m).$ (4.2)

Our finite element scheme is now described in an operator equation as

$$(Q^h): \quad \begin{aligned} F^h(\mu,w^h) &\equiv (I-L^h)w^h + T^h(w^h) - \mu f^h - g^h_\lambda \\ &\equiv (I-P^hL^{(h)})w^h + P^hT^{(h)}(w^h) - \mu P^hf^{(h)} - P^hg^{(h)}_\lambda = 0 \end{aligned} \tag{4.3}$$

Here $T^{(h)}$ is a continuous, compact operator $V \to V$, and of C^p-class (in Fréchet sense, with $p > 3$), such that the q-th Fréchet derivatives (q=1,..,p) are uniformly bounded:

$$\|T^{(h)(q)}_w\|_{V\times\cdots\times V\to V} \le C^{(q)}_o(\|w\|_V) \quad \text{for any } w \in V. \tag{4.4}$$

Also, $T^{(h)}$ is *close* to T in the following sense:

$$\|T(w) - T^{(h)}(w)\|_V \le C^{(o)}_1(\|w\|_W)\, h^{\ell+k} \quad \text{for any } w \in W, \tag{4.5}$$

$$\|T_w^{(q)} - T_w^{(h)(q)}\|_{W\times\cdots\times W\to V} \le C_1^{(q)}(\|w\|_W)\, h^{\ell+k} \quad \text{for any } w \in W. \tag{4.6}$$

$T_w^{(h)'}$ is self-adjoint, and

$$\|T_w' - T_w^{(h)'}\|_{V\to V} \le C_2(\|w\|_W)\, h^{\ell} \quad \text{for any } w \in W. \tag{4.7}$$

$L_\lambda^{(h)}$ is a linear, self-adjoint compact operator $V \to V$, and *close* to L_λ in the following sense:

$$\|L_\lambda - L_\lambda^{(h)}\|_{W\to V} \le C_3 h^{\ell+k} \quad \text{and} \quad \|L_\lambda - L_\lambda^{(h)}\|_{V\to V} \le C_4 h^{\ell}. \tag{4.8}$$

We may also require that $f^{(h)}$ and $g_\lambda^{(h)}$ are some elements of W close to f and g_λ in V, respectively, with the order $h^{\ell+k}$.

The assumptions on (Q) and (Q^h) may be easily checked for the arch problem with C^1-compatible basis. We have $\ell = m = 2$. (Note that in this case, $T^{h)} = T$, $L_\lambda^{(h)} = L$, $f^{(h)} = f$ and $g_\lambda^{(h)} = g_\lambda$.) Concerning shell problems, a smooth domain Ω, or a rectangular domain with simply supported edge ensures the assumptions on (Q) and (Q^h), for compatible C_1-basis, i.e., we have $\ell = m = 2$. For details, see [17], [18].

5. Realization of Ordinary Path

It may be immediately seen that in a neighborhood of an ordinary path of (Q), there exists a unique and smooth ordinary path of (Q^h) in $V^h \equiv R^1 \times V^h$, except in the vicinity of critical points of (Q). In fact, we have the following

Theorem 5.1 : Suppose that $O \equiv (\mu_o, w_o) \in R^1 \times W$ is an ordinary point of (Q). Then, for all $h \le h_o$ (h_o: sufficiently small depending on O), there exists a unique ordinary point $O^h \equiv (\mu_o, w_o^h) \in R^1 \times V^h$ such that $F^h(\mu_o, w_o^h) = 0$, and

$$\|w_o - w_o^h\|_V \le Ch^{\ell}\|w_o\|_W \quad \text{and} \quad \|w_o - w_o^h\|_U \le Ch^{\ell+k}\,\|w_o\|_W. \tag{5.1}$$

Moreover, there is a smooth ordinary path $\mu = \mu_o + \nu$, $w^h = w^h_o + v^h(\nu)$ in $R^1 \times V^h$ which meets (μ_o, w_o^h) at $\nu = 0$.

The proof is essentially due to the implicit function theorem, making use of the following two Lemmas.

Lemma 5.2 : For any $(\mu, w) \in R^1 \times W$ such that $F(\mu, w) = 0$,

$$\|F^h(\mu, P^h w)\|_V \le Ch^{\ell+k}. \tag{5.2}$$

We introduce the following notations for the linearized operators: Let

$$L_w \equiv I + [-L_\lambda + T_w'],$$

and

$$L_{P^h w}^h \equiv I + P^h[-L_\lambda^{(h)} + T_{P^h w}^{(h)'}]. \tag{5.3}$$

Lemma 5.3 : (*Existence and uniform boundedness of the inverse of approximate operator*) Suppose that $(L_w)^{-1}$ exists and is bounded in V for $w \in W$. Then, for all $h \le h_o$ (h_o: sufficiently small depending of this bound), $(L_{P^h w}^h)^{-1}$ exists and is uniformly bounded as an operator in V.

6. A Theorem on Existence of Critical Points in V^h

When $C \equiv (\mu_c, w_c) \in R^1 \times W$ is a *simple, non-degenerate* critical point of (Q), what can we say about the approximate linearized operator $L^h_{Ph_w}$? The following Lemma, which answers this question, is the key to the whole discussion about the behavior of numerical solutions in the vicinity of critical points of (Q).

The results in § 2 show that there is at least one smooth path, say (t)-path, $(\mu(t), w(t))$, with a local parameter t for $t \in \Lambda \subset R^1$, such that $\mu(0) = \mu_c$, and $w(0) = w_c$, and that along this path, there exists a simple, non-degenerate eigenvalue $\zeta(t)$. In other words, on the (t)-path, the eigenproblem

$$(E)_t: \qquad L_{w(t)}\phi(t) = \zeta(t)\ \phi(t) \quad \text{in } V \tag{6.1}$$

has the property that

$$\zeta(0) = 0, \ \frac{\partial \zeta}{\partial t}(0) \neq 0, \text{ and } \phi(0)\text{: simple.} \tag{6.2}$$

We next consider the approximate eigenproblem in V:

$$(E^h)_t: \qquad L^h_{Ph_{w(t)}}\phi^h(t) = \zeta^h(t)\ \phi^h(t). \tag{6.3}$$

Theorem 6.1 : There exists a constant $\tau = \tau(h)$ for all $h \leq h_o$ (h_o: sufficiently small) such that $\zeta^h(\tau) = 0$ and $\frac{\partial \zeta^h}{\partial t}(\tau) \neq 0$; τ is the only zero of $\zeta^h(t)$ for $t \in \Lambda$. Moreover, the eigenstate $(\zeta^h(\tau), \phi^h(\tau))$ is simple, and hold the estimates

$$|\tau| \leq c_1 h^{\ell+k}, \tag{6.4}$$

$$\|\phi(0) - \phi^h(\tau)\|_V \leq c_2 h^{\ell} \quad \text{and} \quad \|\phi(0) - \phi^h(\tau)\|_U \leq c_3 h^{\ell+k}, \tag{6.5}$$

where the constants c_i (i=1,2,3) depend on $\sup_{t\in\Lambda}\|w(t)\|_W$, $\sup_{t\in\Lambda}\|\dot{w}(t)\|_W$ and $\sup_{t\in\Lambda}\|\phi(t)\|_W$.

Let τ be as determined in Theorem 6.1. Then, the kernel of $L^h_{Ph_{w(\tau)}}$ is one dimensional. Let Π^h: $V \to \ker(L^h_{Ph_{w(\tau)}})$ be the orthogonal projection, and let $\omega^h = I - \Pi^h$. There exists the bounded inverse $(L^h_{Ph_{w(\tau)}})^\dagger$ on $\omega^h V$ such that $L^h_{Ph_{w(\tau)}} \cdot L^h_{Ph_{w(\tau)}}{}^{\dagger} \omega^h = \omega^h$. Moreover, we can show the following

Lemma 6.2 : $L^h_{Ph_{w(\tau)}}{}^{\dagger}$ is uniformly bounded on $\omega^h V$.

For the proof of the above Theorem and Lemma, see [17].

7. Snap Point, its Numerical Realization

A non-degenerate snap point $C \equiv (\mu_c, w_c)$ in V is always realized in V^h as itself, say C^h, and the path in a neighborhood of C^h is uniformly convergent to that near C. The numerical buckling load, that is, the value of μ at which (μ, w^h) is critical, converges to μ_c with the order $h^{\ell+k}$. This is the *proposition* we want to show in this section.

Firstly, Theorem 6.1 guarantees the unique existence of $\tau = \tau(h)$, such that $|\tau| \leq Ch^{\ell+k}$, and that

$$\mathcal{L}^h_{P^h w(\tau)} \phi^h_\tau = 0, \quad \text{with } \phi^h_\tau \text{ simple,}$$

where $(\mu(t),w(t)) \in R^1 \times W$ is solutions of (Q) such that $\mu(0)=\mu_c$, and $w(0)=w_c$. Recall that Π^h is the orthogonal projection onto $\ker(\mathcal{L}^h_{P^h w(\tau)})$ and that $\omega^h = I-\Pi^h$.

Letting $w^h = P^h w(\tau) + v^h$ and $\mu = \mu(\tau) + \nu$ in (Q^h), we have

$$(R^h)_\nu: \qquad \mathcal{L}^h_\tau v^h + \frac{1}{2}\mathcal{Q}^h_\tau(v^h,v^h) + \frac{1}{6}\mathcal{C}^h_\tau(v^h,v^h,v^h) + \mathcal{R}^h_\tau(v^h) = -F^h_\tau + \nu f^h. \tag{7.1}$$

Here, we have used the abbreviated symbols:

$$\begin{aligned}
\mathcal{L}^h_\tau &\equiv \mathcal{L}^h_{P^h w(\tau)} = I - L^h_\lambda + T^{h'}_{P^h w(\tau)}, \\
\mathcal{Q}^h_\tau(\cdot,\cdot) &\equiv T^{h''}_{P^h w(\tau)}(\cdot,\cdot), \\
\mathcal{C}^h_\tau(\cdot,\cdot,\cdot) &\equiv T^{h'''}_{P^h w(\tau)}(\cdot,\cdot,\cdot) \\
\mathcal{R}^h_\tau(\cdot) &\equiv R^h(P^h w(\tau),\cdot),
\end{aligned} \tag{7.2}$$

and $\qquad F^h_\tau \equiv F^h(\mu(\tau),P^h w(\tau))$.

Let $\qquad r^h_\tau \equiv \dfrac{1}{h^{\ell+k}} F^h_\tau\,. \tag{7.3}$

It is noted that r^h_τ is uniformly bounded by Lemma 5.2. We first look for the critical point of $(R^h)_\nu$. That is, for given $h > 0$, we seek a triplet $(\nu,v^h,\Phi^h) \in R^1 \times V^h \times V^h$ which simultaneously satisfies $(R^h)_\nu$ and its linearized eigenproblem $(S^h)_\nu$ defined by

$$(S^h)_\nu: \qquad \{\mathcal{L}^h_\tau + \mathcal{Q}^h_\tau(v^h,\cdot) + \frac{1}{2}\mathcal{C}^{h'}_{\tau(v^h)} + \mathcal{R}^{h'}_{\tau(v^h)}\}\Phi^h = 0. \tag{7.4}$$

For this purpose, we introduce a sequence of auxiliary problems $(h \to 0)$:

$$(R^h)_{\nu,\varepsilon}: \qquad \mathcal{L}^h_\tau v^h + \frac{1}{2}\mathcal{Q}^h_\tau(v^h,v^h) + \frac{1}{6}\mathcal{C}^h_\tau(v^h,v^h,v^h) + \mathcal{R}^h_\tau(v^h) = -\varepsilon r^h_\tau + \nu f^h \tag{7.5}$$

with its linearized eigenproblem $(S^h)_{\nu,\varepsilon}$. Note that the auxiliary problems $(R^h)_{\nu,\varepsilon}$ and $(S^h)_{\nu,\varepsilon}$ reduce to $(R^h)_\nu$ and $(S^h)_\nu$, respectively, when $\varepsilon = h^{\ell+k} \equiv \varepsilon(h)$. It turns out that, applying the Lyapunov-Schmidt method to both $(R^h)_{\nu,\varepsilon}$ and $(S^h)_{\nu,\varepsilon}$, and noting the uniform boundedness of $\mathcal{L}^{h\dagger}_\tau$ on $\omega^h V$, the auxiliary problems reduce to the study of solutions $(\alpha,\nu,\varepsilon) \in R^3$ of two scalor equations:

$$\begin{aligned}
\Gamma^h(\alpha,\nu,\varepsilon) \equiv{}& \frac{1}{2}A^h_\tau\alpha^2 + B^h_\tau\alpha\nu + \frac{1}{2}C^h_\tau\nu^2 + X^h_\tau\alpha\varepsilon + Y^h_\tau\nu\varepsilon + Z^h_\tau\varepsilon^2 + \frac{1}{3}D^h_\tau\alpha^3 + \\
&+ \varepsilon\Pi^h r^h_\tau + \nu\Pi^h f^h + (\text{h.o.t.}) = 0,
\end{aligned} \tag{7.6}$$

and $\qquad \Xi^h(\alpha,\nu,\varepsilon) \equiv A^h_\tau\alpha + B^h_\tau\nu + X^h_\tau\varepsilon + (\text{h.o.t.}) = 0, \tag{7.7}$

where (h.o.t.) designates higher order terms. Here, the coefficients A^h_τ, B^h_τ, C^h_τ, D^h_τ, $\cdots$ are the expressions corresponding to, and converging to A_c, B_c, C_c, D_c, $\cdots$, respectively. (See, Eq.(1.3), e.g., $A^h_\tau \equiv \Pi^h\mathcal{Q}^h_\tau(\phi^h_\tau,\phi^h_\tau),\cdots$.) Also, $X^h_\tau \equiv \Pi^h\mathcal{Q}^h_\tau(\phi^h_\tau,\mathcal{L}^{h\dagger}_\tau\omega^h r^h_\tau)$, $Y^h_\tau \equiv \Pi^h\mathcal{Q}^h_\tau(\mathcal{L}^{h\dagger}_\tau\omega^h f^h,\ \mathcal{L}^{h\dagger}_\tau\omega^h r^h_\tau)$ and $Z^h \equiv \Pi^h\mathcal{Q}^h_\tau(\mathcal{L}^{h\dagger}_\tau\omega^h r^h,\mathcal{L}^{h\dagger}_\tau\omega^h r^h_\tau)$. Hence, all the coefficients are uniformly bounded, and especially, A^h_τ and $\Pi^h f^h$ are bounded from both above and below uniformly in h, since A_c and $\Pi f \neq 0$ by hypothesis. These observations imply that $\Gamma^h(\alpha,\nu,\varepsilon) = 0$ and $\Xi^h(\alpha,\nu,\varepsilon) = 0$ exibit a *"fold"* structure uniformly in h, near the origin $(\alpha,\nu,\varepsilon) = (0,0,0)$. We conclude that for each $h \leq h_o$, there exist *uniquely* smooth functions $\nu=\nu^h(\varepsilon)$ and $\alpha=\alpha^h(\nu^h(\varepsilon),\varepsilon)$:

$$\nu = \nu^h(\varepsilon) = \frac{\Pi^h r^h_\tau}{\Pi^h f^h}\varepsilon + O(\varepsilon^2) \tag{7.8}$$

$$\alpha = \alpha^h(\nu^h(\varepsilon),\varepsilon) = -\frac{B^h_\tau}{A^h_\tau}\nu^h(\varepsilon) - \frac{X^h_\tau}{A^h_\tau}\varepsilon + O(\varepsilon^2) \tag{7.9}$$

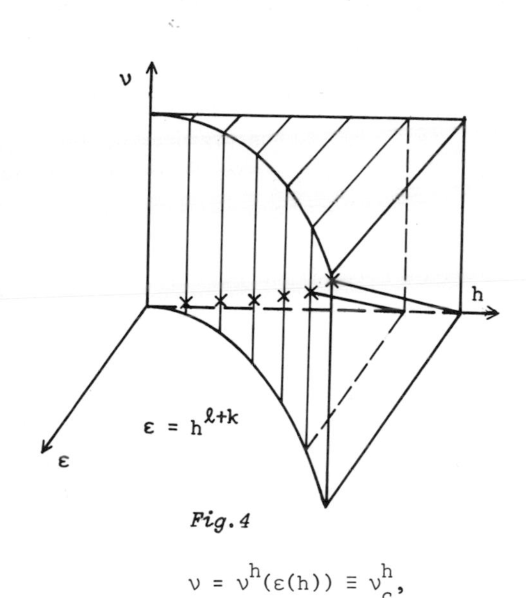

Fig.4

which satisfy $\Gamma^h = \Xi^h = 0$ for all $|\varepsilon| \leq \varepsilon_o$. ($\varepsilon_o$: sufficiently small, but independent of h).

We consider Eq.(7.8) in (ν,ε,h)-space $\subset R^3$, and examine it on the cross-section $\varepsilon = \varepsilon(h)$ $(\equiv h^{\ell+k})$. Then, it follows that on $\varepsilon = \varepsilon(h)$, there is a unique value of

$$\nu = \nu^h(\varepsilon(h)) \equiv \nu^h_c, \tag{7.10}$$

with the bound

$$|\nu^h_c| \leq Ch^{\ell+k}. \tag{7.11}$$

This proves the unique existence of numerical buckling load μ^h_c of (Q^h). Noting that $\mu^h_c = \mu(\tau) + \nu^h_c$, and $|\mu_c - \mu(\tau)| \leq C\tau^2 \leq C'(h^{\ell+k})^2$, we have the error estimate:

$$|\mu_c - \mu^h_c| \leq Ch^{\ell+k}. \tag{7.11}$$

We can also obtain the error estimates of ϕ^h_c, the buckling mode, and of w^h_c, the deflection at the buckled point:

$$\|\phi_c - \phi^h_c\|_V \leq Ch^{\ell}, \qquad \|\phi_c - \phi^h_c\|_U \leq Ch^{\ell+k}, \tag{7.12}$$

$$\|w_c - w^h_c\|_V \leq Ch^{\ell}, \text{ and } \qquad \|w_c - w^h_c\|_U \leq Ch^{\ell+k} \tag{7.13}$$

Existence and convergence of a neighborhood path in V^h is a natural consequence of the above analysis, and we leave the details to [17]. See, Fig.5.

8. Realization of Axi-asymmetric Bifurcation Point

The aim of this section is to show that when the scheme (Q^h) preserves the S-invariance, a simple, non-degenerate, axi-asymmetric, symmetric point of bifurcation, i.e., a cusp bifurcation if $D_c \neq 0$, is realized as itself in V^h. In contrast to this, if the scheme destroys the S-invariance (which (Q) is assumed to have), or if the bifurcation point is of axi-symmetric, we can expect only that the critical point is realized in V^h as a snap point. We study the latter case in the next section. In both cases, however, we can obtain the numerical buckling loads, which exibit a crucial difference between the orders of convergence.

We suppose that (Q) has a S-invariance, and that (Q^h) respects this invariance. Then firstly;

Lemma 8.1 : An ordinary symmetric path of (Q) is always realized as itself in V^h.

Secondly, suppose that $C_+ \equiv (\mu_c, w_{c,+}) \in R^1 \times W_+$ is a simple, non-degenerate, axi-asymmetric, symmetric point of bifurcation with $D_c \neq 0$, i.e., a cusp bifurcation point, on a symmetric path $(\mu, w_+(\mu))$. Then,

Lemma 8.2 : There exists a symmetric path $(\mu, w_+^h(\mu)) \in R^1 \times V_+^h$, for $|\mu - \mu_c| \leq c_o$ (c_o: sufficiently small), which satisfies

$$\|w_+(\mu) - w_+^h(\mu)\|_V \leq Ch^{\ell}, \quad \text{and} \quad \|w_+(\mu) - w_+^h(\mu)\|_U \leq Ch^{\ell+k}, \tag{8.1}$$

where $C = C(\|w_+(\mu)\|_W)$.

This follows from the fact that (Q^h) is decomposable into two components, and that the (+)-component of the linearized operator is uniformly invertible on V_+^h.

Next, we consider the eigenproblem in V_-^h for $|\mu - \mu_c| \leq c_o$:

$$(P_-E^h)_\mu: \qquad P_- L^h_{w_+^h(\mu)} \phi_-^h(\mu) = \zeta^h(\mu) \phi_-^h(\mu). \tag{8.2}$$

Taking notice of $P_-P^h = P^hP_-$, and using a result similar to Theorem 6.1, we obtain the unique existence of a constant $\tau = \tau(h)$, such that $\zeta^h(\mu_c + \tau) = 0$ and $|\tau| \leq Ch^{\ell+k}$. We put $\mu_c^h \equiv \mu_c + \tau$. It is obviously seen that $\Phi_c^h \equiv (0, \phi_-^h(\mu_c^h)) \in V_+^h \times V_-^h$ is the simple eigenfunction of the operator $L^h_{w_+^h}$. The orthogonality $\langle f^h, \Phi_c^h \rangle = 0$ leads to the conclusion that $C_+^h \equiv (\mu_c^h, w_+^h(\mu_c^h))$ is a bifurcation point. By (iii) of Lemma 3.2, C_+^h is also a symmetric point of bifurcation in V^h. Thus, we have the numerical realization of the cusp bifurcation with the $h^{\ell+k}$ convergence of the numerical buckling load. Orders of convergence of $w_+^h(\mu_c^h)$ and of Φ_c^h are as in the snap buckling case. See, Fig.6.

The uniqueness of critical point of (Q^h) in a neighborhood of C_+, and the convergence of paths of (Q^h) near C_+^h to the corresponding paths of (Q) can be also proved. For details, see [17].

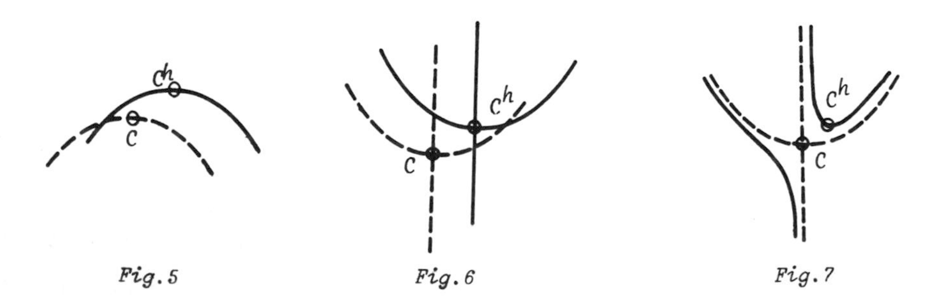

Fig.5 *Fig.6* *Fig.7*

9. When the S-invariance is violated ...

Suppose that C is a cusp bifurcation point as in the previous section. We investigate it in V^h with a different situation, that is, in the case that (Q^h) violates the S-invariance which (Q) actually possesses. One may expect that this leads to a more generic situation as will be seen. In fact, we could see, by virtue of the generalized Morse lemma (See,[19]),that, for a *fixed* $h > 0$, C is never realized numerically as itself, but it separates into two paths around C. The bifurcation point C in V turns to be snap points in V^h. See, Fig.7. Furthermore, under a certain condition, we can prove the unique existence of critical (snap) points in a neighborhood of C. The order of convergence of the numerical buckling load follows the $\frac{2}{3}$ - power law, which reflects the cusp nature of C.

We start the study from Eqs.(7.6) and (7.7). There is an essential difference between the snap and the cusp cases, i.e., in the latter case, A_τ^h and $\Pi^h f^h$ are *not* bounded below uniformly in h. In fact, since $A_c = 0$ and $\Pi f = 0$ by hypothesis, we have

$$|A_\tau^h| \leq Ch^{\ell+k} \quad \text{and} \quad |\Pi^h f^h| \leq Ch^{\ell+k}. \tag{9.1}$$

It holds, however, that D_τ^h and B_τ^h are uniformly bounded from both above and below. This suggests us to let

$$\tilde{A}_\tau^h \equiv \frac{1}{h^{\ell+k}} A_\tau^h \quad \text{and} \quad \widetilde{\Pi^h f^h} \equiv \frac{1}{h^{\ell+k}} \Pi^h f^h, \tag{9.2}$$

and consider the following equations in place of Eqs.(7.6) and (7.7):

$$\begin{aligned}\tilde{\Gamma}^h(\alpha,\nu,\varepsilon) \equiv \frac{1}{3} D_\tau^h \alpha^3 + \frac{1}{2} \tilde{A}_\tau^h \alpha^2 \varepsilon + B_\tau^h \alpha\nu + \cdots \\ + \varepsilon\Pi^h r^h + \nu\varepsilon\widetilde{\Pi^h f^h} + (\text{h.o.t.}) = 0\end{aligned} \tag{9.3}$$

$$\tilde{\Xi}^h(\alpha,\nu,\varepsilon) \equiv D_\tau^h \alpha^2 + \tilde{A}_\tau^h \alpha\varepsilon + B_\tau^h \nu + X^h \varepsilon + (\text{h.o.t.}) = 0. \tag{9.4}$$

Here, we assume that there exists two constants c_o and c_1 such that

$$c_o \leq | \Pi^h r_\tau^h | \leq c_1 \qquad \text{for} \quad 0 < h \leq h_o, \tag{9.5}$$

which corresponds to the assumption that $|\Pi^h F_\tau^h|$ tends to zero actually with the order $h^{\ell+k}$.

Under the assumption (9.5), $\tilde{\Gamma}^h(\alpha,\nu,\varepsilon) = 0$ and $\tilde{\Xi}^h(\alpha,\nu,\varepsilon) = 0$ exibit the *"cusp"* structure uniformly in $h \leq h_o$ at the origin. An analysis similar to that in §7 shows the unique existence of a function $\nu = \nu^h(\varepsilon)$, for each $h \leq h_o$, such that

$$\nu = \nu^h(\varepsilon) = - \frac{D_\tau^h}{B_\tau^h} \left(\frac{3\Pi^h r_\tau^h}{2D_\tau^h} \right)^{\frac{2}{3}} \varepsilon^{\frac{2}{3}} - \frac{X_\tau^h}{B_\tau^h} \varepsilon + O(\varepsilon^{\frac{4}{3}}). \tag{9.6}$$

On the cross-section $\varepsilon = \varepsilon(h)$ $(\equiv h^{\ell+k})$, there is a unique $\nu = \nu^h(\varepsilon(h)) \equiv \nu_c^h$ such that

$$|\nu_c^h| \equiv |\nu^h(\varepsilon(h))| \leq C(h^{\ell+k})^{\frac{2}{3}} \quad \text{for } 0 < h \leq h_o' \ (h_o\text{: sufficiently}$$

small).

In conclusion, we have the orders of convergence of the numerical buckling load μ_c^h, of the numerical buckling mode Φ_c^h and of the numerical deflection w_c^h :

$$|\mu_c - \mu_c^h| \leq Ch^{\frac{2}{3}(\ell+k)},$$
$$\|\phi_c - \phi_c^h\|_V \leq Ch^{\min(\ell, \frac{\ell+k}{3})}, \qquad \|\phi_c - \phi_c^h\|_U \leq Ch^{\frac{\ell+k}{3}}, \tag{9.7}$$
$$\|w_c - w_c^h\|_V \leq Ch^{\min(\ell,\frac{2}{3}(\ell+k))}, \qquad \|w_c - w_c^h\|_U \leq Ch^{\frac{2}{3}(\ell+k)}.$$

It is remarked here that for an axi-symmetric bifurcation, the same conclusion as above holds whether or not the scheme respects the S-invariance.

10. Concluding Remarks

The numerical realization of cusp bifurcation in approximate spaces V^h reveals that " *imperfections* " caused by numerical schemes are very non-generic. From computational viewpoints, the hypothesis that the scheme respects the symmetry of the problem seems natural. Hence, so long as axi-asymmetric bifurcations are concerned, they are generally realized in V^h.

Although we have treated only a compatible class of finite element schemes, and a special class of symmetry $S^2=I$, we hope that our theory may serve as a basis for a more general theory.

It seems that our results justify the numerical computation performed by Endou, Hangai and Kawamata [8], though their space is a hybrid of Galerkin and finite element. See, also [18].

Our numerical experiments for a simplified arch model confirmed the error bounds obtained in §7, §8 and §9. Especially, the 2/3 power law of the convergence order in the numerical buckling load was observed, when the axi-symmetry was destroyed.

A remark about the *algorithm* for solving the scheme (Q^h): For this purpose, Hangai and Kawamata proposed a static perturbation technique [21], which also provides a classification theorem for the discrete nonlinear system (Q^h). See, also Thompson [22]. The key of their method lies in the choice of local parameter to be used. The so-called "displacement control" in the vicinity of critical points corresponds, essentially, to the use of α as the local parameter.

Numerical analysis of secondary bucklings, as well as the "compound" bucklings (i.e., dim ker(L_c) > 1) are the subjects of future research.

Acknowledgements :

The authors wish to express their appreciation to Professor Y. Hangai of the University of Tokyo and Professor T. Nakamura of Kyoto University for invaluable discussions on the nonlinear buckling theory. Special thanks are due Dr. M. Tabata of Kyoto University, with whom they had many stimulating discussions. They also thank to Professor M. Mimura of Konan University, Mr. Y. Nishiura and Mr. Y. Hosono for discussions and reading the manuscript.

A part of this work of the second author has been done during his stay at the Laboratoire d'Analyse Numérique of the University of Paris VI, France.

References :

[1] Th.von Karman and H.S.Tsien, The buckling of spherical shells by external pressure, J. Aero. Sci., 7 (1939).

[2] E.L.Reiss,H.J.Greenbergand H.B.Keller, Nonlinear deflections of shallow spherical shells, J. Aero. Sci. 24 (1957).

[3] B.Budiansky, Buckling of clamped shallow spherical shells,*Proc. IUTAM Symp. on the Theory of Thin Elastic Shells,* Delft 1959.

[4] N.C.Huang, Unsymmetric buckling of thin shallow spherical shells, J. appl. Mech. 31, (1964).

[5] J.R.Fitch and B.Budiansky, Buckling and post-buckling behavior of spherical caps under axi-symmetric load, AIAA J. 8 (1970).

[6] M.Yamada, Effect of initial imperfection on the buckling of spherical thin shells under external pressure load, Thesis, Tohoku University 1973.

[7] J.A.Stricklin, Geometrically nonlinear static and dynamic analysis of shells of revolution, *Proc. IUTAM Symp. on High Speed Computing,* Liege 1970.

[8] A.Endou,Y.Hangai and S.Kawamata, Post-buckling analysis of elastic shells of revolution, Rep. Inst. Ind. Sci., 26 (1976), The University of Tokyo.

[9] M.Berger, On von Karman's equations and the buckling of a thin elastic plate I, Comm. Pure Appl. Math. 20 (1967).

[10] S.N.Chow,J.Hale and J.Mallet-Paret, Applications of generic bifurcation I, Arch. Rat. Mech. Anal., 59 (1975).

[11] G.H.Knightly and D.Sather, Nonlinear buckled states of rectangular plates, Arch. Rat. Mech. Anal., 54 (1974)

[12] G.H.Knightly, Some mathematical problems from plate and shell theory, *Nonlinear Functional Analysis and Differential Equations,* M.Dekker, 1976.

[13] D.H.Sattinger, Group representation theory and branch points of nonlinear functional equations, SIAM J. Math. Anal., 8 (1977).

[14] F.Kikuchi, An iterative finite element scheme for bifurcation analysis of semilinear elliptic equations, Rep. Inst. Space Aero. Sci., 542 (1976), The University of Tokyo.

[15] R.Weiss, Bifurcation in difference approximations to two-point boundary value problems, Math. Comp., 29 (1976).

[16] K.Marguerre, Zur Theorie der gekrummten Platte grosser Formanderung, *Proc. 5th Int. Congr. Appl. Mech.,* 1938.

[17] H.Fujii and M.Yamaguti, Structure of singularities and its numerical deformation in nonlinear elasticity, (to appear).

[18] H.Fujii and M.Tabata, On numerical analysis of snap and bifurcation bucklings of arches and shells, (in preparation).

[19] L.Nirenberg, *Topics in nonlinear functional analysis*, Courant Inst., Ne York University, 1974.

[20] T.Miyoshi, A mixed finite element method for the solutions of the von Karman equations, Numer. Math. 26 (1976).

[21] Y.Hangai and S.Kawamata, Analysis of geometrically nonlinear and stability problems by static perturbation method, Rep. Inst. Ind. Sci., 22 (1973), The University of Tokyo.

[22] J.M.T.Thompson, A general theory for the equilibrium and stability of discrete conservative systems, Z.A.M.P., 20 (1969).

NUMERICAL METHODS FOR FREE SURFACE PROBLEMS BY MEANS OF PENALTY

H. Kawarada
University of Tokyo, Tokyo, JAPAN
and
IRIA LABORIA, Le Chesnay, FRANCE

Introduction

The penalty method[1),2),3)] is applied to the free surface problems. In Section 1, we show what meaning the additional penalty term has in the penalized problem of the elliptic boundary value problem defined in a bounded domain in R^2 and we introduce the concept of the integrated penalty.[4)] This integrated penalty leads to some approximation models for the free surface problems including the problems of hydraulics[5)] and plasma physics[6)] in Sections 2 and 5. By means of the approximation theories[7),8)] of the variational inequalities, the mathematical foundation of these approximation models is given in Section 3. Finally we give an algorithm to solve the approximation model for the problem of hydraulics and its numerical result in Section 4.

1. Penalty method and integrated penalty

For example we shall choose the following boundary value problem:

(1.1a) $\quad -\Delta u = f \quad$ in Ω ,

(1.1b) $\quad u|_S = 0$.

To simplify our discussions, we assume that Ω is bounded in R^2 and S, the boundary of Ω, is smooth. (1.1) has a unique solution $u \in H^2(\Omega) \cap H_0^1(\Omega)$ for given f in $L^2(\Omega)$. Take B large enough so as to contain Ω.

We introduce:

$$\tilde{u} = u \quad \text{in } \Omega , \qquad 0 \quad \text{in } B \setminus \Omega ,$$

$$\tilde{f} = f \quad \text{in } \Omega , \qquad 0 \quad \text{in } B \setminus \Omega ,$$

χ = the characteristic function in B of $B \setminus \overline{\Omega}$

and we now compute $\Delta\tilde{u} + \tilde{f}$ in the sense of distribution in B.

Then we have (for any $\phi \in \mathcal{D}(B)$)

(1.2) $$\langle \Delta\tilde{u}+\tilde{f}, \phi \rangle = \iint_{\Omega} u\Delta\phi dxdy + \iint_{\Omega} f\phi dxdy$$

$$= \int_S \left(u \frac{\partial \phi}{\partial n} - \frac{\partial u}{\partial n} \phi\right) dS + \iint_\Omega (\Delta u + f) \phi \, dx dy$$

(by Green's formula)

$$= -\int_S \frac{\partial u}{\partial n} \phi \, dS \qquad \text{(by (1.1a) and (1.1b))}$$

where n = normal to S directed towards the exterior of Ω. Let us choose some portion S_1 of S which is a manifold. S_1 may be represented by (cf. Fig.1)

$$y = h(x) \quad (x_1 < x < x_2)$$

or

$$x = g(y) \quad (y_1 < y < y_2) \ .$$

Let

$$V_1 = \{(x,y) \mid d_1 < x < 1,\ y_1 < y < y_2\}$$

and

$$V_2 = \{(x,y) \mid x_1 < x < x_2,\ 0 < y < d_2\}$$

and let

$$\frac{1}{n_1} = \sqrt{1 + (g'(y))^2}$$

and

$$\frac{1}{n_2} = \sqrt{1 + (h'(x))^2}$$

If we introduce

$$\frac{1}{n_1} \cdot \frac{\partial u}{\partial n}(g(y), y) \frac{\partial \chi}{\partial x}$$

and

$$\frac{1}{n_2} \cdot \frac{\partial u}{\partial n}(x, h(x)) \frac{\partial \chi}{\partial y} \quad ,$$

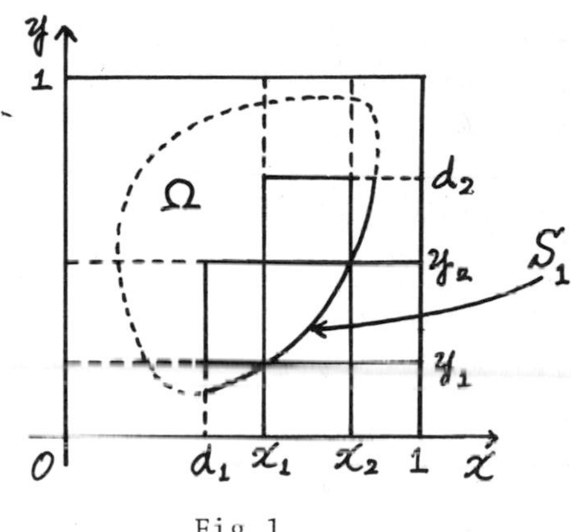

Fig.1.

then we have

$$(1.3) \qquad \left\langle \frac{1}{n_1} \cdot \frac{\partial u}{\partial n} \cdot \frac{\partial \chi}{\partial x}, \phi \right\rangle = \int_S \frac{\partial u}{\partial n} \phi \, dS \quad \text{for any } \phi \in \mathcal{D}(B \cap V_1)$$

and

$$(1.4) \qquad \left\langle \frac{1}{n_2} \cdot \frac{\partial u}{\partial n} \cdot \frac{\partial \chi}{\partial y}, \phi \right\rangle = -\int_S \frac{\partial u}{\partial n} \phi \, dS \quad \text{for any } \phi \in \mathcal{D}(B \cap V_2) \ .$$

Therefore we have

$$(1.5) \qquad \Delta \tilde{u} + \tilde{f} = \frac{1}{n_1} \cdot \frac{\partial u}{\partial n} \cdot \frac{\partial \chi}{\partial x} \quad \text{in} \quad B \cap V_1 \ .$$

$$(1.6) \qquad = \frac{1}{n_2}\cdot\frac{\partial u}{\partial n}\cdot\frac{\partial \chi}{\partial y} \quad \text{in} \quad B\cap V_2 \quad .$$

Let us describe the basic concept of the integrated penalty. Now we approximate (1.1) by the penalized problem for every $\varepsilon>0$:

$$(1.7a) \qquad -\Delta u_\varepsilon + \frac{1}{\varepsilon}\chi\, u_\varepsilon = \tilde{f} \quad \text{in} \quad B \quad ,$$

$$(1.7b) \qquad u_\varepsilon\big|_{\partial B} = 0 \; .$$

We easily check that as $\varepsilon \to 0$,

$$(1.8) \qquad u_\varepsilon \to \tilde{u} \quad \text{in} \quad L^2(B) \quad .$$

Put

$$p_\varepsilon = p_\varepsilon(x,y) = (\frac{1}{\varepsilon}\chi u_\varepsilon)(x,y)$$

and

$$q_\varepsilon^{(1)} = q_\varepsilon^{(1)}(x,y) = \int_x^1 p_\varepsilon(\xi,y)\,d\xi \quad ,$$

$$q_\varepsilon^{(2)} = q_\varepsilon^{(2)}(x,y) = \int_0^y p_\varepsilon(x,\eta)\,d\eta \quad .$$

Then we have

Theorem 1.1. Let $\varepsilon \to 0$, then one has

$$(1.9) \qquad p_\varepsilon \longrightarrow -\frac{1}{n_1}\cdot\frac{\partial u}{\partial n}\cdot\frac{\partial \chi}{\partial x} \quad \text{in} \quad \mathcal{D}'(B\cap V_1) \quad ,$$

$$(1.10) \qquad p_\varepsilon \longrightarrow \frac{1}{n_2}\cdot\frac{\partial u}{\partial n}\cdot\frac{\partial \chi}{\partial y} \quad \text{in} \quad \mathcal{D}'(B\cap V_2) \quad ,$$

and

$$(1.11) \qquad p_\varepsilon \longrightarrow 0 \quad \text{locally uniformly} \quad \text{in } B\backslash\overline{\Omega} \quad .$$

Proof If we use (1.8), then we have

$$(1.12) \qquad p_\varepsilon = \Delta u_\varepsilon + \tilde{f} \longrightarrow \Delta\tilde{u} + \tilde{f} \quad \text{in} \quad \mathcal{D}'(B) \quad \text{as} \quad \varepsilon \to 0 \quad .$$

From (1.12) and (1.5) , (1.6) follow (1.9) and (1.10). We easily check (1.11).

Theorem 1.2. Let $\varepsilon \to 0$, then one has

$$(1.13) \qquad q_\varepsilon^{(1)} \longrightarrow -\frac{1}{n_1}\cdot\frac{\partial u}{\partial n}\cdot(1-\chi) \quad \text{in} \quad \mathcal{D}'(B\cap V_1) \quad ,$$

$$(1.14) \qquad q_\varepsilon^{(2)} \longrightarrow -\frac{1}{n_2}\cdot\frac{\partial u}{\partial n}\cdot(1-\chi) \quad \text{in} \quad \mathcal{D}'(B\cap V_2) \quad .$$

Henceforce we call $i_\varepsilon^{(1)} = q_\varepsilon^{(1)}(d_1, y)$ and $i_\varepsilon^{(2)} = q_\varepsilon^{(2)}(x, d_2)$ the integrated penalty. By virtue of Theorem 1.2 we conclude that $-i_\varepsilon^{(1)}$ and $-i_\varepsilon^{(2)}$ approximate $\frac{1}{n_1}\frac{\partial u}{\partial n}$ and $\frac{1}{n_2}\frac{\partial u}{\partial n}$ in some sense, respectively. These $i_\varepsilon^{(k)}$ $(k = 1,2)$ furnish numerical algorithms which are rather efficient for the free surface problems.[10)]

Remark 1.3. Let $\delta > 0$ be any number. Then we denote ω_δ by the boundary strip of Ω with the width 2δ, that is, $\omega_\delta = \{x \in B \mid \text{dis.}(x,S) < \delta\}$. If δ is sufficiently small, we also check that :

$$(1.15) \quad \Delta\tilde{u} + \tilde{f} = -\frac{\partial u}{\partial n}\cdot\frac{\partial \chi}{\partial \rho} \quad \text{in} \quad \omega_\delta \ ,$$

where $\frac{\partial \chi}{\partial \rho} = \frac{\partial \chi}{\partial \rho}(\xi + \rho n(\xi))$ ($\xi \in S$, $n(\xi)$ = unitary normal to S directed towards the extrior of Ω). Therefore let $\varepsilon \to 0$, then one has

$$(1.16) \quad p_\varepsilon \to -\frac{\partial u}{\partial n}\cdot\frac{\partial \chi}{\partial \rho} \quad \text{in} \quad \mathcal{D}'(\omega_\delta) \ .$$

2. A free surface problem in hydraulics

2.1. The problem[5)]

Two water reservoirs of different levels, say y_1, y_2 ($y_1 > y_2 \geq 0$) are separated (cf. Fig.2) by an earth-dam, G, with the width a and the height b. Suppose the porous media (earth-dam) to be homogeneous and the flow to be steady, incompressible and irrotational. Let Ω be the region of the flow and let

$h(x) = \sup\{y \in (0,b) \mid (x,y) \in \Omega\}$ $(0 < x < a)$. Denote by $\pi = \pi(x,y)$ the pressure at point (x,y) of the dam. This problem (FBP) is to find $\{\pi,\ y = h(x)\}$ such that

$$(2.1) \quad \Delta\pi = 0 \quad \text{in} \quad \Omega \ ,$$

$$(2.2) \quad \pi = y_1 \quad \text{on} \quad [AF] \ ,$$

$$(2.3) \quad \pi = y_2 \quad \text{on} \quad [BC] \ ,$$

$$(2.4) \quad \pi = y \quad \text{on} \quad [CC_h] \ ,$$

$$(2.5) \quad \frac{\partial \pi}{\partial y} = 0 \quad \text{on} \quad [AB] \ .$$

On the free surface $S = \{(x,y) \mid y = h(x),\ 0 < x < a\}$,

$$(2.6) \quad \pi = y \ ,$$

$$(2.7) \quad \frac{\partial \pi}{\partial n} = 0 \quad (n = \text{normal to } S) \ .$$

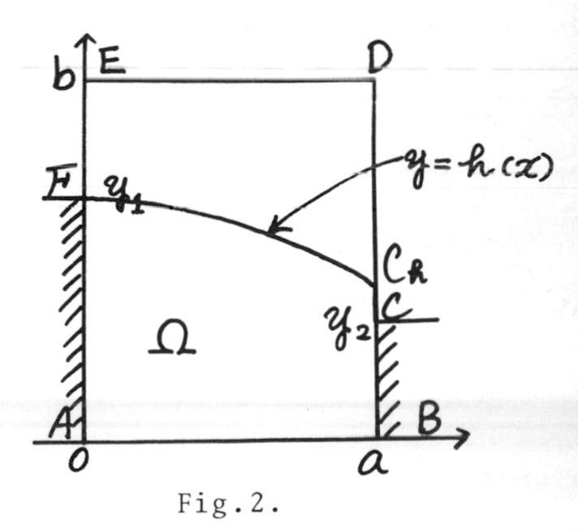

Fig.2.

2.2 Transformation of the problem(FBP)

We want first to give an equivalent formulation of (FBP) which uses in an essential manner the characteristic function χ in G of $G\setminus\overline{\Omega}$.

We introduce:

$$\tilde{\pi} = \pi \quad \text{in} \quad \Omega \quad , \quad y \quad \text{in} \quad G\setminus\Omega$$

and we compute $\Delta\tilde{\pi} = \Delta(\tilde{\pi}-y)$ in the sense of distribution.
Then we have[5)]

$$(2.8) \qquad \Delta\tilde{\pi} = -\frac{1}{n_2}\frac{\partial}{\partial n}(\pi - y)\frac{\partial\chi}{\partial y} = (\text{by } (2.7) \text{ and } (1.6)) = \frac{\partial\chi}{\partial y} \quad \text{in } G.$$

2.3. Introduction of an approximation model

In what follows, for every $\varepsilon>0$, we introduce the penalized problem:

$$(2.9) \qquad -\Delta\pi_\varepsilon + \frac{1}{\varepsilon}\chi(\pi_\varepsilon - y) = 0 \qquad \text{in} \quad G ,$$

$$(2.10) \qquad \pi_\varepsilon = y \qquad \text{on} \quad [EF] \quad ,$$

$$(2.11) \qquad \pi_\varepsilon = y \qquad \text{on} \quad [C_hD] \quad ,$$

$$(2.12) \qquad \frac{\partial\pi_\varepsilon}{\partial y} = 1 \qquad \text{on} \quad [DE]$$

and on the remaining part of ∂G , π_ε satisfies the same boundary condition as π . We easily check that as $\varepsilon \to 0$,

$$\pi_\varepsilon \to \tilde{\pi} \qquad \text{in} \qquad L^2(G)$$

Put

$$p_\varepsilon = p_\varepsilon(x,y) = \frac{1}{\varepsilon}\chi(\pi_\varepsilon - y) \; ,$$

$$q_\varepsilon = q_\varepsilon(x,y) = \int_y^b p_\varepsilon(x,\eta)\,d\eta \quad .$$

By an application of Theorems 1.1, 1.2 and by use of (2.8), we have

$$(2.13) \qquad p_\varepsilon \to \frac{\partial\chi}{\partial y} \qquad \text{in} \quad \mathcal{D}'(G) \quad \text{as} \quad \varepsilon \to 0 \; ,$$

$$(2.14) \qquad q_\varepsilon \to 1 - \chi \quad \text{in} \quad \mathcal{D}'(G) \quad \text{as} \quad \varepsilon \to 0 \; .$$

Considering the above results we present an approximation model $(FBP)_\varepsilon$ which is to find $\{\pi_\varepsilon \; , \; y = h_\varepsilon(x)\}$ such that

$$(2.15) \qquad -\Delta\pi_\varepsilon + \frac{1}{\varepsilon}\chi_\varepsilon(\pi_\varepsilon - y) = 0 \qquad \text{in} \quad G ,$$

$$(2.16) \qquad i_\varepsilon = \int_0^b \frac{1}{\varepsilon}\chi_\varepsilon(\pi_\varepsilon - \eta)\,d\eta = 1 \qquad (0 < x < a) \; ,$$

$$(2.17) \qquad \chi_\varepsilon = \text{the characteristic function in G of } G\setminus\overline{\Omega}_\varepsilon, \text{ where } \Omega_\varepsilon = \{(x,y) \mid 0 < x < a \, , \; 0 \lessdot y < h_\varepsilon(x)\} \; ,$$

$$(2.18) \qquad \pi_\varepsilon \text{ satisfies the same boundary conditions as stated above.}$$

3. Existence and convergence Theorems of $(FBP)_\varepsilon$ and error estimates

3.1. Transformation of approximation model $(FBP)_\varepsilon$

Following Baiocchi,[5)] we introduce the following transformation:

$$u_\varepsilon = u_\varepsilon(x,y) = \int_y^b (\pi_\varepsilon(x,\eta) - \eta)\,d\eta - \varepsilon \; .$$

Then we obtain from (2.15) and (2.12)

$$(3.1)\qquad \Delta u_\varepsilon - \frac{1}{\varepsilon}\chi_\varepsilon u_\varepsilon = 1 \quad \text{in } G.$$

By an elementary use of maximum principle and (2.16), we easily have the following a prio'ri estimates:

(i) $\frac{\partial u_\varepsilon}{\partial y} \leqq 0$ in G; (ii) $\frac{\partial u_\varepsilon}{\partial y} < 0$ in Ω_ε ;

(iii) $u_\varepsilon = 0$ on $y = h_\varepsilon(x)$ $(0 < x < a)$.

Therefore we see that $u_\varepsilon > 0$ in Ω_ε and $u_\varepsilon \leqq 0$ in $G\backslash\overline{\Omega}_\varepsilon$. By use of the notation $u_\varepsilon^- = \sup(-u_\varepsilon, 0)$, we are able to rewrite (3.1) as follows:

$$(3.2)\qquad \Delta u_\varepsilon + \frac{1}{\varepsilon}u_\varepsilon^- = 1 \quad \text{in } G .$$

Corresponding to the transformation stated above, the boundary value of u_ε follows:

$$(3.3)\qquad u_\varepsilon = \frac{1}{2}(y_1 - y)^2 - \varepsilon \quad \text{on } [AF] ,$$

$$(3.4)\qquad u_\varepsilon = \frac{1}{2}(y_2 - y)^2 - \varepsilon \quad \text{on } [BC] ,$$

$$(3.5)\qquad u_\varepsilon = \frac{1}{2}(y_1^2 - \varepsilon) - \frac{x}{2a}(y_1^2 - y_2^2) \quad \text{on } [AB] .$$

$$(3.6)\qquad u_\varepsilon = -\varepsilon \quad \text{on } [CD] \cup [DE] \cup [EF] .$$

Using (2.15) and (2.16), we have easily $\lim_{y\to 0} \frac{\partial^2}{\partial x^2} u_\varepsilon(x,y) = 0$ $(0 < x < a)$, which implies (3.5) . Henceforce we denote by $\overline{(FBP)_\varepsilon}$ this transformed boundary value problem (3.2)∿(3.6). Here we should note that $\overline{(FBP)_\varepsilon}$ is an approximation problem of the variational inequality of stationary type introduced by Lions.[7)] Lions proved the existence of the solution of $\overline{(FBP)_\varepsilon}$ and its convergence to the solution of the variational inequality as $\varepsilon \to 0$.

3.2. Existence theorems

Following Lions,[7)] we have

Theorem 3.1.

$\overline{(FBP)_\varepsilon}$ has a unique solution $u_\varepsilon \in H^2(G)$.

By virtue of the Sobolev imbedding theorems we see $u_\varepsilon \in C^\circ(\overline{G})$. Let

$$\pi_\varepsilon = -\frac{\partial u_\varepsilon}{\partial y} + y ,$$

$$\Omega_\varepsilon = \{(x,y) \in G \mid u_\varepsilon(x,y) > 0\} ,$$

and $h_\varepsilon(x) = \sup\{y \in (0,b) \mid (x,y) \in \Omega_\varepsilon\}$ $(0 < x < a)$.

Theorem 3.2. (Existence Theorem)

$(FBP)_\varepsilon$ has a unique solution $\pi_\varepsilon \in H^1(G)$ and $y = h_\varepsilon(x) \in C^\circ([0,a])$.

Let $g = g(x,y)$ be the function defined on ∂G by:

$$g = \frac{1}{2}(y_1 - y)^2 \text{ on } [AF], \quad g = \frac{1}{2}(y_2 - y)^2 \text{ on } [BC], \quad g = \frac{1}{2}y_1^2 - \frac{x}{2a}(y_1^2 - y_2^2)$$

on [AB] and $g = 0$ on $[CD] \cup [DE] \cup [EF]$.

Set now

$$K = \{u \in H^1(G) \mid u|_{\partial G} = g, \quad u \geq 0 \text{ a.e. in } G\} .$$

Then we introduce the variational inequality (V)[5] equivalent to (FBP) :

Find $u \in K$ such that

$$\text{(V)} \qquad \iint_G \operatorname{grad} u \cdot \operatorname{grad}(v-u)\,dxdy + \iint_G (v-u)\,dxdy \geq 0 , \qquad \text{for any } v \in K .$$

It is well known that (V) has a unique solution $u_0 \in H^2(G)$. Therefore (FBP) has a unique solution $\pi_0 \in H^1(G)$ and $y = h_0(x) \in C^\circ([0,a])$ corresponding to u_0 . In fact, $(FBP)_\varepsilon$ is an approximation problem of (V).

3.3. Convergence theorem

Theorem 3.3. (Lions)[8)]

One has

$$(3.7) \qquad \|u_\varepsilon - u_0\|_{H^1(G)} \leq C_1\sqrt{\varepsilon} ,$$

where C_1 is a positive constant independent of ε.

Lemma 3.4. (Mignot and Puel)[9)] Let $0 < \varepsilon_1 \leq \varepsilon_2$, then one has

$$(3.8) \qquad u_{\varepsilon_1} \geq u_{\varepsilon_2} \quad \text{in } \overline{G}.$$

From Lemma 3.4, we see

Lemma 3.5. Let $0 < \varepsilon_1 \leq \varepsilon_2$, then one has

$$(3.9) \qquad \Omega_{\varepsilon_1} \supseteq \Omega_{\varepsilon_2} , \quad \text{i.e.,}$$

$$(3.10) \qquad h_{\varepsilon_1}(x) \geq h_{\varepsilon_2}(x) \quad \text{in } [0,a] .$$

Using Theorem 3.3, Lemmas 3.4, 3.5 and Dini's theorem, we have

Theorem 3.6. (Convergence Theorem) Let $\varepsilon \to 0$, then one has

$$(3.11) \qquad u_\varepsilon \to u_0 \quad \text{uniformly in } \overline{G} ,$$

and

$$(3.12) \qquad h_\varepsilon \to h_0 \quad \text{uniformly in } [0,a] .$$

3.4. Error estimates

Using Theorems 3.3, 3.6 and Lemma 3.5, then we have the following error estimates:

Theorem 2.7. One has

$$(3.13) \qquad \|\pi_\varepsilon - \pi_0\|_{L^2(G)} \leq C_1\varepsilon^{\frac{1}{2}} ,$$

(3.14) $\|h_\varepsilon - h_0\|_{L^2(I_1)} \leqq C_2\varepsilon^{\frac{1}{4}}$ $(I_1 = \{x|\ 0 < x < a_1,\ 0 < a_1 < a\})$,

where C_1, C_2 are positive constants independent of ε.

4. Algorithm to solve $(FBP)_\varepsilon$

We fix $\varepsilon > 0$ sufficiently small.

Step 1 : choose $h_\varepsilon = h_\varepsilon^{(0)}$.

Step 2 : compute $\pi_\varepsilon^{(0)}$.

Step 3 : compute $i_\varepsilon^{(0)}$.

Step 4 : define $h_\varepsilon^{(1)} = h_\varepsilon^{(0)} - k(i_\varepsilon^{(0)} - 1)$.

$k > 0$ is sufficiently small constant.

Numerical result is shown in Fig.3.

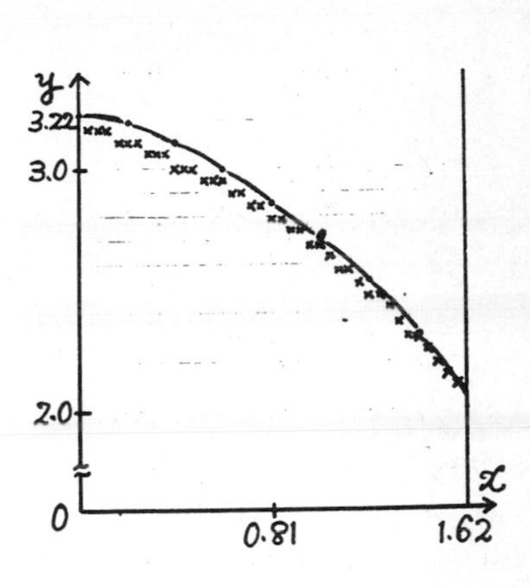

Fig.3.

5. A free surface problem in plasm physics[6)]

5.1. The problem

Let Ω be a bounded domain in R^2 and let Ω_p be a domain occupied by the plasm $(\overline{\Omega}_p \subset \Omega)$. This problem is to find $\{u, \lambda \in \mathbb{R}\}$ such that

(5.1) $-\Delta u = \lambda u$ in Ω_p ,

(5.2) $-\Delta u = 0$ in $\Omega_v = \Omega \setminus \overline{\Omega}_p$ (the vacuum) ,

(5.3) $u|_{\partial\Omega} = 1$,

and on the free surface S which separates Ω_p from Ω_v

(5.4) $u = 0$

(5.5) $\frac{\partial u}{\partial n}$ continuous (n = normal to S directed towards Ω_v) .

By an application of the maximum principle, we have[6)]

$$\Omega_p = \{(x,y) \in \Omega |\ u < 0\} ,$$

$$\Omega_v = \{(x,y) \in \Omega |\ u > 0\} .$$

Thus (5.1) and (5.2) are combined into

(5.6) $\Delta u = \lambda u^-$ in Ω .

5.2. Transformation of the problem by the integrated penalty

We introduce

$u^+ = \sup(u,0) = u + u^-$,

χ = the characteristic function of Ω_v in Ω

and we compute Δu^+ and $\Delta u + \lambda u$. By virtue of Remark 1.3 we have

$$(5.7)\quad \Delta u^+ = -\frac{\partial u^+}{\partial n}\cdot\frac{\partial}{\partial \rho}(1-\chi) \quad \text{in} \quad \omega_\delta ,$$

$$(5.8)\quad \Delta u^- + \lambda u^- = -\frac{\partial u^-}{\partial n}\cdot\frac{\partial \chi}{\partial \rho} \quad \text{in} \quad \omega_\delta .$$

From (5.7) and (5.8) follows

$$(5.9)\quad \Delta u - \lambda u^- = \left(\frac{\partial u^+}{\partial n} + \frac{\partial u^-}{\partial n}\right)\frac{\partial \chi}{\partial \rho} \quad \text{in} \quad \omega_\delta .$$

(5.5) implies

$$(5.10)\quad \int_{-\delta}^{\delta}\left(\frac{\partial u^+}{\partial n} + \frac{\partial u^-}{\partial n}\right)\frac{\partial \chi}{\partial \rho}\,d\rho = 0 \quad .$$

5.3. Introduction of an approximation model

Using the integrated penalty, we introduce the penalized problem which is to find $\{\alpha_\varepsilon, \beta_\varepsilon, \lambda_\varepsilon, S_\varepsilon\}$ such that

$$(5.11)\quad -\Delta\alpha_\varepsilon + \frac{1}{\varepsilon}(1-\chi_\varepsilon)\alpha_\varepsilon = 0 \quad \text{in} \quad \Omega ,$$

$$(5.12)\quad \alpha_\varepsilon\big|_{\partial\Omega} = 1 ,$$

$$(5.13)\quad -\Delta\beta_\varepsilon + \frac{1}{\varepsilon}\chi_\varepsilon\beta_\varepsilon = \lambda_\varepsilon\beta_\varepsilon \quad \text{in} \quad \Omega ,$$

$$(5.14)\quad \beta_\varepsilon\big|_{\alpha\Omega} = 0 ,$$

$$(5.15)\quad i_\varepsilon = \int_{-\delta}^{\delta}\{\frac{1}{\varepsilon}(1-\chi_\varepsilon)\alpha_\varepsilon + \frac{1}{\varepsilon}\chi_\varepsilon\beta_\varepsilon\}(\xi + \rho n(\xi))\,d\rho = 0 \quad (\xi \in S_\varepsilon) ,$$

(5.16) χ_ε = the characteristic function in Ω of $\Omega_{v,\varepsilon}$,

(5.17) S_ε = the free surface which separates $\Omega_{v,\varepsilon}$ from $\Omega_{p,\varepsilon}$ $(= \Omega\setminus\bar{\Omega}_{v,\varepsilon})$.

5.4. Algorithm to solve the approximation model

Step 1 : choose $S_\varepsilon = S_\varepsilon^{(0)}$.

Step 2 : compute $\alpha_\varepsilon^{(0)}$, $\beta_\varepsilon^{(0)}$ and $\lambda_\varepsilon^{(0)}$.

Step 3 : compute $i_\varepsilon^{(0)}(S_\varepsilon^{(0)})$.

Step 4 : define $S_\varepsilon^{(1)} = \{\xi - k\, i_\varepsilon^{(0)}(S_\varepsilon^{(0)})n(\xi) \mid \xi \in S_\varepsilon^{(0)}\}$

Here k and ε are sufficiently small positive constants.

REFERENCES

1. R. Courant, Bull. Amer. Math. Soc., 49(1943), 1-23.

2. J.L. Lions, Revue Roumaine de Math. Pures et Appliquées., 9(1964), 11-18.

3. H. Fujita nad N. Sauer, Bull. Amer. Math. Soc., 75(1969), 465-468.

4. H. Kawarada and M. Natori, Japan-France Seminar on Functional Analysis and Numerical Analysis, Tokyo, September 1976.

5. C. Baiocchi, V. Comincioli, E. Magenes and G.A. Pozzi, Annali di Mat. 96(1972), 1-82.

6. R. Temam, Arch. Rat. Mech. Anal., (1976), 51 - 73.

7. J.L. Lions, Int. Conference on Functional Analysis, Tokyo, Avril 1969.

8. J.L. Lions, W.Eckhaus(ed.), New Developments in Differential Equations, North-Holland Publ. Co., (1976).

9. F.Mignot and J.P. Puel, (1975) C. R. Ac. Sc. Paris.

10. I. Yanagisawa, Annal Reports of the 31th Scientific Lectures of JSCE, Part 3(1976) 19.

A FAMILY OF MODEL PROBLEMS IN PLASTICITY

Gilbert Strang
Massachusetts Institute of Technology

ABSTRACT

We look for the simplest configuration which will lead to the problems most characteristic of plasticity--the contrast between incremental flow laws and deformation theory, and between the collapse of a perfectly plastic structure and its continued stresses and strains when hardening is allowed. The problems are to be continuous rather than discrete, and not one-dimensional; they will be governed by linear partial differential equations with inequality constraints. Each has a primal form in which the stresses are the unknowns, and a dual in terms of velocities or displacements. Our main goal is to clarify these different possibilities, in the case known as _anti-plane shear_.

The preparation of this paper was supported by the National Science Foundation (MCS 76-22289).

The torsion of a rod has been a valuable model for plasticity, both in its mathematical analysis--it reduces to the obstacle problem in the theory of variational inequalities--and in the testing of numerical algorithms [1]. Nevertheless, there are important features it does not possess. If a circular rod is sufficiently twisted, a plastic region will appear at the boundary and gradually extend into the interior. No angle of twist, however, will produce collapse; there is always an elastic core at the center. Our goal is to describe a closely related problem, again with a minimum of unknown stress and velocity components, in which collapse <u>can</u> occur. This new problem requires not only a description of the way plasticity develops, but also a limit analysis--to find the time and the mechanism of collapse Or alternatively, it allows the introduction of a hardening law which is fundamental to the behavior of a real material.

The "new" problem is known to engineers as anti-plane (or longitudinal) shear [2-4], and it describes the behavior of an infinitely long pipe with constant cross-section Ω. The pipe is subject to a body force f defined over Ω, or a surface traction defined on its boundary; both f and g are independent of z but in the direction of z, along the pipe. The forces will vary with time--they may be proportional to the time--but the only displacements they can produce will be in the z-direction. Therefore there is a single displacement component $u(x,y)$. The stresses will be the two shears $\sigma_1 = \sigma_{xz}$ and $\sigma_2 = \sigma_{yz}$, also independent of z.

When the material inside the pipe is isotropic, it is the quantity $|\sigma| = (\sigma_1^2 + \sigma_2^2)^{\frac{1}{2}}$ which appears in the laws of plasticity. The yield conditions of von Mises and Tresca coincide, bounding the admissible stresses by

$$\|\sigma\| = \sup_{\Omega} |\sigma(x,y)| \leq 1 .$$

For stresses below this limit, the problem is elastic and Hooke's law is in force; we take the shear modulus to be unity, so that stress equals strain. In the perfectly plastic case, stresses may reach the yield surface but may not go beyond it; in the case of hardening they are allowed to be arbitrarily large but the stress-strain law is changed.

We could also imagine a non-isotropic material, in which the rotationally invariant quantity $|\sigma|$ is changed to one with preferred directions; the yield surface can become a polygon, for example a square in the case $-1 \leq \sigma_1 \leq 1$, $-1 \leq \sigma_2 \leq 1$.* With linear constraints our model would fit exactly into the framework of quadratic and linear programming, and a numerical algorithm of "simplex type" has been proposed by Maier, De Donato, Cottle, and Kaneko [5]. It differs entirely from the established systems such as MARC and ADINA, which do not begin with a linear approximation of the convex surface $|\sigma| = 1$, and are not based on the ideas of mathematical programming; they are more flexible and more general. We stay with the constraint $|\sigma| \leq 1$.

Our goal is to describe several different laws of plasticity, and to write them out explicitly for the problem of anti-plane shear. We include four possibilities:

1) Deformation theory for perfect plasticity
2) The incremental "flow law"
3) Limit analysis
4) Kinematic strain-hardening

In each case there is a choice between the equations for the stresses and the dual problem in terms of velocities or displacements. Mathematically, there are advantages on the side of the stresses; uniqueness is easier to decide, and the constraint $||\sigma|| \leq 1$ is simple and straightforward. Computationally, however, the favorites are the velocities--going back to the choice in elasticity of minimum potential energy over minimum complementary energy, and the displacement method over the equilibrium method. The former allows an easier construction of finite elements and a routine assembly of the stiffness matrix. In calculations with plasticity, these steps will remain at the heart of the program; the nonlinearity presents an important complication, which can increase the cost enormously, but it has not meant a completely new approach. In fact the first step away from the displacement form is more likely to be a "mixed method," in which stresses also appear and are approximated independently--as in the minimax expressions which we will meet in going from primal to dual in the four problems above.

For simplicity we allow in all these problems a body force f

* The introduction of edges in the yield surface should be reflected by discontinuities (slip lines) in the solution. Several examples are known, but a regularity theory which distinguishes between smooth constraints and corners remains an open problem.

but no surface traction g, and impose boundary conditions only on the displacement.

1. DEFORMATION THEORY

This is the most immediate extension of ordinary elasticity theory. In terms of stresses, we always have the equilibrium equations, which take an especially convenient form in the problem of anti-plane shear; there is only the single equation

$$(1) \qquad -\mathrm{div}\,\sigma = -\frac{\partial \sigma_1}{\partial x} - \frac{\partial \sigma_2}{\partial y} = f \quad \text{in} \quad \Omega .$$

By itself this cannot determine the stresses, since there are plenty of solutions to the homogeneous equation. (To construct a "self-stress" in balance with the force $f = 0$, take $\sigma_1 = \phi_y$ and $\sigma_2 = -\phi_x$ for any choice of ϕ.) A minimum principle is used in elasticity to choose the correct stress, and in plastic deformation theory we simply add the constraint $||\sigma|| \leq 1$:

PRIMAL: Minimize $\displaystyle\int_\Omega \frac{|\sigma|^2}{2}\,dxdy$, subject to $-\mathrm{div}\,\sigma = f$ and $||\sigma|| \leq 1$.

We remark first that if the minimizing σ happens to satisfy $||\sigma|| < 1$, so that the new constraint is inactive, then the problem is still purely elastic. This happens whenever f is sufficiently small; the minimizing σ is a gradient ($\sigma = \mathrm{grad}\, u$ with $-\Delta u = f$) and we are back to equations rather than inequalities.

At the other extreme, the new constraint may be incompatible with the equilibrium equation $-\mathrm{div}\,\sigma = f$; then no stresses are even admissible in the minimization. This will happen if f is sufficiently large, and is the concern of limit analysis. In between, we expect a solution which is partly elastic and partly plastic. In one region of Ω we find $|\sigma| < 1$ and in another $|\sigma| = 1$; topologically the principal condition seems to be that the elastic region should be connected, since otherwise we are at the point of collapse: the material has nothing secure to hold to, and falls through the pipe.

To construct the dual problem we introduce a Lagrange multiplier u, replacing the equilibrium equation by an inner maximization:

$$\min_{\substack{-\mathrm{div}\,\sigma = f \\ ||\sigma|| \leq 1}} \int \frac{|\sigma|^2}{2} = \min_{||\sigma|| \leq 1} \; \max_{u} \int \frac{|\sigma|^2}{2} + u(\mathrm{div}\,\sigma + f) \; .$$

The maximum will be $+\infty$ unless $-\text{div}\,\sigma = f$, so the ensuing minimum is concerned only with stresses that are in equilibrium, and agrees with the minimum on the left. There is a natural boundary condition $u = 0$ to be imposed on the multiplier (which is destined to be the displacement). A formal integration by parts produces its gradient ∇u and leaves

$$\min_{\|\sigma\| \leq 1} \quad \max_{u} \int \left(\frac{|\sigma|^2}{2} - \sigma \cdot \nabla u + uf \right) dxdy \quad .$$

This is the Lagrangian form of the problem. To reach the dual, we exchange minimum and maximum and then carry out the minimization: at each point of Ω,

$$\left(\frac{|\sigma|^2}{2} - \sigma \cdot \nabla u \right) \quad \text{is minimized at} \quad \begin{cases} \sigma = \nabla u, & \text{if} \quad |\nabla u| \leq 1 \\ \sigma = \dfrac{\nabla u}{|\nabla u|}, & \text{if} \quad |\nabla u| \geq 1. \end{cases}$$

In the first case the minimum is $-\frac{1}{2}|\nabla u|^2$; in the second it is $\frac{1}{2} - |\nabla u|$. Substituting these minimum values, we find the

$$\underline{\underline{\text{DUAL}}}. \quad \underline{\text{Maximize}} \quad \int_{\Omega} \left(uf + \begin{cases} -\dfrac{|\nabla u|^2}{2}, & \text{if} \quad |\nabla u| \leq 1 \\ \frac{1}{2} - |\nabla u|, & \text{if} \quad |\nabla u| \geq 1 \end{cases} \right) dxdy.$$

The last expression, which is a function θ of the strain $\varepsilon = |\nabla u|$, involves a continuously differentiable θ; it is a parabola up to $\varepsilon = 1$ and afterwards linear, with value $-\frac{1}{2}$ and slope -1 at the point $\varepsilon = 1$.

The chief difficulty with these equations of deformation theory is that they are not correct. They produce a stress σ and a displacement u, given the present value of the load f--and in general that is impossible. Instead the history of the loads has to be taken into account, and the correct physical law is an incremental one; at each instant it is the rates of change which are to be computed, and σ and u are the integrals of these rates. Of course it is possible that the incremental law leads to the same results as deformation theory; this happens in the exceptional circumstance that, whenever the stress at a point reaches the yield surface $|\sigma| = 1$, it stays there. Otherwise there is unloading, and the two theories will differ; the discrete case is discussed in

[6], and the general case in [7]. In our problem, the exceptional circumstance does seem to occur with proportional loading $f(t) = tF$, for a fixed $F \geq 0$. (This conjecture remains to be verified.) But for a general loading the incremental law must be given priority, and in any case it is very likely the more useful in computations.

2. INCREMENTAL THEORY FOR PERFECT PLASTICITY

This law has several forms. The most direct is a minimum principle for the instantaneous stress rate $\dot{\sigma}$, with an inequality constraint that is imposed only at those points where the stress itself has reached $|\sigma| = 1$:

PRIMAL: Minimize $\int \frac{|\dot{\sigma}|^2}{2}$, subject to $-\mathrm{div}\, \dot{\sigma} = \dot{f}$ and $\dot{\sigma}\cdot\sigma \leq 0$ where $|\sigma| = 1$.

The last condition keeps σ in the admissible region $|\sigma| \leq 1$, since

$$\frac{d}{dt}|\sigma|^2 = 2\dot{\sigma}_1\sigma_1 + 2\dot{\sigma}_2\sigma_2 = 2\dot{\sigma}\cdot\sigma$$

is non-positive when the boundary $|\sigma| = 1$ has been reached; there is no danger of going outside. The stresses change as little as possible, under this local constraint of admissibility and the global constraint of equilibrium. We add the initial condition $\sigma(0) = 0$ (assuming $f(0) = 0$).

To find the dual, we introduce the multiplier $\dot{u}$ (the velocity), integrate by parts as before, and assume that duality justifies the exchange of minimum and maximum; we reach

$$\max_{\dot{u}} \min_{\substack{\dot{\sigma}\cdot\sigma \leq 0 \\ \text{if } |\sigma| = 1}} \int \frac{|\dot{\sigma}|^2}{2} - \dot{\sigma}\cdot\nabla\dot{u} + \dot{u}\dot{f} .$$

At each point of Ω, the minimization has two possibilities:

A: if $|\sigma| < 1$ or $\nabla\dot{u}\cdot\sigma \leq 0$, the minimum is at $\dot{\sigma} = \nabla\dot{u}$

B: if $|\sigma| = 1$ and $\nabla\dot{u}\cdot\sigma > 0$, the minimum is at $\dot{\sigma} = \nabla\dot{u} - (\nabla\dot{u}\cdot\sigma)\sigma$.

Substituting these minimizing values into the integral yields the

$$\text{DUAL. Maximize } \int \left(\dot{u}\dot{f} + \begin{cases} \dfrac{-|\nabla \dot{u}|^2}{2} & \text{in case A} \\ \dfrac{(\nabla \dot{u} \cdot \sigma)^2 - |\nabla \dot{u}|^2}{2} & \text{in case B} \end{cases} \right).$$

At any instant this requires the current stress σ, which is kept up to date by integrating the rate $\dot{\sigma}$. We hope the reader, remembering that the shear modulus was taken as unity--removing a factor G from our σ^2 and $\dot{\sigma}^2$--will not be disturbed by the unusual combination of dimensions.

Both formulations suggest a way to make the problem discrete, with finite increments of load instead of infinitesimals. Mathematically, however, they have a substantial drawback: It is comparatively easy to study at any instant the existence of a minimizing $\dot{\sigma}$, but comparatively difficult to show that these rates can be integrated to give a reasonable stress σ. The problem is posed in too disconnected a way. Therefore analysts have considered instead a form which connects σ more closely to $\dot{\sigma}$ (Duvaut-Lions [8], Moreau [9], Johnson [10]). The incremental law is rewritten as a variational rather than an extremal principle; it corresponds to an equation of virtual work but with inequalities.

VARIATIONAL INEQUALITY. Find $\sigma \in K$ so that $\int_\Omega \dot{\sigma} \cdot (\tau - \sigma)\, dxdy \geq 0$ for all $\tau \in K$.

The inequality is to hold at every time $t \leq T$, with $\sigma(0) = 0$. K is the set of plastically admissible stress histories:

$$K = \{\tau \mid -\operatorname{div} \tau = f(t) \text{ and } \|\tau\| \leq 1 \text{ at all } t \leq t\}.$$

The solution to this problem gives the complete evolution in time of the stress $\sigma(t)$; nothing is affected by the choice of a final time T, except that it must come before collapse or the set K would be empty.

The equivalence of this incremental form with the previous one is sketched in [6] for the discrete (finite-dimensional) case, with the loading $f(t) = tF$ and a piecewise linear yield condition; the stress is in that case also piecewise linear, with a change in $\dot{\sigma}$ only when a constraint is activated. In the continuous case, with a new $\dot{\sigma}$ at every instant, the equivalence requires a much more careful study of the differentiability in t.

3. LIMIT ANALYSIS

Collapse occurs when there is no admissible stress; the external loads cannot be kept in equilibrium without exceeding the yield condition. If we think of proportional loading, $f(t) = tF$ with increasing t, then the collapse time t^* can be found either from deformation theory or the incremental flow law; the former is easier. In fact it is exactly the possibility of computing t^* without requiring the whole history that makes limit analysis attractive, but also less general; it demands a rather idealized situation and supplies only one number t^* (possibly with an indication of the final state, or collapse mechanism). Furthermore, it depends more heavily than the other problems on mathematical programming techniques, and their comparative unfamiliarity has been a serious drawback to widespread adoption. A 1977 conference at the University of Waterloo [11] did make a strong attempt to establish programming as an efficient tool, and in our problem it provides the one case found so far in which, without assuming circular symmetry, an exact solution is possible.

In terms of stresses, the question is this: At what point does the plastically admissible set (the feasible set in deformation theory) become empty?

PRIMAL. Maximize t, subject to $-\text{div}\,\sigma = tF$ and $||\sigma|| \leq 1$.

This is the static form, or the lower bound theorem--so named because any t for which there exists an admissible σ is automatically a lower bound to t^*. Therefore an approximation based on equilibrium finite elements provides a t_h^* which is safe; if the structure is not required to support a load greater than $t_h^* F$, it will not let you down. It is unfortunate that this safe result, which is the desirable one, comes from the equilibrium method whose elements seem less easy to construct; the more familiar displacement elements go into the dual, and produce an upper bound.

To find the dual, we rewrite the problem as

$$\max_{||\sigma|| \leq 1} \; \min_{\int vF = 1} \; -\int v \,\text{div}\,\sigma \,dxdy ,$$

recognizing that the inner minimum will be $-\infty$ unless $\text{div}\,\sigma$ is proportional to F. (For vectors the same is true: the constraint $a \cdot b = 1$ allows $a \cdot c$ to take any value at all, unless c is a multiple of b.) When div σ is proportional to F, say $-\text{div}\,\sigma = tF$, the inner minimum is exactly t and then the maximization over such σ is exactly the primal.

As before, we impose zero boundary conditions on the Lagrange multiplier v, integrate by parts, and exchange max and min:

$$\min_{\int vF = 1} \ \max_{\|\sigma\| \leq 1} \int \sigma \cdot \nabla v \, dxdy \ .$$

At each point of Ω, the maximum occurs when the stress points in the same direction as ∇v and has unit length, the largest allowed: $\sigma = \nabla v / |\nabla v|$. The integrand becomes $\sigma \cdot \nabla v = |\nabla v|$.

DUAL. Minimize $\int |\nabla v|$ subject to $\int vF = 1$.

This is the kinematic form of limit analysis and it must be connected to a breakdown of the dual problem in deformation theory, just as the static problem is linked to infeasibility of the primal. Recalling that there was an integral $I = \int uf + \frac{1}{2} - |\nabla u|$, and that $f = tF$, the connection is not difficult to find: if t exceeds $\int |\nabla v|$, with $\int vF = 1$, then we can let u be a larger and larger multiple of v and find that I is unbounded. If t exceeds the collapse factor, the maximum in the dual problem of deformation theory is $+\infty$.

We return to the present problem, and try to solve it. Because everything is homogeneous in v, it can be rewritten as a minimization of the ratio $\int |\nabla v| / \int vF$; the numerator is an internal energy dissipation rate, and the denominator is the external work rate. Similar expressions appear in the limit load theory of three-dimensional bodies and of plates and beams, but it is the special simplicity of $\int |\nabla v|$ which allows an exact solution for anti-plane shear.

Suppose we consider all those velocities for which $\int |\nabla v| \leq 1$. This is a convex set--a "unit ball" in velocity space, with the integral of the gradient as the norm--and our problem is to find the v which maximizes the denominator $\int vF$. Since this expression is linear in v, its maximum must occur at an extreme point of the convex set. Here we are anticipating that the unit ball has extreme points, and that all other points are combinations of them; otherwise only a supremum can be expected, and a maximizing v (a collapse mechanism) cannot be guaranteed. This depends on the precise definition of the class of admissible v, and the right choice seems to be $BV(\Omega)$: the velocities v should be functions of bounded variation. Roughly speaking, this is the largest class for which $\int|\nabla v|$ is finite; v is allowed to jump in passing from one part of Ω to another, and the gradient of v is a measure.

For this space, Fleming [12] has found all the extreme points: they are multiples of characteristic functions $v = \chi(E)$, with $v = 1$ in the subset $E \subset \Omega$ and $v = 0$ elsewhere. The boundary ∂E is a simple closed curve of finite length. In this case the gradient of χ is a "line of δ-functions" along this boundary, and $\int|\nabla\chi|$ is exactly the perimeter of E. Physically, the set E describes the part of Ω which falls through the pipe when the limit load t^*F is reached. This collapse mechanism is a particularly simple one--it might be called a hinge--and with uniform load $F = 1$ it is frequently possible to determine E explicitly.

With square cross-section Ω, we computed t^* and E in [13]. In general, the dual problem with $F = 1$ minimizes

$$\frac{\int|\nabla v|}{\int v} = \frac{\text{perimeter (E)}}{\text{area (E)}} ,$$

which is a classical isoperimetric problem with the extra constraint $E \subset \Omega$. Without the constraint, E would be circular; with it, E will still be circular whenever it is not pushed all the way out to $\partial\Omega$. Furthermore, when ∂E meets $\partial\Omega$ the two are tangent; otherwise a little shortcut would reduce the ratio of perimeter to area [13]. Therefore ∂E is a smooth curve formed from circular arcs and pieces of $\partial\Omega$, and the remarkable thing is that all the circular arcs have the same radius. For a triangle, as for a convex quadrilateral and a number of other polygons [14], this leads to an exact expression for the minimum ratio: if the polygon has perimeter P and it circumscribes a circle of circumference C, then

$$t^* = 2\pi(C^{-1} + (PC)^{-\frac{1}{2}}) .$$

Note: We have quietly abandoned the boundary condition $v = 0$, on which the earlier integration by parts was based. In fact our collapse set E will always reach the boundary of Ω if $F = 1$, and for a circular pipe we actually have $v = 1$ throughout the whole of Ω. We are very grateful to Roger Témam for demonstrating that this deceit can be justified; he constructs a "relaxed problem"

$$\text{Minimize} \int_\Omega |\nabla v| + \int_{\partial\Omega} |v|, \quad \text{subject to} \int_\Omega vF = 1,$$

and proves that it too is in duality with our primal. The extreme points are still the characteristic functions.

In problems more general than anti-plane shear, the collapse mechanism may be much more complicated than a hinge, but we believe the right admissible space for the velocities is still BV. This was discovered by E. Christiansen [15]. The full theory is not yet in an elegant form--we seem to need a Korn's inequality in L_1--but convex analysis does give a rigorous proof of inf-sup duality between the static and kinematic problems; we refer to Koiter [7], Nayroles [16], Debordes [17], Mercier [18], and to forthcoming articles of Christiansen, Matthies, and the author. For plates and beams, the required inequality seems to hold and there exists an extremal whose <u>second</u> derivatives are measures.

4. KINEMATIC HARDENING

In a real material the yield condition $\|\sigma\| \leq 1$ is not inflexible, and hardening occurs. It can take many forms, but we prefer the <u>internal parameters</u> adopted by Halphen-Nguyen [19] and Johnson [20]. In anti-plane shear with kinematic hardening, the yield set remains a unit circle in stress space, but it is shifted around as the stresses change; the internal parameters $\xi = (\xi_1, \xi_2)$ locate the center of the circle, and (using deformation theory for simplicity) the stress formulation is

<u>PRIMAL</u>. <u>Minimize</u> $\displaystyle\iint\left(\frac{|\sigma|^2}{2} + \frac{\alpha}{2}|\xi|^2\right)$, <u>subject to</u> $-\text{div}\,\sigma = f$, $\|\sigma-\xi\| \leq 1$.

Limit analysis does not arise, because the constraints are now always compatible--but there is a cost in leaning too heavily on ξ. The material constant α is related to the slope of the stress-strain curve beyond the elastic region; it is a hardening parameter.

To construct the dual, we enforce equilibrium with the multiplier u as before, and introduce a nonnegative multiplier $\lambda(x,y)$ to maintain the yield condition; the Lagrangian form of the problem is

$$\min_{\sigma,\xi}\ \max_{\lambda \geq 0, u} \iint\left(\frac{|\sigma|^2}{2} + \frac{\alpha}{2}|\xi|^2 + u(\text{div}\,\sigma + f) - \lambda(1 - |\sigma|^2 + 2\sigma\cdot\xi - |\xi|^2)\right).$$

Integrating by parts and presupposing duality, we have

$$\max_{\lambda \geq 0, u} \; \min_{\sigma, \xi} \int \left(\frac{|\sigma|^2}{2} + \frac{\alpha}{2} |\xi|^2 - \sigma \cdot \nabla u + uf - \lambda(1 - |\sigma|^2 + 2\sigma \cdot \xi - |\xi|^2) \right).$$

At each point, this is a quadratic to be minimized in σ and ξ; it is just like $\frac{1}{2}x^T A x - x^T b$, with minimum value $-\frac{1}{2}b^T A^{-1} b$ attained where $Ax = b$. For our quadratic $Ax = b$ becomes

$$\begin{bmatrix} 1 + 2\lambda & -2\lambda \\ -2\lambda & \alpha + 2\lambda \end{bmatrix} \begin{bmatrix} \sigma \\ \xi \end{bmatrix} = \begin{bmatrix} \Delta u \\ 0 \end{bmatrix},$$

and the minimizing values are

$$\sigma = \frac{\alpha + 2\lambda}{\alpha + 2(1+\alpha)\lambda} \nabla u, \qquad \xi = \frac{2\lambda}{\alpha + 2(1+\alpha)\lambda} \nabla u .$$

Substituting into the integral, we have

$$\max_{\lambda \geq 0, u} \int -\frac{1}{2} \frac{\alpha + 2\lambda}{\alpha + 2(1+\alpha)\lambda} |\nabla u|^2 - \lambda + uf .$$

The maximum over λ occurs at $\lambda = 0$, if $|\nabla u| \leq 1$; this is the elastic case, when $\sigma = \nabla u$ and $\xi = 0$. Otherwise, it occurs where the derivative with respect to λ is zero: $\alpha|\nabla u| = \alpha + 2(1+\alpha)\lambda$. In this case we find

$$\sigma = \left(\frac{|\nabla u| + \alpha}{1 + \alpha} \right) \frac{\nabla u}{|\nabla u|}, \qquad \xi = \left(\frac{|\nabla u| - 1}{1 + \alpha} \right) \frac{\nabla u}{|\nabla u|} .$$

Notice that $|\sigma - \xi| = 1$; the stress is at yield, lying on the boundary of the unit circle around ξ. And each increment in $|\nabla u|$ increases σ by $\frac{1}{1+\alpha}$; this is the slope of the stress-strain curve in the hardening range. Substituting for λ in the integral, we have a displacement formulation:

DUAL. Maximize $\displaystyle \int \left(uf - \begin{cases} \dfrac{|\nabla u|^2}{2} & \text{if } |\nabla u| \leq 1 \\[2mm] \dfrac{|\nabla u|^2 + 2\alpha|\nabla u| - \alpha}{2(1 + \alpha)} & \text{if } |\nabla u| > 1 \end{cases} \right).$

The admissible space H_0^1, and its finite element subspaces, are the same as in linear elasticity.

We add a final note on the relation of anti-plane shear to St. Venant torsion. They look nearly identical, because in the elastic region both are governed by Poisson's equation with constant right hand side, and in the plastic region both have a sum of squares equal to the square of the yield stress, say unity. Nevertheless, there is a difference. In torsion the equilibrium equation is homogeneous, div $\sigma = 0$, and the compatibility equation is not. (It comes from the assumptions $u = -y\theta z$ and $v = x\theta z$ for the horizontal displacements; these were zero in anti-plane shear.) In our problem it was the equilibrium equation which was inhomogeneous, $-\text{div}\ \sigma = f$. Since in both cases it is equilibrium which remains in force in the plastic region, the stresses there will not generally coincide. In fact, it is the impossibility of maintaining equilibrium for large f which produces a collapse mechanism in the shear problem which was absent in torsion.

We are very grateful to John Hutchinson for recognizing our original equations as anti-plane shear. We hope they have value as a simple test case in plasticity, and in their own right as a starting point for fracture mechanics.

REFERENCES

1. R. Glowinski, J.L. Lions, and R. Trémolières, Analyse Numérique des Inéquations Variationelles, Dunod, Paris, 1976.

2. J. Hult and F.A. McClintock, Elastic-plastic stress and strain distribution around sharp notches under repeated shear, Ninth Congress of Applied Mechanics, 8, 1956.

3. J.R. Rice, Contained plastic deformation near cracks and notches under longitudinal shear, Int. J. of Fracture Mechanics, 2 (1966), 426-447.

4. J.R. Rice, On the theory of perfectly plastic anti-plane straining, Report GK 286, Brown University, 1966.

5. I. Kaneko, Complete solutions for a class of elastic-plastic structures, Dept. of Industrial Engineering, Univ. of Wisconsin, 1977.

6. G. Strang, Some recent contributions to plasticity theory, J. Franklin Institute 302 (1976), 429-441.

7. W. Koiter, General theorems for elastic-plastic solids, Progress in Solid Mechanics, 165-221, North Holland, Amsterdam, 1960.

8. G. Duvaut and J.L. Lions, Inequalities in Mechanics and Physics, Springer, New York, 1976. (French original: Dunod, Paris, 1972).

9. J.J. Moreau, Application of convex analysis to the treatment of elastoplastic systems, Springer Lecture Notes 503, 1976.

10. C. Johnson, Existence theorems for plasticity problems, J. Math. Pures et Appl., 55 (1976), 431-444.

11. M.Z. Cohn, ed., Engineering Plasticity by Mathematical Programming, University of Waterloo, Canada, 1977.

12. W.H. Fleming, Functions with generalized gradient and generalized surfaces, Annali di Matematica 44 (1957), 93-103.

13. G. Strang, A minimax problem in plasticity theory, AMS Symposium on Functional Analysis Methods in Numerical Analysis, Springer Lecture Notes, to appear.

14. T.P. Lin, Maximum area under constraint, Math. Magazine 50 (1977), 32-34.

15. E. Christiansen, MIT Thesis, 1976.

16. B. Nayroles, Essai de théorie fonctionelle des structures rigides plastiques parfaites, J. de Mécanique 9 (1970), 491-506.

17. O. Debordes, Dualité des théorèmes statique et cinématique, C.R. Acad. Sc. Paris 282 (1976), 535-537.

18. B. Mercier, Une méthode de résolution du problème des charges limites utilisant les fluides de Bingham, C.R. Acad. Sci. Paris 281 (1975), 525-527.

19. B. Halphen and Nguyen Quoc Son, Sur les matériaux standards généralisés, J. de Mécanique 14 (1975), 39-63.

20. C. Johnson, On plasticity with hardening, to appear.

HOMOGENIZATION
HOMOGENEISATION

The Computational Aspects of the

Homogenization Problem

I. Babuska

Institute for Physical Sciences and Technology

University of Maryland

College Park, Maryland 20742

Abshock. The homogenization is discussed as an (numerical) approximation problem. It is shown that it is closely related to a treatment of a pseudo-differential operator. Its approximation leads then to the usual homogenization approach. The paper discusses the Fourier transformation and finite element approach for solving the problem with highly oscilatory coefficients.

1. Introduction

We are interested in finding numerically the solution u of a differential equation with highly oscilatory coefficients, e.g.,

$$\sum_{i=1}^{n} \frac{\partial}{\partial x_i} a\left(\frac{x}{H}\right) \frac{\partial u}{\partial x_i} + b\left(\frac{x}{H}\right) u = f \quad , \tag{1}$$

where

$0 < C_1 \leq a(x),\ b(x) \leq C_2 < \infty$ and $0 < H \leq 1$. Here $a(x)$, $b(x)$ are are periodic functions with period $(2\pi)^n$.

We will explain the main problems and ideas related to the numerical solution in the simplest case of (1), namely in one dimension, but the generalization is almost straightforward. Problems of the mentioned type arise in many fields, e.g., in mechanics, in chemistry, in computation of nuclear problems, etc. For the survey of the basic results and applications in different fields we refer to [1] [2] [3], where an extensive list of references is given. Recently many additional theoretical results have been achieved by Lions and his co-workers. Also, other lectures with these proceedings are related to these results, see e.g. [4] [5] .

2. Formulation of the problem and its numerical aspects

As a model problem we will be interested in the numerical solution of the differential equation

$$- \frac{d}{dx} \; a\left(\frac{x}{H}\right) \frac{du^H}{dx} + b\left(\frac{x}{H}\right) u^H = f \; , \; - \infty < x < + \infty \qquad (2)$$

where

$$0 < C_1 \le a(x) \; , \; b(x) \le C_2 < \infty \; , \; 0 < H \le 1 \qquad (3)$$

and a and b are 2π-periodic functions.

By H^1, resp H^0, we denote the usual Sobolev spaces, and we seek the solution $u^H \in H^1$. Equation (2) describes, e.g., the behavior of a periodic composite material, with $a(x)$, $b(x)$ being the properties of the material cell of size $2\pi H$. This means that H and $f(x)$ are physically given and cannot be arbitrarily changed. On the other hand H is typically so small that the usual direct numerical solutions, e.g., by finite element method, is impossible, because of the cost. The above formulation of the problem has a unique solution for any $f \in H^0$. More precisely

THEOREM 1. For every $0 < H \le 1$ and $f \in H^0$ there exists a unique (weak) solution $u^H(x) \in H^1$ such that

$$B_H(u^H,v) = \int_{-\infty}^{+\infty} fv \, dx \; , \forall v \in H^1 \qquad (4)$$

where

$$B_H(u,v) \equiv \int_{-\infty}^{+\infty} \left[a\left(\frac{x}{H}\right) u'(x) \, v'(x) + b\left(\frac{x}{H}\right) uv \right] dx \qquad (5)$$

In addition

$$||u^H||_{H^1} \le \frac{1}{C^1} ||f||_{H^0} \qquad (6)$$

where C_1 is given in (3).

If H is sufficiently small one can expect that u^H is close to the limiting solution U, i.e., $\lim_{H \to 0} u^H = U$. We can ask in what sense (if at all) u^H converges to U as $H \to 0$ and how to find this function U. The situation is described by the following well known theorem of the homogenization approach.

THEOREM 2. For fixed smooth f, there exists $U \in H^1$ so that

$$||u^H - U||_{H^0} \to 0 \; , \; H \to 0 \qquad (7)$$

$$||u^H - U - HU'\chi\left(\frac{x}{H}\right)||_{H^1} \to 0 \ , \ H \to 0 \tag{8}$$

where U is the weak solution of the homogenized differential equation

$$L(u) \equiv \frac{d}{dx} \ A \frac{dU}{dx} + BU = f \tag{9}$$

with

$$A^{-1} = \frac{1}{2\pi} \int_{-\pi}^{+\pi} a^{-1}(x) \ dx \ , \ B = \frac{1}{2\pi} \int_{-\pi}^{+\pi} b(x) \ dx \ . \tag{10}$$

The function $\chi(x)$ is periodic with period 2π, independent of f and H, and $\int_{-\pi}^{+\pi} \chi(x) \ dx = 0$.

The main questions which arise with respect to the use of the homogenized solution U have the following characteristics:

a) Is (the physically given) H small enough to lead to the homogenized solution U such that

$$||u^H - U||_{H^0} \leq \tau \tag{11}$$

for an a-priori given tolerance $\tau > 0$?

b) How should one proceed if H is not small enough with respect to the given tolerance?

We mention that H has a completely relative character. For any H we can make the obvious transformation of coordinates, $\xi = \frac{x}{H}$, so that we get $H = 1$. So we have to ask the question with respect to what should H be small. As an illustration of the ideas and approaches we will use the following concrete examples:

$$a(x) = 5 + 2 \sin x \ , \ b(x) = 1 \ , \ H = 1 \tag{12a}$$

$$a(x) = 5 + 2 \sin x \ , \ b(x) = 1 + .25 \cos x \ , \ H = 1 \tag{12b}$$

3. The method based on the Fourier transform.

<u>THEOREM 3.</u> There exists a function $\phi(x,t,H) \in H^1_{PERIODIC}$, which is 2π-periodic in x and is defined for any $-\infty < t < \infty$ and $0 < H \leq 1$ (depending only on a,b) so that

$$u^H(x) = \frac{1}{2\pi} \int_{-\infty}^{+\infty} F_f(t) \ e^{itx} \ \phi\left(\frac{x}{H} \ , \ t, \ H\right) \ dt \tag{13}$$

where $F_f(t) \in H^0$ is the Fourier transform of f.

<u>THEOREM 4.</u> The function $\phi(x,t,H) \in H^1_{PERIODIC}$ is uniquely determined by the condi-

tion

$$\int_{-\pi}^{+\pi} \left[a(x) \frac{d}{dx} \left(\phi e^{itHx}\right) \frac{d}{dx} \left(\overline{ve^{itHx}}\right) + H^2 \; b(x) \; \phi \bar{v} \right] dx \tag{14}$$

$$= H^2 \int_{-\pi}^{+\pi} \bar{v} \; dx \; , \forall \; v \in H^1_{PERIODIC}$$

and has (among others) the following properties:

a) $\phi(x,t,H)$ is analytic in t,H , $-\infty < t,H < \infty$,

b) $\phi(x,t,0)$ is constant in x ,

c) $\phi(x,0,H) \equiv \phi_0(x,H) \neq 0$ and is real.

Theorems 3 and 4 have important corrolaries.

We define

$$v(x,y,H) = \frac{1}{2\pi} \int_{-\infty}^{+\infty} F_f(t) \; e^{itx} \; \phi(y,t,H) \; dt \tag{15}$$

for all $-\pi < y \leq \pi$; then we have

$$u^H(x) = v(x,[x]_H \; ,H) \tag{16}$$

where $\frac{x}{H} = [x]_H + 2\pi k$, k is integral, and $-\pi < [x]_H \leq \pi$. For any y (fixed), v (as a function of x) is the solution of a convolution type of equation

$$\Lambda(y) \; v = f \tag{17}$$

with $\phi^{-1}(y,t,H)$ (as a function of t) being the symbol of the operator $\Lambda(y)$.

Using Theorem 2 we see that

$$u^H(x) - U(x) = \frac{1}{2\pi} \int_{-\infty}^{+\infty} F_f(t) \; e^{itx} \left[\phi\left(\frac{x}{H} \; ,t \; ,H\right) - \frac{1}{At^2 + B}\right] dt \tag{18}$$

where A and B are given in (10).

Figure 1 shows the symbol for both the homogenized operator L of (9) and of the convolution operator Λ of (17) for a few values of y in the case (12a). We note that the symbol of the homogenized operator L is real and the symbol of Λ is complex.

Let us mention that in the case (12a) we have $\phi(x,0,1) = 1$ but in the case (12b) this is not so and the curves will not go through the same point when $t = o$ as they do in Figure 1.

Coming back to (18) we can define

$$\kappa(x,t,H) = \left| \; \phi(x,t,H) - \frac{1}{At^2 + B} \; \right| \tag{19}$$

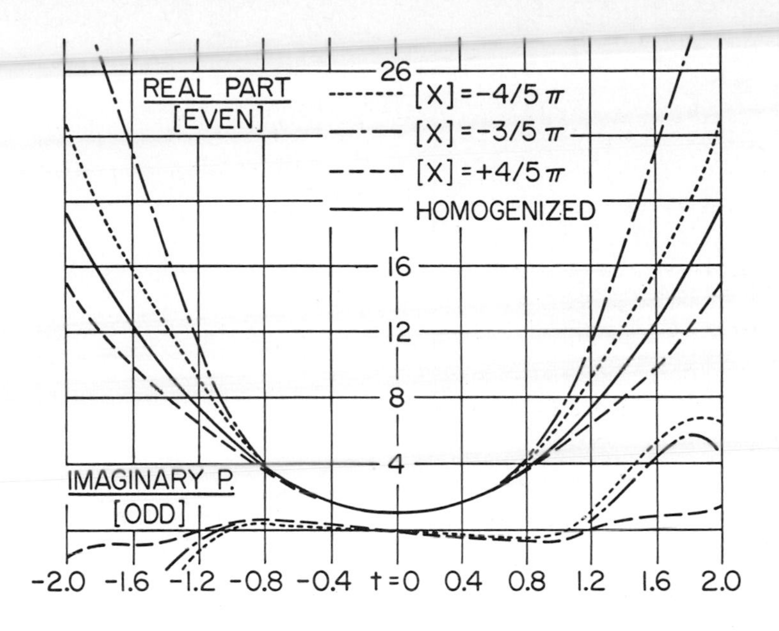

Figure 1. Symbols of the equation based on (12a)

and

$$\rho(t,H) = \sup_{-\pi < x \leq \pi} \kappa(x,t,H) \tag{20}$$

The function ρ allows one to decide whether or not H is sufficiently small.

Figure 2 shows, for the case (12a) the graph of the function $\kappa(x,t,H)$ - (which is even in t) - for some values $[x]$ and for $0 \leq t \leq 2.0$. This fig. shows well for which f the homogenized solution is acceptable.

Theorem 4 yields immediately

$$\phi(x,t,H) = g(t,H) \left[\phi_0(x,H) + \psi(x,t,H)\right] \tag{21}$$

$$g(t,H) = \frac{\int_{-\pi}^{+\pi} \phi(x,t,H)\ \phi_0(x,H)\ dx}{\int_{-\pi}^{+\pi} \phi_0^{\,2}(x,H)\ dx} \tag{22}$$

The functions ψ and g are analytic in t and $g(t,H) > 0$ in a neighborhood of $t = 0$. In addition $\psi(x,0,H) = 0$ and with

$$\psi(x,t,H) = it\ \phi_1(x,H) - t^2\ \phi_2(x,H) + \ldots \tag{23}$$

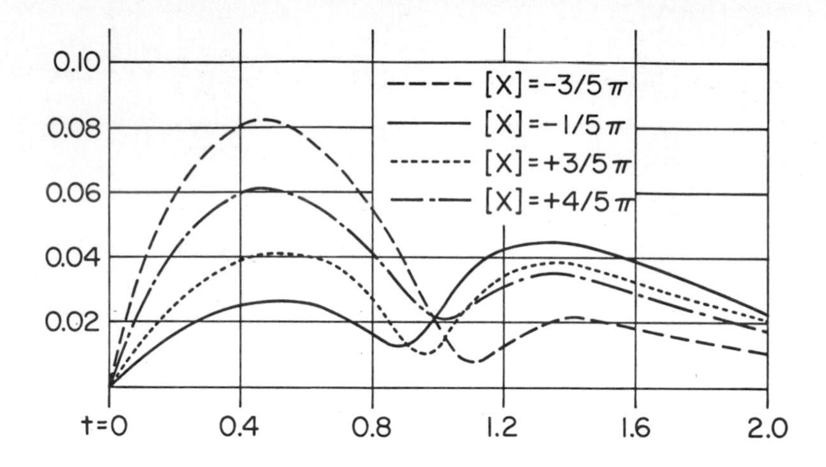

Figure 2. Accuracy of the Homogenized solution. Graph of the function $\kappa(x,t,H)$

we can write

$$u^H(x,H) = W\,\phi_0\left(\frac{x}{H}\,,\,H\right) + W'\,\phi_1\left(\frac{x}{H}\,,\,H\right) + W''\,\phi_2\left(\frac{x}{H}\,,\,H\right) + \ldots \quad (24)$$

where

$$W(x) = \frac{1}{2\pi}\int_{-\infty}^{+\infty} F_f(t)\; e^{itx}\; g(t,H)\; dt\;. \quad (25)$$

Displacing $g^{-1}(t,H)$ by the first few terms of Taylor series, W can be replaced by $\hat{W}$ satisfying a differential equation. For $H \to 0$ we will get the form which is in agreement with the form related to Theorem 2 (see (8)).

Equation (15) and arguments analogous to those which lead to (24) and (25), show how to proceed if H is not sufficiently small. We underline that from the computational point of view to compute the functions ϕ_0, ϕ_1, g, it is not more laborious than implementing the "classical" homogenization procedure. This also gives one the possibility to find the solution with arbitrary accuracy.

The ideas explained above can be generalized in a very broad way. For more see [6]. By the above outlined approach the edge effect for composed material and combined effects (e.g. stresses and humidity) of composite material etc. can be studied and many classical application of Fourier transform can be used here. In general, the computation of $\phi(x,t,H)$ is not a simple one when a and b are not smooth functions. Here, the mixed version of finite element is an effective approach. For more see [7].

4. The finite element method

As before we will deal only with our model problem (2) and will, for simplicity assume that we are using uniform elements with length h . This shows that we have two parameters which have to be considered simultaneously. We define $\mu(h) = \frac{1}{2\pi}\frac{h}{H}$ as the relation between these two parameters. Let us denote by $S(h)$ the usual set of continuous functions, which are linear on every $I_k(h) \equiv \{kh < x < (k+1)h$, $k = \ldots,1,0,1,\ldots\}$. Now we can find the usual finite element solution u_h using $S(h)$ as the trial and test function sets. Now the following holds.

<u>THEOREM 5.</u> Let $\mu(h) = K, K \geq 1$ integral, then

$$||u_h - V||_{H^1} \to 0 \quad \text{as} \quad h \to 0 \ , \tag{26}$$

and V satisfies the differential equation

$$-\frac{d}{dx} A^* \frac{dV}{dx} + BV = f \tag{27}$$

where

$$A^* = \frac{1}{2\pi}\int_{-\pi}^{+\pi} a(x)\, dx \tag{28}$$

and B is given in (10.

Comparing Theorems 5 and 2 we see that the usual finite element approach cannot be used when the linear elements with a size larger than the cell size are used. Of course $||u_h - u^H||_{H^1} \to 0$ when $\mu(h) \to 0$ simultaneously with $h \to 0$.

Denote now by $\psi_i(x,H)$, $i = 0,1, 2\pi$-periodic functions such that $V_i(x,H) = x^i + \psi_i\left(\frac{x}{H}, H\right)$ satisfies (weakly) the equation

$$-\frac{d}{dx}\, a\left(\frac{x}{H}\right)\frac{d}{dx}\left(x^i + \psi_i\left(\frac{x}{H}, H\right)\right) + b\left(\frac{x}{H}\right)\left(x^i + \psi_i\left(\frac{x}{H}, H\right)\right) - x^i \tag{29}$$

The functions $\psi_i(x,H)$ are uniquely determined by (29).

Let us take now $\mu = K_0$, with $K_0 \geq 1$ is integral and let $S_{K_0}(h) = \{\rho \mid \rho = \phi_{K_0}(h)\omega,$ $\omega \in S(h)\}$ with $\phi_{K_0}(h)\omega$ defined so that for $\omega = C_1 + C_2 x$ on $I_k(h)$, $\rho = \phi_{K_0}(h)\omega =$ $C_1 + C_2 x + C_1\psi_0\left(\frac{x}{H}, H\right) + C_2\psi_1\left(\frac{x}{H}, H\right)$. Now of course $S_{K_0}(h) \not\subset H^1$ so that with $S_{K_0}(h)$ we are using in fact the nonconforming elements. Denoting, as usual $||\cdot||^2_{H^1_h} = \sum_{k=\infty}^{\infty}||\cdot||^2_{H^1(I_k)}$ we can prove

THEOREM 6. Let ρ_h be the finite element solution based on $S_{K_0}(h)$ as the sets of trial and test functions. Then

$$||\rho_h - u^H||_{H_h^1} \to 0 \ , \ h \to 0 \tag{30}$$

The approach leading to Theorem 6 can be easily generalized and the rate of convergence can be studied. Because the elements depend on (a given) H the accuracy is usually a good one. Nevertheless, Theorem 6 still does not give an estimate of the accuracy in the given concrete case, but an posteriori estimates technique can be obtained by a generalization of the a-posteriori technique developed in [8] [9] [10].

References

[1] I. Babuska, Solution of problems with interfaces and singularities, Mathematical Aspects of Finite Elements in Partial Differential Equations, Academic Press, 1974, 213-277.

[2] I. Babuska, Homogenization and its application. Mathematical and computational problems, Numerical Solution of Partial Differential Equations-III, Academic Press, 1976, 89-116.

[3] I. Babuska, Homogenization approach in engineering, Lecture Notes in Economics and Mathematical Systems, M. Beckman and H. P. Kunzi (eds.), Springer-Verlag, 1975, 137-153.

[4] G. Papanicolaou, Homogenization and related problems in linear transport theory, proceedings from the third international symposium on computing methods in applied sciences and engineering, December, 1977.

[5] E. Larsen, Homogenization in neutron transport problems, proceedings from the third international symposium on computing methods in applied sciences and engineering, December, 1977.

[6] R. Morgan Thesis.

[7] I. Babuska, and W. Rheinboldt, Numerical treatment of eigenvalue problems for differential equations with discontinuous coefficients, Tech. Note 853, Institute for Physical Science and Technology, University of Maryland, College Park, April, 1977.

[8] I. Babuska and J. E. Osborn, Computational aspects of finite element analysis, Technical Report TR-518, Computer Science Technical Report Series, University of Maryland, College Park, Maryland, April, 1977.

[9] I. Babuska and W. C. Rheinboldt, Error estimates for adaptive finite element computations, Technical Note BN-854, Institute for Physical Science and Technology, University of Maryland, College Park, Maryland, May, 1977. To appear in SIAM J Num. Math, 1978.

[10] I. Babuska and W. C. Rheinboldt, A-posteriori error estimates for the finite element method, Institute for Physical Science and Technology, University of Maryland, College Park, Maryland, May, 1977. To appear in Int. Journ. Num. Math Eng., 1978.

ASYMPTOTICS FOR BRANCHING TRANSPORT PROCESSES

A. Bensoussan, J. L. Lions and G. C. Papanicolaou

IRIA and U. of Paris IX, IRIA and Collège de France, Courant Inst., N.Y.U.

INTRODUCTION

In [1] we analyzed the asymptotic behavior, including boundary layers and homogenization, of a class of transport processes without branching (i.e. fission in neutron transport). We shall present here extensions of the results in [1] to problems with branching. We state only two theorems here and prove one completely. To avoid lengthy preliminary considerations of a detailed probabilistic nature we have abridged our results to their essential analytical content. The full treatment is given in [2] including problems with boundaries (simple extension of present calculations) and problems with boundary layers but no branching on the boundary (extensions of present calculations and results in [1]). We have not considered yet problems in which the boundaries induce branching.

Our methods are similar to the ones in [1] and [3 Chapter III]. By specializing the present results to the asymptotic behavior of the average density of particles we recover results previously obtained by Larsen and Keller [4], Larsen [5] and Williams [6].

1. BRANCHING TRANSPORT PROCESSES

The physical theory of such processes, in neutron or radiative transport, has been known for some time; cf. for example [7], [8], [9]. The general theory of [10] applies here also but we follow more closely the setup of point processes and random measure-valued processes [11], [12], [13], [14], [15].

Let $S \subset R^m$ be fixed and let $F(x,y)$ be an R^n-valued smooth and bounded function on $E \equiv R^n \times S$. Let $\xi(t) = \xi(t;x,y)$ be defined by

(1.1) $$\frac{d\xi(t)}{dt} = F(\xi(t),y) , \quad \xi(0) = x \in R^n .$$

Let $0 \leq q(x,y) \leq C < \infty$ be given and smooth. For each $k = 0,1,2,\ldots$ and each $(x,y) \in E$ let $\pi_k(x,y,A_1,\ldots,A_k)$ be a bounded measure on $S \times \ldots \times S$ (k times); here $A_1,A_2,\ldots,A_k$ are Borel subsets of S and π_k depends smoothly on (x,y). It is assumed that

(1.2) $$\sum_{k=0}^{\infty} \pi_k(x,y,S,\ldots,S) = 1 , \qquad \sum_{k=1}^{\infty} k\pi_k(x,y,S,\ldots,S) < \infty .$$

We shall define a <u>branching transport process</u> (BPT) on E as follows. A <u>single particle</u> (particle is used synonymously with point, neutron, photon, etc.) begins at (x,y) and moves along the trajectories of (1.1) for a random length of time τ_1. At the instant τ_1- the particle is at $(\xi(\tau_1),y)$ and at τ_1+ it disappears. In its place k particles appear at position $\xi(\tau_1)$ (in (x,y), x stands for "position", y for "velocity") with random velocities $y_1(\tau_1),\dots,y_k(\tau_1)$ where

$$\text{(1.3)}\quad P\Big\{\tau_1 \in [0,t] \text{ and k particles are created with velocities}$$

$$y_1(\tau_1)\in A_1,\dots,y_k(\tau_1)\in A_k\Big\} = \int_0^t \pi_k\big(\xi(s),y,A_1,A_2,\dots,A_k\big)q\big(\xi(s),y\big)e^{-\int_0^s q(\xi(\gamma),y)d\gamma}ds$$

After time τ_1, either the process stops (extinction) if zero particles are created or each new particle moves independently of the others (branching property) according to (1.1) and splits according to (1.3) etc.

Let $N(t) = 0,1,2,\dots$ be the number of particles at time t and let $\big(x_k(t),y_k(t)\big)$, $k = 1,2,\dots,N(t)$, denote their positions and velocities. Under the condition $q\le C<\infty$ and (1.2), $N(t) < \infty$ with probability one. Let $f(x,y)$ be a measurable function with $0 \le f(x,y) \le 1$, $(x,y) \in E$. Consider

$$\text{(1.4)}\quad u(t,x,y;f) = E\{f(x_1(t),y_1(t)) \dots f(x_{N(t)}(t),y_{N(t)}(t))\},$$

with the integrand on the right interpreted as one when $N(t) = 0$.

By the usual renewal argument we obtain the following nonlinear integral equation for $u(t,x,y)$ (omit f for simplicity).

$$\text{(1.5)}\quad u(t,x,y) = f\big(\xi(t),y\big)\exp\Big\{-\int_0^t q\big(\xi(s),y\big)\,ds\Big\}$$

$$+\int_0^t \Phi(u)(t-s,\xi(s),y)q(\xi(s),y)\exp\Big\{-\int_0^s q\big(\xi(\gamma),y\big)d\gamma\Big\}ds.$$

Here Φ is the nonlinear operator

$$\text{(1.6)}\quad \Phi(u)(t,x,y) = \sum_{k=0}^{\infty}\underbrace{\int_S\cdots\int_S}_{k\text{ times}} u(t,x,y_1)\dots u(t,x,y_k)\ \pi_k(x,y,dy_1,\dots,dy_k).$$

Note that Φ acts on u as a function of y only, x is a parameter while t enters only because u depends on t. To emphasize parametric dependence on x we write $\Phi(\cdot)=\Phi_x(\cdot)$.

Starting from (1.5) one can show analytically that under $0 \le q \le C < \infty$, (1.2) and $0 \le f \le 1$, (1.5) has a unique solution such that $0 \le u(t,x,y;f) \le 1$.

Let $\lambda(x,y) \ge 0$ be bounded and measurable. On replacing f by $\exp(-\lambda)$ in (1.4) we obtain

$$\text{(1.7)}\quad u(t,x,y;\lambda) = E\Big(\exp\Big\{-\sum_{k=1}^{N(t)}\lambda(x_k(t),y_k(t))\Big\}\Big)$$

$$= E_{x,y}\Big(\exp\Big\{-\int_E \lambda(\bar{x},\bar{y})\ \nu_t(d\bar{x}\ d\bar{y})\Big\}\Big)$$

Here we have introduced the random measure $\nu_t(B)$, $B \subset E$ such that $\nu_t(B) = 0,1,\dots$ (a random point measure) is the random number of particles in B (a subset of phase space E) at time t. Expectation given that a single particle starts at (x,y) is written $E_{x,y}$. In the following we use the notation

$$(1.8)\qquad (\lambda,\nu_t) = \int_E \lambda(\bar{x},\bar{y})\ \nu_t(d\bar{x}\ d\bar{y}) .$$

Let $M^+(E)$ be the set of Radon measures on E with the topology of vague convergence (cf. [13]); it is a separable, locally compact, metrizable space. Let $M_p^+(E) \subset M^+(E)$ be the closed subset of point measures on E. We denote by $X = D([0,\infty);\ M^+(E))$ (or $X_p = D([0,\infty);M_p^+(E))$ the space of right continuous functions on $[0,\infty) \to M^+(E)$ (or $M_p^+(E)$), having left-hand limits and endowed with the Skorohod metric. X and X_p are complete separable metric spaces.

Let (Ω,F,P) be the probability space on which the process ν_t above is constructed. Here $\nu_\bullet$ is considered as a measurable mapping from Ω into X_p. The image of P under this mapping is denoted by Q and is the probability measure on X_p which interests us. It is the measure of a right continuous strong Markov process on the state space $M_p^+(E)$.

On account of the branching property if $\mu \in M_p^+(E)$ and $\mu < \infty$ (μ is not random), we have

$$(1.9)\qquad E^{Q_\mu}\left\{e^{-(\lambda,\nu_t)}\right\} \equiv E_\mu\left\{e^{-(\lambda,\nu_t)}\right\} = \exp\left\{\int_E \log u(t,x,y;\lambda)\ \mu(dx\ dy)\right\}$$

Here Q_μ is the measure on X_p induced by ν_t with $\nu_0 = \mu$ and expectation is denoted for short by E_μ. When $\mu = \delta_{(x,y)}$, the delta measure at (x,y), we write $E_{x,y}$.

In conclusion, the Markov measures Q_μ , $\mu \in M_p^+(E)$, $\mu < \infty$, on X_p are completely determined by $u(t,x,y;\lambda)$ which solves (1.5) with $f = \exp\{-\lambda\}$ and it suffices to take $\lambda \in C_0^+(E)$ (nonnegative continuous functions on E with compact support). The function $u(t,x,y;\lambda)$ is called the Laplace functional of the process ν_t.

The infinitesimal generator of ν_t on functions of the form $\exp\{-(\lambda,\mu)\}$, $\lambda \in C_0^+(E)$, $\mu \in M_p^+(E)$, has the form

$$(1.10)\qquad Le^{-(\lambda,\mu)} \equiv \frac{d}{dt} E_\mu\left\{e^{-(\lambda,\nu_t)}\right\}\Big|_{t=0} = e^{-(\lambda,\mu)}\left(N(\lambda),\mu\right) ,$$

where $N(\cdot)$ is the nonlinear operator

$$(1.11)\qquad N(\lambda)(x,y) = -F(x,y)\cdot\frac{\partial\lambda(x,y)}{\partial x} + q(x,y)\left[e^{\lambda(x,y)}\ \Phi_x(e^{-\lambda})(x,y) - 1\right] .$$

2. MOMENTS

The differential form of (1.5) satisfied by the Laplace functional $u(t,x,y;\lambda)$ is (formally)

$$(2.1)\qquad \frac{\partial u(t,x,y)}{\partial t} = F(x,y)\cdot\frac{\partial u(t,x,y)}{\partial x} + q(x,y)\left[\Phi(u)(t,x,y) - u(t,x,y)\right] ,$$
$$t > 0,\qquad (x,y) \in E,\qquad u(0,x,y) = e^{-\lambda(x,y)} ,\qquad \lambda \in C_0^+(E) .$$

Even though we write (2.1), we always take it to hold in the integral form. One frequently wants only partial information about statistics of ν_t and not u itself which determines Q_μ completely. Because of the branching property moments of ν_t

satisfy linear equations as follows.

Let B be a bounded subset of E and let $\chi_B(x,y)$ denote its indicator function. Define

$$u^{(1)}(t,x,y;B) = E_{x,y}\{\nu_t(B)\} \equiv -\frac{d}{d\alpha} u(t,x,y;\alpha\chi_B)\Big|_{\alpha=0} \tag{2.2}$$

which will be seen to exist. Replacing λ by $\alpha\chi_B$ in (2.1), differentiating with respect to α and setting $\alpha = 0$ we obtain

$$\frac{\partial u^{(1)}(t,x,y)}{\partial t} = F(x,y)\cdot\frac{\partial u^{(1)}(t,x,y)}{\partial x} + M_x u^{(1)}(t,x,y) , \tag{2.3}$$

$$t > 0 , \quad (x,y) \in E , \quad u^{(1)}(t,x,y) = \chi_B(x,y) .$$

The linear operator M_x acts on the y variables only and depends parametrically on x. Acting on $v = v(y)$ it has the form

$$(M_x v)(y) = q(x,y)\left[\left(\Phi_x^{(1)}(1)v\right)(y) - v(y)\right] \tag{2.4}$$

where

$$\Phi_x^{(1)}(1)v = \int_S v(y_1)\ \pi^{(1)}(x,y,dy_1) \tag{2.5}$$

$$\pi^{(1)}(x,y,A) = \sum_{k=1}^{\infty}\sum_{j=1}^{k} \pi_k(x,y,S,\ldots,\overset{\longleftarrow j \longrightarrow}{A},\ldots,S) , \qquad A \subset S . \tag{2.6}$$

Equation (2.3) is the adjoint or backward version of the usual linear equation of transport theory. The kernel $\pi^{(1)}$ ($q\pi^{(1)}$ is the differential scattering cross section) is not in general a probability kernel but is finite by (1.2). If $u^{(1)}(t,x,y;B)$, which is a measure in the last variable has a density $u^{(1)}(t,x,y;\bar{x},\bar{y})$ relative to Lebesgue measure then $u^{(1)}$ satisfies the formal adjoint of (2.3) as a function of $(t,\bar{x},\bar{y})$ with initial conditions $\delta(x-\bar{x})\delta(y-\bar{y})$.

Let us assume that

$$\sum_{k=2}^{\infty} k(k-1)\ \pi_k(x,y,S,\ldots,S) < \infty . \tag{2.7}$$

Then we can find an equation for second moments: $E_{x,y}\{\nu_t(B_1)\nu_t(B_2)\}$, $B_1,B_2 \subset E$. Omitting a few simple details we have the following. If $u^{(2)}$ is defined by

$$E_\mu\{\nu_t(B_1)\nu_t(B_2)\} - E_\mu\{\nu_t(B_1)\}\ E_\mu\{\nu_t(B_2)\} \tag{2.8}$$

$$= \int_E u^{(2)}(t,x,y;B_1,B_2)\ \mu(dx\ dy) - \int_E u^{(1)}(t,x,y;B_1)\ u^{(1)}(t,x,y;B_2)\ \mu(dx\ dy) ,$$

then $u^{(2)}$ satisfies the equation

$$\frac{\partial u^{(2)}(t,x,y)}{\partial t} = F(x,y)\cdot\frac{\partial u^{(2)}(t,x,y)}{\partial x} + M_x u^{(2)}(t,x,y) + q(x,y)\left(\Phi^{(2)}(1)u^{(1)},u^{(1)}\right)\cdot(t,x,y) \tag{2.9}$$

$$t > 0, \quad u^{(2)}(0,x,y;B_1,B_2) = \chi_{B_1}(x,y)\chi_{B_2}(x,y) , \qquad (x,y) \in E.$$

Here $\Phi^{(2)}(1)$ denotes the bilinear operator defined by

$$(2.10)\qquad \left(\Phi^{(2)}(1)u,v\right) = \int_S\int_S u(y_1)v(y_2)\ \pi^{(2)}(x,y;dy_1,dy_2)\ ,$$

and where with $A_1, A_2 \subset S$

$$(2.11)\qquad \pi^{(2)}(x,y,A_1,A_2) = \sum_{k=2}^{\infty}\ \sum_{\substack{j=1\\ j\neq \ell}}^{k}\ \sum_{\ell=1}^{k}\ \pi_k(x,y,S,\ldots,\overset{\longleftarrow\, j\,\longrightarrow}{A_1},\ldots,\overset{\longleftarrow\,\ell\,\longrightarrow}{A_2},\ldots,S)\ .$$

3. ERGODICITY HYPOTHESES AND THE EXPECTATION PROCESS

We shall be interested in the asymptotic behavior of BTP under various scalings discussed in Section 4. We need for this the following hypotheses which are not, however, optimal in any sense.

First we assume that the set of velocities S is a compact subset of R^m ($m = n$ usually). Next we assume that there is a finite reference measure γ on S such that $\pi^{(1)}$ (cf. (2.6)) has a density $\tilde{\pi}^{(1)}(x,y,\bar{y})$ relative to γ and

$$(3.1)\qquad 0 < \pi_\ell \le \tilde{\pi}^{(1)}(x,y,\bar{y}) \le \pi_u < \infty\ .$$

We also strengthen $0 \le q \le C < \infty$ to

$$(3.2)\qquad 0 < q_\ell \le q(x,y) \le q_u < \infty\ .$$

Under these hypotheses the operator M_x of (2.4) has an isolated maximal eigenvalue with right and left eigenfunctions $\phi(y;x)$ and $\tilde{\phi}(y;x)$, respectively ($x \in R^n$ is a parameter) which are strictly positive [11, p. 67]. We assume that this eigenvalue is equal to zero identically in $x \in R^n$. This means that locally the BTP is assumed to be critical. More explicitly ϕ and $\tilde{\phi}$ satisfy

$$(3.3)\qquad M_x\phi = q(x,y)\int_S \phi(y_1;x)\ \tilde{\pi}^{(1)}(x,y,y_1)\ \gamma(dy_1) - q(x,y)\ \phi(y;x) = 0$$

$$(3.4)\qquad \tilde{\phi}M_x = \int_S q(x,y_1)\ \tilde{\phi}(y_1;x)\ \tilde{\pi}^{(1)}(x,y_1,y)\ \gamma(dy_1) - q(x,y)\ \tilde{\phi}(y;x) = 0\ .$$

Define an operator Q_x acting on functions $v(y)$, with x a parameter, by

$$(3.5)\qquad Q_x v = \frac{1}{\phi(y;x)}\left(M_x(\phi v)\right)(x,y)\ .$$

It is called the expectation operator. It is the infinitesimal generator of an ergodic Markov process on S. The unique invariant measure of Q_x is

$$(3.6)\qquad \bar{p}(dy;x) = \tilde{\phi}(y;x)\ \phi(y;x)\ \gamma(dy)$$

We denote by $\psi(y,dy_1;x)$ the recurrent potential kernel of Q_x, i.e., the kernel of the operator $-Q_x^{-1}$. This operator is well defined on functions orthogonal to $\bar{p}$ and maps such bounded functions to bounded functions.

If one is interested only in the analysis of $E_\mu\{\nu_t\}$, i.e. of $u^{(1)}(t,x,y;B)$ of (2.3), then this can be done by replacing $u^{(1)}$ by $u^{(1)}(t,x,y;B)/\phi(y;x)$ and studying it. The resulting analysis, including boundary layers and homogenization, reduces to the one carried out in [1].

4. SCALING

We are interested in the diffusion limit of transport theory, also called the small mean free path or high density limit. In properly defined dimensionless variables, this amounts at first to replacing F and q by F/ε and q/ε^2, $0 < \varepsilon << 1$ in (2.1) (the equation for the Laplace functional of ν_t). Call the $M_p^+(E)$-valued process so scaled $\tilde{\nu}_t^\varepsilon$.

For fixed $\mu \in M_p^+(E)$, $\mu < \infty$ it is easily seen that the processes $\tilde{\nu}_t^\varepsilon$, $\tilde{\nu}_0^\varepsilon = \mu$, (i.e. the measures Q_μ^ε) do not converge in any way; they must be scaled further. An appropriate scaling is the analog in the present case of Lamperti's [16] and Jiřina's [17] continuous state branching limit as follows.

For each $\mu \in M^+(E)$, $\mu < \infty$ let $\mu^\varepsilon \in M_p^+(E)$ be such that $\varepsilon^2\phi\mu^\varepsilon \to \mu$ weakly as $\varepsilon \to 0$ i.e. for each $\lambda \in C(E)$

$$(\varepsilon^2\phi\lambda,\mu^\varepsilon) \longrightarrow (\lambda,\mu) \ , \qquad \varepsilon \to 0 \ . \tag{4.1}$$

There are many ways for doing this; fix a particular one. Let ν_t^ε be the process defined by

$$\nu_t^\varepsilon(dx\ dy) = \varepsilon^2\phi(y;x)\ \tilde{\nu}_t^\varepsilon(dx\ dy) \ , \qquad \tilde{\nu}_0^\varepsilon = \mu^\varepsilon \tag{4.2}$$

It is easily seen that the Laplace functional of ν_t^ε is given by

$$E_{\mu^\varepsilon}\left\{e^{-(\lambda,\nu_t^\varepsilon)}\right\} = \exp\left\{\int_E \log u^\varepsilon(t,x,y;\varepsilon^2\phi\lambda)\ \mu^\varepsilon(dx\ dy)\right\} \tag{4.3}$$

Here $u^\varepsilon(t,x,y;\varepsilon^2\phi\lambda)$ satisfies

$$\frac{\partial u^\varepsilon}{\partial t} = \frac{1}{\varepsilon} F \cdot \frac{\partial u^\varepsilon}{\partial x} + \frac{q}{\varepsilon^2}\left[\Phi(u^\varepsilon) - u^\varepsilon\right] \ , \qquad t > 0 \ , \tag{4.4}$$

$$u^\varepsilon(0,x,y;\varepsilon^2\phi\lambda) = \exp\left\{-\ \varepsilon^2\phi(y;x)\ \lambda(x,y)\right\} .$$

There are many other scalings that are interesting. Some of them are discussed in [2] and lead to asymptotic results different from the ones that follow.

5. THE MAIN CONVERGENCE THEOREM

Let Q_μ^ε be the measures induced by ν_t^ε of the last section on $X = D([0,\infty),M^+(E))$. Because of the scaling, ν_t^ε is not in $M_p^+(E)$ any longer. We show in [2] that under the hypotheses of Theorem 1 below, the Q_μ^ε converge weakly as $\varepsilon \to 0$. Here we show only that for each t the Laplace functionals converge but our proof contains all the essential steps in general.

Assume now the following conditions:

$$\int_S \tilde{\phi}(y;x)\ \phi(y;x)\ F(x,y)\ \gamma(dy) = 0\ , \tag{5.1}$$

$$\int_S \tilde{\phi}(y;x)\ F(x,y)\cdot \frac{\partial\phi(y;x)}{\partial x}\ \gamma(dy) = 0\ , \qquad x \in R^n. \tag{5.2}$$

If $F(x,y) = y$ and $\phi = \phi(y)$ only, (5.2) is automatic and (5.1) follows usually from elementary symmetry considerations. Assume further that

$$\sum_{k=1}^{\infty} k^3 \pi_k(x,y;S,S,\ldots,S) \le C < \infty\ . \tag{5.3}$$

<u>Theorem 1</u>. <u>For each</u> $t \ge 0$ <u>fixed and each</u> $\lambda(x) \in C_0^+(R^n)$, <u>the logarithm of the Laplace functionals</u> (4.3), <u>given by</u>

$$\int_E \log u^\varepsilon(t,x,y;\varepsilon^2\phi\lambda)\ \mu^\varepsilon(dx\ dy)\ ,$$

<u>converges as</u> $\varepsilon \to 0$ <u>to</u>

$$-\int_E v(t,x;\lambda)\ \mu(dx\ dy)\ .$$

<u>Here</u> $v(t,x;\lambda)$ <u>is the unique solution of the nonlinear diffusion equation</u>

$$\frac{\partial v(t,x;\lambda)}{\partial t} = Lv(t,x;\lambda) - g(x)v^2(t,x;\lambda)\ , \qquad t > 0 \tag{5.4}$$

$$v(0,x;\lambda) = \lambda(x) \ge 0\ , \qquad x \in R^n\ .$$

<u>For</u> $f(x) \in C_0^\infty(R^n)$ <u>the diffusion operator</u> L <u>is given by</u>

$$Lf(x) = \int_S\int_S \tilde{\phi}(y;x)F(x,y)\cdot \frac{\partial}{\partial x}\left[\frac{\phi(y;x)\psi(y,dz;x)}{\phi(z;x)}\, F(x,z)\cdot\frac{\partial}{\partial x}\big(\phi(z;x)f(x)\big)\right]\gamma(dy)\ , \tag{5.5}$$

<u>with</u> ψ <u>defined below</u> (3.6). <u>The function</u> $g(x) \ge 0$ <u>is defined by</u>

$$g(x) = \frac{1}{2}\int_S\int_S\int_S \tilde{\phi}(y;x)\ \gamma(dy)\ q(y,x)\ \pi^{(2)}(x,y,dy_1,dy_2)\psi(y_1;x)\phi(y_2;x)\ . \tag{5.6}$$

<u>The kernel</u> $\pi^{(2)}$ <u>in</u> (5.6) <u>is defined by</u> (2.11).

<u>Remark 1</u>. Let

$$v^\varepsilon(t,x,y;\lambda) = -\frac{\log u^\varepsilon(t,x,y;\varepsilon^2\phi\lambda)}{\varepsilon^2\phi(y;x)}\ , \qquad \lambda = \lambda(x) \ge 0. \tag{5.7}$$

In view of (4.1), the theorem says in effect, and we shall prove in Section 7, that for each $t \ge 0$,

$$\sup_{(x,y)\in E} |v^\varepsilon(t,x,y;\lambda) - v(t,x;\lambda)| \to 0\ , \qquad \text{as } \varepsilon \to 0\ . \tag{5.8}$$

The fact (5.8) is the major analytical content of the theorem. It is easily seen that v^{ε} satisfies the nonlinear equation

$$\frac{\partial v^{\varepsilon}}{\partial t} = \frac{1}{\varepsilon}\frac{1}{\phi} F \cdot \frac{\partial}{\partial x}(\phi v^{\varepsilon}) - \frac{q}{\varepsilon^2}\left[\frac{e^{\varepsilon^2 \phi v^{\varepsilon}} \Phi(e^{-\varepsilon^2 \phi v^{\varepsilon}}) - 1}{\varepsilon^2 \phi}\right], \tag{5.9}$$
$$v^{\varepsilon}(0,x,y;\lambda) = \lambda(x) .$$

We may take, as we will do, $\lambda(x) \geq 0$, $\lambda \in C_0^{\infty}(R^n)$.

Remark 2. From (5.8) and by simple computations similar to the ones in Section 2, we deduce equations for the mean and covariance of the limit of ν_t^{ε}. If $A_1, A_2 \subset R^3$, then

$$E_{\mu^{\varepsilon}}\left\{\nu_t^{\varepsilon}(A_1 \times S)\nu_t^{\varepsilon}(A_2 \times S)\right\} - E_{\mu^{\varepsilon}}\left\{\nu_t^{\varepsilon}(A_1 \times S)\right\} E_{\mu^{\varepsilon}}\left\{\nu_t^{\varepsilon}(A_2 \times S)\right\} \tag{5.10}$$
$$\rightarrow \int_E v^{(2)}(t,x;A_1,A_2)\,\mu(dx\,dy), \quad \text{as} \quad \varepsilon \to 0 ,$$

where

$$\frac{\partial v^{(2)}}{\partial t} = L v^{(2)} + 2g v_1^{(1)} v_2^{(1)} , \qquad t > 0 , \tag{5.11}$$
$$v^{(2)}(0,x;A_1,A_2) = 0$$

and

$$v_i^{(1)}(t,x) = v^{(1)}(t,x,A_i) , \qquad i = 1,2 .$$

Moreover, we have that

$$\frac{\partial v_i^{(1)}}{\partial t} = L v_i^{(1)} , \quad t > 0 , \quad v_i^{(1)}(0,x) = \chi_{A_i}(x) \tag{5.12}$$

and

$$E_{\mu^{\varepsilon}}\left\{\nu_t^{\varepsilon}(A_i \times S)\right\} \longrightarrow \int_E v_i^{(1)}(t,x)\,\mu(dx\,dy) . \tag{5.13}$$

The results (5.12), (5.13) constitute the usual diffusion approximation of transport theory (one speed).

6. HOMOGENIZATION

We now assume that:

(6.1) as functions of x, $q(x,y)$ and $\pi_k(x,y;A_1,\ldots,A_k)$, $k = 0,1,2,\ldots$ are periodic functions of period one in each direction, and

(6.2) $F(x,y) \equiv y$, S = Unit ball in R^n, i.e. S^{n-1}.

We replace q and $\Phi_x(\cdot)$ by $q(x/\varepsilon,y)$ and $\Phi_{x/\varepsilon}(\cdot)$ in (4.4) and call the resulting process $\tilde{\nu}_t^{H,\varepsilon}$; we must next scale the initial measure and the $\tilde{\nu}_t^{H,\varepsilon}$ measures as in (4.2). For this we need some facts proved in [1, section 5.2] as follows.

Consider the operator

(6.3) $$M_\zeta^H = y\cdot\frac{\partial}{\partial\zeta} + M_\zeta\ , \quad (M_\zeta \text{ given by (2.4)})\ ,$$

acting on functions on $T^n \times S^{n-1}$, the n-dimensional unit torus cross the unit n-dimensional ball. It is shown in [1] that M_ζ^H has an <u>isolated</u> maximal eigenvalue and <u>strictly positive</u> right and left eigenvectors $\phi_H(\zeta,y)$, $\tilde{\phi}_H(\zeta,y)$ (reference measure is $d\zeta\, dy$, Lebesgue measure). We assume that <u>the maximal eigenvalue is zero</u> as in Section 3.

Given $\mu \in M^+(E)$, $\mu < \infty$ pick $\mu^\varepsilon \in M_p^+(E)$, $\mu^\varepsilon < \infty$ such that for each $\lambda \in C(E)$

(6.4) $$\int_E \lambda(x,y)\ \varepsilon^2\phi_H(\tfrac{x}{\varepsilon},y)\ \mu^\varepsilon(dx\ dy) \longrightarrow \int_E \lambda(x,y)\ \mu(dx\ dy)\ .$$

Now define $\nu_t^{H,\varepsilon}$, the scaled BTP, as follows

(6.5) $$\nu_t^{H,\varepsilon}(dx\ dy) = \varepsilon^2\phi_H(\tfrac{x}{\varepsilon},y)\ \tilde{\nu}_t^{H,\varepsilon}(dx,dy)\ , \qquad \tilde{\nu}_0^{H,\varepsilon} = \mu^\varepsilon.$$

Assume further that (5.3) holds and that

(6.6) $$\int_{T^n\times S^{n-1}} \tilde{\phi}_H(\zeta,y)\ \phi_H(\zeta,y) y\ d\zeta\ dy = 0\ .$$

<u>Theorem</u> 2. <u>For each</u> $t \geq 0$ <u>and each</u> $\lambda(x) \in C_0^+(R^n)$, <u>the logarithm of the Laplace functionals of</u> $\nu_t^{H,\varepsilon}$

$$\log E_{\mu^\varepsilon}\left\{e^{-(\lambda,\nu_t^{H,\varepsilon})}\right\} = \int_E \log u_H^\varepsilon(t,x,y;\varepsilon^2\phi_H\lambda)\ \mu^\varepsilon(dx\ dy)\ ,$$

<u>converges as</u> $\varepsilon \to 0$ <u>to</u>

$$-\int_E v_H(t,x;\lambda)\ \mu(dx\ dy)\ ,$$

<u>where</u> $v_H(t,x;\lambda)$ <u>is the unique solution of the nonlinear diffusion equation</u>

(6.7) $$\frac{\partial v_H}{\partial t} = L_H v_H - g_H v_H^2\ , \qquad t > 0\ ,$$

$$v_H(0,x;\lambda) = \lambda(x) \geq 0\ , \qquad x \in R^n\ .$$

<u>Here the homogenized diffusion operator</u> L_H <u>is defined on</u> $C_0^\infty(R^n)$ <u>by</u>

(6.8) $$L_H f(x) = \sum_{i,j=1}^n a_{ij}\ \frac{\partial^2 f(x)}{\partial x_i \partial x_j}\ ,$$

<u>with the constant matrix</u> (a_{ij}) <u>equal to the symmetric part of</u> $(\tilde{a}_{ij})$

(6.9) $$\tilde{a}_{ij} = \iint_{(T^n\times S^{n-1})^2} \tilde{\phi}_H(\zeta,y)\phi_H(\zeta,y)\psi_H(y,\zeta;d\bar{y}\ d\bar{\zeta})\ \phi_H(\bar{\zeta},\bar{y})\ y_i\bar{y}_j\ dy\ d\zeta\ ,$$

<u>and</u> $\psi_H(y,\zeta;d\bar{y}\,d\bar{\zeta})$ <u>is the kernel of the integral operator</u> $-(M_\zeta^H)^{-1}$ (cf. (6.3)) <u>acting on functions orthogonal to</u> $\tilde{\phi}_H$. <u>The constant</u> $g_H \geq 0$ <u>is given by</u>

$$(6.10) \quad g_H = \iiiint_{(T^n\times S^{n-1})\times(S^{n-1})^2} \tilde{\phi}_H(\zeta,y)q(\zeta,y)\,\pi^{(2)}(\zeta,y,dy_1\,dy_2)\,\phi_H(\zeta,y_1)\phi_H(\zeta,y_2)\,dy\,d\zeta.$$

7. PROOF OF THEOREM 1

The proof of Theorem 2 is nearly identical to one for Theorem 1 so only the latter one will be given in detail. We shall actually prove (5.8) where v^ε satisfies (5.9) and v satisfies (5.4) and $\lambda \geq 0$ is in $C_0^\infty(R^n)$.

First of all we note that $v(t,x;\lambda)$ is smooth since coefficients and data in (5.4) are smooth by hypothesis. Define $v_1(t,x,y)$ by

$$(7.1) \qquad Q_x v_1 + \frac{1}{\phi} F \cdot \frac{\partial}{\partial x}(\phi v) = 0$$

which is well defined by (5.1), (5.2) and the Fredholm alternative for Q_x. Define also $v_2(t,x,y)$ by

$$(7.2) \qquad Q_x v_2 + \frac{1}{\phi} F \cdot \frac{\partial}{\partial x}(\phi v_1) + V(v,v) - \frac{\partial v}{\partial t} = 0$$

where

$$(7.3) \quad V(v,v) = \frac{q}{2\phi}\left[\Phi^{(2)}(1)\phi v,\phi v\right] - \frac{q\phi}{2}v^2 - qv\,\Phi^{(1)}(1)\phi v + \frac{q}{\phi}\Phi^{(1)}(1)\left(\frac{(\phi v)^2}{2}\right) .$$

Again, v_2 is well defined because the solvability condition for it holds; it is in fact (5.4).

Define (suppressing the λ in the notation)

$$(7.4) \qquad w^\varepsilon(t,x,y) = v^\varepsilon(t,x,y) - v(t,x) - \varepsilon v_1(t,x,y) - \varepsilon^2 v_2(t,x,y) .$$

We find by direct computation that w^ε satisfies the nonlinear equation (its integral equation version)

$$(7.5) \quad \frac{\partial w^\varepsilon}{\partial t} = \frac{1}{\varepsilon}\frac{1}{\phi} F \cdot \frac{\partial}{\partial x}(w^\varepsilon\phi) + \frac{1}{\varepsilon^2} Q_x w^\varepsilon - V(2h^\varepsilon+w^\varepsilon,w^\varepsilon) + \varepsilon^2 N^\varepsilon(w^\varepsilon+h^\varepsilon) + \varepsilon g_1^\varepsilon ,$$

$$w^\varepsilon(0,x,y) = \varepsilon g_2^\varepsilon(x,y) .$$

Here $g_1^\varepsilon(t,x,y)$ is some complicated but <u>known</u> function such that for any $T < \infty$,

$$(7.6) \qquad \sup_{0\leq t\leq T}\ \sup_{(x,y)\in E} |g_1^\varepsilon(t,x,y)| \leq C_1 , \quad \text{independently of } \varepsilon .$$

Similarly for g_2^ε,

$$(7.7) \qquad \sup_{(x,y)\in E} |g_2^\varepsilon(x,y)| \leq C_2 \qquad \text{independently of } \varepsilon .$$

The function $h^\varepsilon = v + \varepsilon v_1 + \varepsilon^2 v_2$ and the operator $N^\varepsilon(\cdot)$ is given by

$$(7.8)\qquad \varepsilon^2 N^\varepsilon(v) = -\frac{q}{\varepsilon^2}\,\frac{e^{\varepsilon^2\phi v}\,\Phi_x(e^{-\varepsilon^2\phi v}) - 1}{\varepsilon^2\phi} - \frac{1}{\varepsilon^2}Q_x v + V(v,v)$$

It is easily seen that if

$$(7.9)\qquad \|g\| = \sup_{0\le t\le T}\ \sup_{(x,y)\in E} |g(t,x,y)|,$$

then

$$(7.10)\qquad \|N^\varepsilon(g)\| \le C_3\|g\|^3\, e^{C_4\varepsilon^2\|g\|},$$

with C_3, C_4 constants independent of ε. At this point we know that (7.5) has a solution (its integral equation form) but we have no estimates. We shall show that

$$(7.11)\qquad \|w^\varepsilon\| = O(\varepsilon)\ , \quad \text{for}\quad \varepsilon > 0 \quad \text{sufficiently small}\ ,$$

which proves Theorem 1; we shall automatically get existence for (7.5) as a byproduct. The following lemma is basic.

Lemma 1. Let U^ε be the solution of

$$(7.12)\qquad \frac{\partial U^\varepsilon}{\partial t} = \frac{1}{\varepsilon}\frac{1}{\phi}F\cdot\frac{\partial}{\partial x}(\phi U^\varepsilon) + \frac{1}{\varepsilon^2}Q_x U^\varepsilon \equiv L^\varepsilon U^\varepsilon\ ,$$
$$U^\varepsilon(0,x,y) = g(x,y)\ .$$

There exist constants C_5 and C_6 independent of g and $\varepsilon\in(0,\varepsilon_0]$, ε_0 sufficiently small, such that

$$(7.13)\qquad |U^\varepsilon(t,x,y)| \le C_5\, e^{C_6 t}\|g\|, \qquad t\ge 0\ ,\quad (x,y)\in E.$$

Proof of Lemma 1: Let $\chi(y;x)$ be defined by

$$(7.14)\qquad Q_x\chi + \frac{1}{\phi}F\cdot\frac{\partial\phi}{\partial x} = 0\ ;\quad \chi \text{ well defined by (5.2) and bounded,}$$

and put

$$(7.15)\qquad U^\varepsilon = \phi(1+\varepsilon\chi)Z^\varepsilon$$

Then we find by direct calculation that for ε sufficiently small

$$(7.16)\qquad \frac{\partial Z^\varepsilon}{\partial t} = \frac{1}{\varepsilon}F\cdot\frac{\partial Z^\varepsilon}{\partial x} + \frac{1}{\varepsilon^2}Q^\varepsilon Z^\varepsilon + \big(\phi(1+\varepsilon\chi)\big)^{-1}F\cdot\frac{\partial(\phi\chi)}{\partial x}\,Z^\varepsilon,$$
$$Z^\varepsilon(0,x,y) = \big(\phi(1+\varepsilon\chi)\big)^{-1}g\ ,$$

with

$$(7.17)\qquad Q^\varepsilon Z^\varepsilon = \big(\phi(1+\varepsilon\chi)\big)^{-1}\left[M_x\big(\phi(1+\varepsilon\chi)Z^\varepsilon\big) + \varepsilon F\cdot\frac{\partial\phi}{\partial x}Z^\varepsilon\right].$$

The operator Q^ε is a Markov generator, i.e. it generates a positivity preserving, contraction semigroup. It follows that for ε sufficiently small Z^ε and hence by (7.15) U^ε satisfy (7.13).

We return to (7.5) and prove the following Lemma 2 which immediately finishes the proof of Theorem 1.

Lemma 2. *For $T < \infty$ fixed and $\varepsilon > 0$ sufficiently small there exists a constant C independent of ε such that*

$$\sup_{0\le t\le T} \sup_{(x,y)\in E} |w^{\varepsilon}(t,x,y)| \le \varepsilon C . \tag{7.18}$$

The constant C depends on $\lambda \ge 0$, $\lambda \in C_0^{\infty}(R^n)$ which is fixed here.

Proof of Lemma 2: Let $w_0 \equiv 0$ and define successively $w_n^{\varepsilon}(t,x,y)$, $n = 1,2,\ldots,$ by

$$\frac{\partial w_n^{\varepsilon}}{\partial t} = L^{\varepsilon} w_n^{\varepsilon} - V(2h^{\varepsilon}+w_{n-1}^{\varepsilon}, w_n^{\varepsilon}) + \varepsilon^2 N^{\varepsilon}(w_{n-1}^{\varepsilon}+h^{\varepsilon}) + \varepsilon g_1^{\varepsilon} , \tag{7.19}$$

$$w_n^{\varepsilon}(0,x,y) = \varepsilon g_2^{\varepsilon} .$$

The semigroup $e^{L^{\varepsilon}t}$ satisfies (7.13); hence (7.19) has the integral form

$$w_n^{\varepsilon}(t) = \varepsilon\, e^{L^{\varepsilon}t} g_2^{\varepsilon} - \int_0^t e^{(t-s)L^{\varepsilon}} \Big[V(2h^{\varepsilon}+w_{n-1}^{\varepsilon}, w_n^{\varepsilon}) - \varepsilon^2 N^{\varepsilon}(w_{n-1}^{\varepsilon}+h^{\varepsilon}) - \varepsilon g_1^{\varepsilon} \Big] ds .$$

Hence for $0 \le t \le T < \infty$ we have

$$|w_n^{\varepsilon}(t)| \le \varepsilon C_7 + C_8\big(1 + \|w_{n-1}^{\varepsilon}\|\big) \int_0^t |w_n^{\varepsilon}(s)|\, ds + \varepsilon^2 C_9 e^{\varepsilon^2 C_{10}\|w_{n-1}^{\varepsilon}\|} + \varepsilon C_{11} \tag{7.20}$$

(cf. (7.9)).

Here we use (7.10), (7.6), (7.7) and of course (7.13). By Gronwall's inequality we have

$$\|w_n^{\varepsilon}\| \le \varepsilon C_{12} \exp\Big\{C_{13}\|w_{n-1}^{\varepsilon}\|\Big\} , \tag{7.21}$$

where C_{12} and C_{13} are constants independent of ε and of $n = 1,2,\ldots$.

Inequality (7.21) implies that there is a constant C such that

$$\|w_n^{\varepsilon}\| \le \varepsilon C \quad \text{for all} \quad n = 0,1,2,\ldots \tag{7.22}$$

if $\varepsilon > 0$ is sufficiently small.

It is quite easy now to show that we also have

$$\|w_{n+1}^{\varepsilon} - w_n^{\varepsilon}\| \le \rho \|w_n^{\varepsilon} - w_{n-1}^{\varepsilon}\| \quad \text{for all } n = 1,2,\ldots , \tag{7.23}$$

with $0 < \rho < 1$ and $\varepsilon > 0$, sufficiently small.

This proves that $w_n^{\varepsilon} \to w^{\varepsilon}$ as $n \to \infty$, $\varepsilon > 0$ sufficiently small fixed, where w^{ε} is the unique solution of the integral equation version of (7.5) and (7.18) holds. The proof of Theorem 1 is complete.

REFERENCES

[1] A. Bensoussan, J. L. Lions, and G. C. Papanicolaou, Boundary layers and homogenization of transport processes, J. Publ. RIMS, Kyoto Univ., to appear.

[2] To appear.

[3] A. Bensoussan, J. L. Lions and G. C. Papanicolaou, Asymptotic Methods in Periodic Structures, North-Holland, Amsterdam, 1978.

[4] E. Larsen and J. B. Keller, Solution of the steady, one-speed neutron transport equation for small mean free paths, J. Math. Phys. 15 (1974), pp. 299-305.

[5] E. Larsen, Neutron transport and diffusion in inhomogeneous media I, J. Math. Phys. 16 (1975), pp. 1421-1427.

[6] M. Williams, Ph.D. dissertation, New York Univ., 1976.

[7] G. I. Bell, Stochastic formulations of neutron transport, in SIAM-AMS Proceedings, Vol. 1, Transport Theory, R. Bellman, G. Birkhoff and I. Abu-Shumays, editors, Providence, R. I., 1969, pp. 181-197.

[8] J. E. Moyal, The general theory of stochastic population processes. Acta Math. 108 (1962) pp. 1-31. See also article in the same volume as [7], pp. 198-212.

[9] T. W. Mullikin, Branching processes in neutron transport theory, in: Probabalistic Methods in Appl. Math., Vol. 1, A. T. Bharucha-Reid, editor, Academic Press, New York, 1968, pp. 199-281.

[10] N. Ikeda, M. Nagasawa and S. Watanabe, Branching Markov processes, J. Math. Kyoto Univ. 8 (1968), pp. 233-278, 356-410, and 9 (1969), pp. 95-160.

[11] T. Harris, The Theory of Branching Processes, Springer, Berlin, 1963.

[12] J. Kerstan, K. Matthes and J. Mecke, Unbegrenzt teilbare Punktprozesse, Akademie-Verlag, Berlin, 1974.

[13] O. Kallenberg, Random measures, Akademie-Verlag, Berlin, 1976.

[14] J. Neveu, Processes ponctuels, École d'été de Saint Flour, 1976.

[15] D. A. Dawson, Stochastic evolution equations and related measure processes, J. Multiv. Anal. 5 (1975), pp. 1-52.

[16] J. Lamperti, Continuous state branching processes, BAMS 73 (1967) pp. 382-386.

[17] M. Jiřina, Diffusion branching processes with several types of particles, Zeitschrift für Wahr. 18 (1971) pp. 34-46.

NUMERICAL EXPERIMENTS OF THE HOMOGENIZATION METHOD FOR OPERATORS WITH PERIODIC COEFFICIENTS

J.F. BOURGAT
IRIA-LABORIA
Domaine de Voluceau
Rocquencourt
78150 LE CHESNAY

ABSTRACT

A boundary value problem in an heterogeneous medium is numerically impossible to solve if the number of heterogeneities is very large. The homogenization consists in replacing the heterogeneous medium by an "equivalent" homogeneous one. In the particular case of a periodical distribution of heterogeneities, results on the mathematical aspects of the homogenization have been obtained recently. We present here some numerical experiments on the computation of the homogenized operator, the approximation of the real solution by the homogenized one and the efficiency of "correctors" to approximate the periodic behaviour of the solution.

INTRODUCTION

Let us consider a domain Ω , bounded open set of $\mathbb{R}^n$ with n=1,2,3, which consists in a composite material with a periodic structure. Let us denote by ε the period of the structure (we take here the same length ε of the period in all directions for the sake of clearness but this need not be the case in general). We suppose that ε is "small" that is to say the length of the period is small with respect to the size of Ω. In each cell we have two regions in which the physical constants are different (see fig. 1).

Let us make the scaling $\frac{1}{\varepsilon}$ on a given cell, we obtain the "unit cell" $Y =]0,1[^n$ and in Y we have two regions Y_1 and Y_o (see fig. 1).
If the material which corresponds to Y_o (resp. Y_1) is modelled by the operator

$$- \sum_{i,j=1}^{n} a_{ijo} \frac{\partial^2}{\partial y_i \partial y_j} \quad (\text{resp. } - \sum_{i,j=1}^{n} a_{ij1} \frac{\partial^2}{\partial y_i \partial y_j}) ,$$

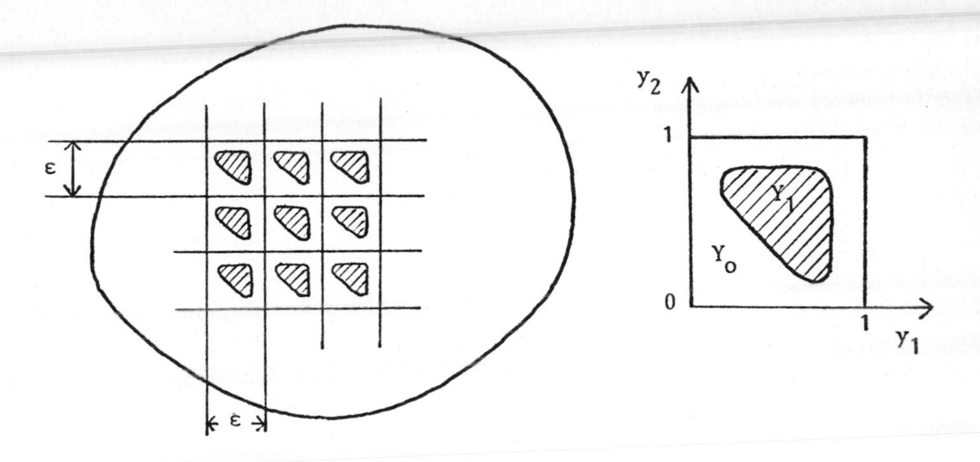

Figure 1 - A cell structure and the representative cell

we are led to introduce

$$a_{ij}(y) = a_{ijo} \text{ in } Y_o, \; a_{ij1} \text{ in } Y_1$$

and then the state of the system is given by the solution of

$$- \sum_{i,j=1}^{n} \frac{\partial}{\partial x_i} \left(a_{ij}\left(\frac{x}{\varepsilon}\right) \frac{\partial u_\varepsilon}{\partial x_j} \right) = f \text{ in } \Omega,$$

where f is a given function and where u_ε is subject to some boundary condition on $\Gamma = \partial\Omega$.

We now have the following questions :

- How to compute an operator with constant coefficients which models the macroscopic behaviour of the material.
- How to compute u_ε when the number of cells is very large.
- How to obtain an approximation of the microperiodic behaviour of u_ε .

In this paper we present some numerical results obtained in these directions. Recents results on the mathematical aspects of the homogenization have been given by BABUSKA [2], BENSOUSSAN-LIONS-PAPANICOLAOU [1], [2], DE GIORGI-SPAGNOLO [1], SANCHEZ PALENCIA [1] and TARTAR [1].

For this report we use theoretical results of BENSOUSSAN-LIONS-PAPANICOLAOU [2].

1. - THE HOMOGENIZED OPERATOR

We shall use the convention of summation on the repeated index.

1.1. Statement of the basic result.

We consider functions $a_{ij}(y)$ defined in $\mathbb{R}^n$, with real values, such that

$$a_{ij} \in L^\infty(\mathbb{R}^n),\ a_{ij} \text{ is Y-periodic, (i.e. takes equal values on opposite sides of Y).} \tag{1}$$

and

$$a_{ij}(y)\xi_i\xi_j \geq \alpha\,\xi_i\xi_i\ ,\ \alpha>0\ ,\ \forall \xi \in \mathbb{R}^n. \tag{2}$$

We define, for $\varepsilon > 0$, the operator A^ε by

$$A^\varepsilon \phi = -\frac{\partial}{\partial x_i}\left(a_{ij}\left(\frac{x}{\varepsilon}\right)\frac{\partial \phi}{\partial x_j}\right) \tag{3}$$

and we consider the equation

$$\begin{cases} A^\varepsilon u_\varepsilon = f \text{ in } \Omega,\ f \in L^2(\Omega) \\ u_\varepsilon = 0 \text{ on } \Gamma. \end{cases} \tag{4}$$

The operators A^ε are "uniformly elliptic in ε" in the following sense. For u,v in $H_0^1(\Omega)$ we set

$$a^\varepsilon(u,v) = \int_\Omega a_{ij}\left(\frac{x}{\varepsilon}\right)\frac{\partial u}{\partial x_j}\frac{\partial v}{\partial x_i}\,dx\ , \tag{5}$$

then

$$a^\varepsilon(v,v) \geq \alpha\|v\|^2_{H_0^1(\Omega)}\ . \tag{6}$$

So, as $\varepsilon \to 0$, one has

$$\|u_\varepsilon\|_{H_0^1(\Omega)} \leq \frac{\|f\|_{L^2(\Omega)}}{\alpha} \tag{7}$$

and therefore we can extract a subsequence, still denoted by u_ε , such that

$$u_\varepsilon \to u \text{ in } H_0^1(\Omega) \text{ weakly.} \tag{8}$$

Then one has the following result (see BENSOUSSAN-LIONS-PAPANICOLAOU [2], TARTAR [1])

THEOREM 1 : It exists a second order elliptic operator $\mathcal{A}$ with constant coefficients q_{ij} such that u is the solution of the equation

$$(9)\qquad \begin{cases} -\, q_{ij} \dfrac{\partial^2 u}{\partial x_i \partial x_j} = f \text{ in } \Omega \\ u = 0 \text{ on } \Gamma . \end{cases}$$

$\mathcal{A}$ is the "homogenized operator", q_{ij} the "homogenized coefficients", (9) the "homogenized problem" and u the "homogenized solution".

Construction of $\mathcal{A}$

We introduce the space W defined by

$$(10)\qquad W = \{\psi \in H^1(Y), \psi \text{ is Y-periodic (i.e. takes equal values on opposite sides of Y)}\}$$

and the "reduced" operator A_Y defined by

$$(11)\qquad A_Y \phi = - \frac{\partial}{\partial y_i} \left(a_{ij}(y) \frac{\partial \phi}{\partial y_j}\right).$$

The bilinear form associated to A_Y is

$$(12)\qquad a_Y(\phi,\psi) = \int_Y a_{ij}(y) \frac{\partial \phi}{\partial y_j} \frac{\partial \psi}{\partial y_i} dy$$

and we consider the function $\chi^i \in W$ defined by

$$(13)\qquad a_Y(\chi^i - y_i, \psi) = 0 \qquad \forall \psi \in W.$$

We remark that (13) defines χ^i up to an additive constant. We choose χ^i such that the mean value of χ^i is zero (i.e. $\int_Y \chi^i(y)\, dy = 0$). This choice will be justified in section 3.

Then the "homogenized coefficients" q_{ij} are obtained by

$$(14)\qquad q_{ij} = \frac{1}{|Y|} a_Y(\chi^j - y_j, \chi^i - y_i),$$

where $|Y|$ denotes the measure of Y.

1.2. Asymptotic expansion

Let us briefly recall the multiple scale method (see BENSOUSSAN-LIONS-PAPANICOLAOU [2]) which gives formulas (14).

We set

(15) $$y_i = x_i/\varepsilon \quad i=1,\dots,n.$$

Applied to a function of x and y the operator $\frac{\partial}{\partial x_i}$ becomes $\frac{\partial}{\partial x_i} + \frac{1}{\varepsilon}\frac{\partial}{\partial y_i}$ and we can write

(16) $$A^{\varepsilon} = \varepsilon^{-2}A_1 + \varepsilon^{-1}A_2 + A_3 ,$$

with

(17) $$A_1 = -\frac{\partial}{\partial y_i}\left(a_{ij}(y)\frac{\partial}{\partial y_j}\right) ,$$

(18) $$A_2 = -\frac{\partial}{\partial y_i}\left(a_{ij}(y)\frac{\partial}{\partial x_j}\right) - \frac{\partial}{\partial x_i}\left(a_{ij}(y)\frac{\partial}{\partial y_j}\right) ,$$

(19) $$A_3 = -a_{ij}(y)\frac{\partial^2}{\partial x_i \partial x_j} .$$

We look for u_ε in the form

(20) $$u_\varepsilon(x) = w_o(x,y) + \varepsilon\, w_1(x,y) + \varepsilon^2 w_2(x,y) + \dots$$

where

$w_j(x,y)$ is Y-periodic in y.

Then we can write (4) in the form

$$(\varepsilon^{-2}A_1 + \varepsilon^{-1}A_2 + A_3)(w_o + \varepsilon w_1 + \varepsilon^2 w_2 + \dots) = f$$

and if we identify in ε we obtain

(21) $$A_1 w_o = 0,$$

(22) $$A_1 w_1 + A_2 w_o = 0,$$

(23) $$A_1 w_2 + A_2 w_1 + A_3 w_o = f .$$

If we notice that the only solution in Y of $A_1\theta = 0$, θ periodic, is θ = constant, we see that

(24) $$w_o(x,y) = u(x).$$

Then equation (22) becomes

(25) $$A_1 w_1 = \frac{\partial}{\partial y_i}\left(a_{ij}(y)\frac{\partial u}{\partial x_j}(x)\right).$$

Therefore if we introduce χ^j by (13), (i.e. $A_1\chi^j = -\frac{\partial}{\partial y_i}(a_{ij}(y))$, we obtain

$$w_1(x,y) = -\chi^j(y)\frac{\partial u}{\partial x_j}(x) + \tilde{w}_1(x). \tag{26}$$

Using the Fredholm's alternative (i.e. a necessary and sufficient condition, for $A_1\theta = F$, θ periodic, to have a solution is $\int_Y F(y)dy = 0$) equation (23) in w_2 has a solution iff

$$\int_Y (f - A_2w_1 - A_3w_o)dy = 0 \quad \forall x. \tag{27}$$

Using (18) and (19) we obtain

$$-\frac{\partial}{\partial x_i}\int_Y a_{ij}(y)\frac{\partial w_1}{\partial y_i}d_y - [\int_Y a_{ij}(y)dy]\frac{\partial^2 u}{\partial x_i \partial x_j} = |Y|\ f$$

and replacing w_1 by its value, we obtain

$$(-\int_Y a_{ij}(y)dy + \int_Y a_{ik}(y)\frac{\partial \chi^j}{\partial y_k}(y)dy)\frac{\partial^2 u}{\partial x_i \partial x_j} = |Y|\ f. \tag{28}$$

Then the coefficient of $\frac{\partial^2 u}{\partial x_i \partial x_j}$ equals

$$q_{ij} = a_Y(\chi^j - y_j, -y_i) \tag{29}$$

and since χ^i is periodic we have

$$a_Y(\chi^j - y_j, \chi^i) = 0 \tag{30}$$

and therefore we obtain (14).

1.3. Computation of the homogenized coefficients

The homogenized operator coefficients are computed only on the unit cell Y. For solving the problem (13) which gives the functions χ^i we use the finite element method with Lagrange finite elements of degree 1 on triangles. This computation need not a priori a fine mesh however, as the homogenized coefficients q_{ij} are obtained from the first derivatives of the functions χ^i, we have to use a sufficiently fine mesh of Y.

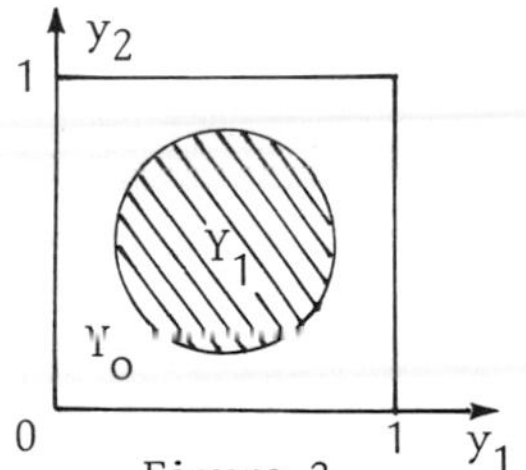

Figure 2.

Thus, if we consider the unit cell representation as in Fig. 2, with

Y_1 = disk of area 0.25

$a_{ijo}(y) = \delta_{ij}$ (the Kronecker's index)

$a_{ij1}(y) = 10\ \delta_{ij}$

and a triangulation such as the discontinuities of the coefficients a_{ij} coincide with sides of triangles, we obtain the results shown in table 1.

Number of triangles	q_{11}
32	1.494
128	1.512
512	1.516
2048	1.516

Table 1 - Accuracy of the homogenized coefficients computations.

For solving the linear system associated to the discretized problem, it is very difficult to use direct methods because the periodicity conditions disturb the band structure of the matrix. So we have used the overrelaxation method with an optimal parameter.

In several examples studied here, the coefficients a_{ij} have some properties of symmetry, as it is often the case in physical examples. Then the question is : does the homogenized coefficients have the same properties ?

One has (see BOURGAT-DERVIEUX [1]) the

PROPOSITION 1

If the coefficients a_{ij} are such as

(31) a_{ii} is symmetric with respect to the middle hyperplans $\Delta_i (i=1,\dots,n)$ of the representative cell (see fig. 3),

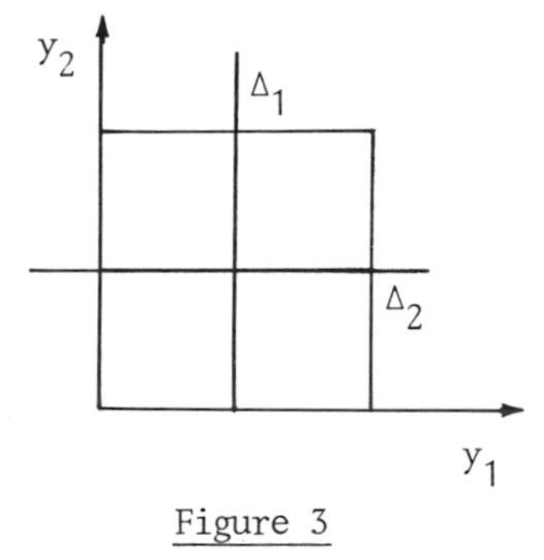

Figure 3

Then

(32) $a_{ij} = 0$ for $i \neq j \Longrightarrow q_{ij} = 0$ for $i \neq j$.

If we consider the representative cell as in fig. 4, with

$$a_{ijo} = \delta_{ij} \, , \, a_{ij1} = 10 \, \delta_{ij},$$

we have $a_{ij} = 0$ if $i \neq j$ but the assumption (31) is not satisfied. The homogenized coefficients obtained

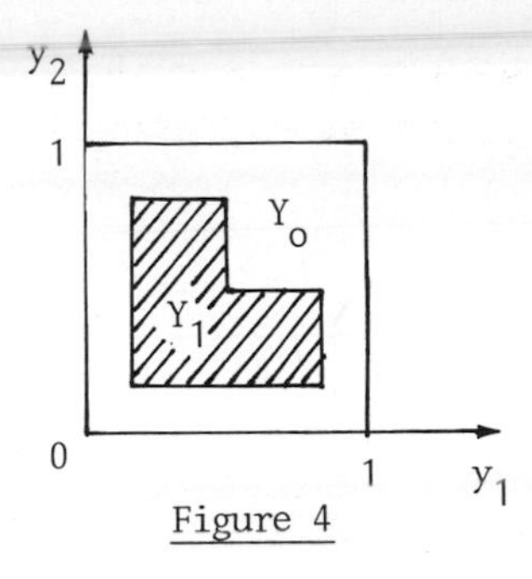

Figure 4

$$q_{11} = q_{22} = 1.915, \quad q_{12} = q_{21} = -0.101,$$

show that $q_{ij} \neq 0$ for $i \neq j$. Therefore an heterogeneous medium made of two isotropic materials can be homogenized into an effective non isotropic medium.

We present now various examples showing some properties of the homogenized operator.

Example 1 : Homogenized coefficients and mean values

Let us consider the representative cell as in fig. 5, with

$$(33) \quad \begin{cases} a_{ijo} = \delta_{ij} \\ a_{ij1} = \lambda\delta_{ij}, \ \lambda > 0. \end{cases}$$

In this case one has

$$q_{11} = q_{22} \, , \, q_{12} = q_{21} = 0.$$

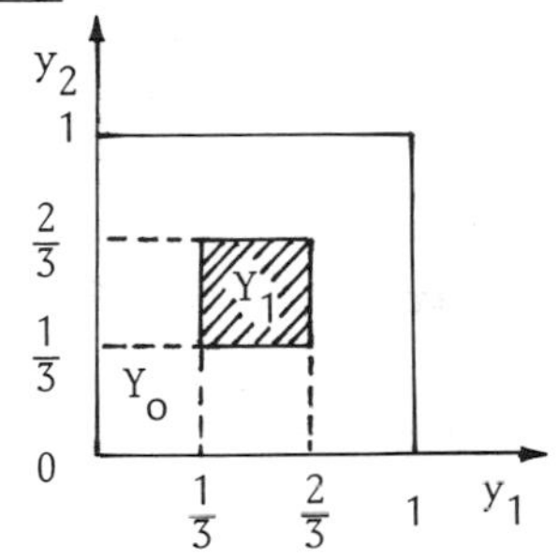

Figure 5

The values of q_{11} as λ varies from 0 to $+\infty$ are represented on fig. 6 and compared to

$$(34) \qquad \mathcal{m}(a_{11}) = \frac{1}{|Y|} \int_Y a_{11}(y)dy \ , \text{ the mean value of } a_{11}(y)$$

and also to

$$(35) \qquad (\mathcal{m}(\frac{1}{a_{11}}))^{-1} = \left[\frac{1}{|Y|} \int \frac{1}{a_{11}(y)}\right]^{-1}$$

which is the homogenized coefficient if $a_{11}(x)$ is periodic in the direction x_1 only.

We remark in the figure 6 that q_{11} is very different from $\mathcal{m}(a_{11})$ or $(\mathcal{m}(\frac{1}{a_{11}}))^{-1}$ and that q_{11} converges as $\lambda \to 0$ or as $\lambda \to \infty$. We remark also that one has (see TARTAR [1])

$$(36) \qquad \left[\mathcal{m}(\frac{1}{a_{11}})\right]^{-1} < q_{11} < \mathcal{m}(a_{11}) \qquad \forall \lambda > 0.$$

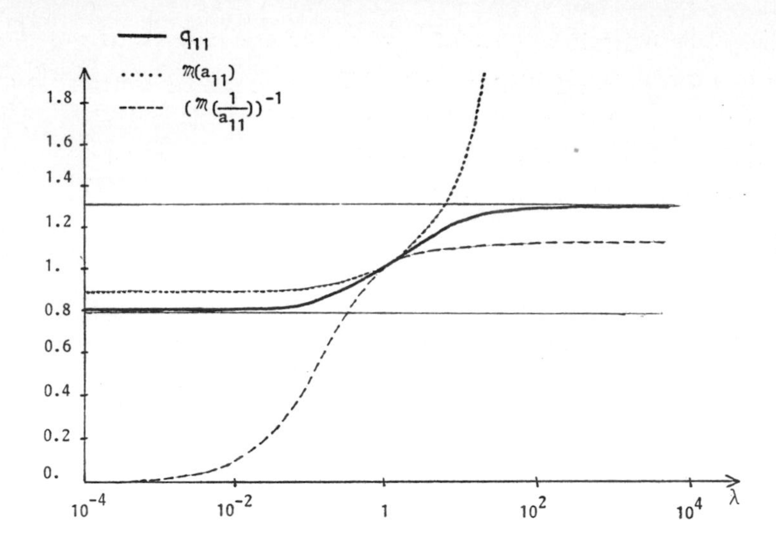

Figure 6 - Comparison of q_{11}, $\mathcal{M}(a_{11})$, $[\mathcal{M}(\frac{1}{a_{11}})]^{-1}$.

Example 2 : Torsion of fiber-reinforced bars (see BOURGAT-LANCHON [1])

The equation (4) models the elastic torsion of a cylindrical fiber-reinforced bar. Ω is the cross section of the bar, Y_1 and Y_o are the image in the scaling $\frac{1}{\varepsilon}$ of the fiber and of the matrix respectively. u_ε is called the "stress function" and gives the components of the stress tensor.
The operator A^ε is modelled by

$$a_{ijo} = \delta_{ij} \text{ , } a_{ij1} = \lambda\delta_{ij} \text{ ,}$$

with $\lambda = \dfrac{\mu_1}{\mu_o}$, where μ_1 and μ_o are the shear modulus of the fiber and of the matrix respectively. The representative cell is chosen as in fig. 5.
For a "glass-epoxy" composite (glass in Y_1) with

(37) $$\mu_o = 2.2 \; 10^5 \text{ psi, } \mu_1 = 4 \; 10^6 \text{ psi (i.e. } \lambda = \frac{1}{18})$$

we obtain the homogenized shear modulus $\overline{\mu} = 2.68 \; 10^5$ psi while the law of mixture would have given $2.46 \; 10^5$ psi.
For a "boron-epoxy" composite (boron in Y_1) with

(38) $$\mu_o = 2.2 \; 10^5 \text{ psi, } \mu_1 = 2.5 \; 10^7 \text{ psi (i.e. } \lambda = \frac{1}{114} \text{)}$$

we obtain $\bar{\mu} = 2.75\ 10^5$ psi in place of $2.47\ 10^5$ by the law of mixtures.
To be able to compare our results with the experimental ones given in GARG-SVATBONAS-GURTMAN [1], we make the same computation as above for a "glass epoxy" composite with $\frac{\text{meas}(Y_1)}{\text{meas}\ (Y)} = 0.7$. We obtain $\bar{\mu} = 0.965\ 10^6$ psi in place of $0.65\ 10^6$ psi by the mixture law and $0.95\ 10^6$ psi in GARG-SVATBONAS-GURTMAN [1].

Example 3 :

Let us consider the three representative cells of the figure 7. The area of Y_1 has the value 1/4 in each cell. One can notice that each cell satisfies the symmetry property (31).

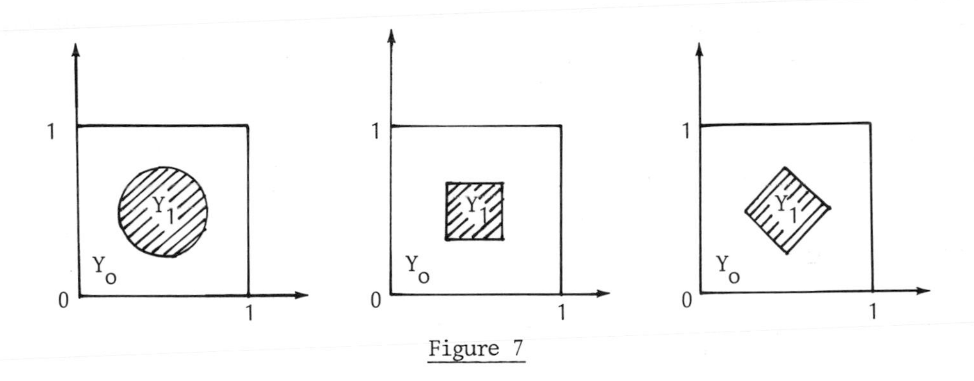

Figure 7

For the following choice of a_{ij}

$$a_{ijo} = \delta_{ij}\ ,\ a_{ij1} = 10\ \delta_{ij},$$

we have obtained with a very fine triangulation (2048 triangles)

$$q_{11} = \begin{array}{ll} 1.516 & \text{for the disk} \\ 1.548 & \text{for the square} \\ 1.573 & \text{for the lozenge} \end{array}$$

Therefore, in an application, if the shape of Y_1 is not well defined, the results of the homogenization computations depend strongly on the chosen shape.

Example 4 :

Let us consider the representative cell $Y =]0,a[\times]0,1[$ as in fig. 8, with

$a_{ijo} = \delta_{ij}\ ,\ a_{ij1} = 10\ \delta_{ij}$.

The two materials are in the same pro-

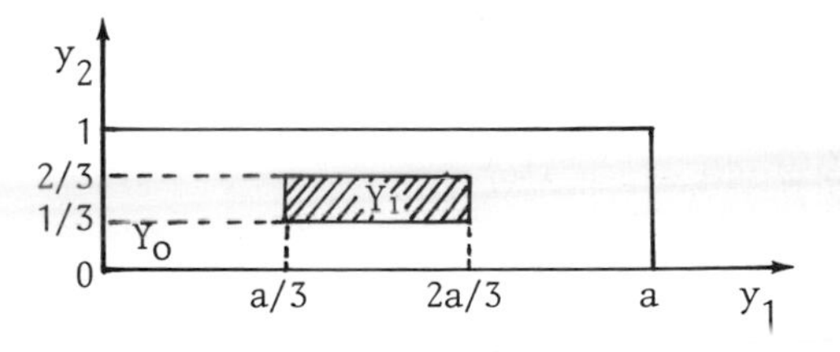

Figure 8

portion for every value of a.

Figure 9 shows how the homogenized coefficient q_{11} varies as a varies from 0 to ∞.

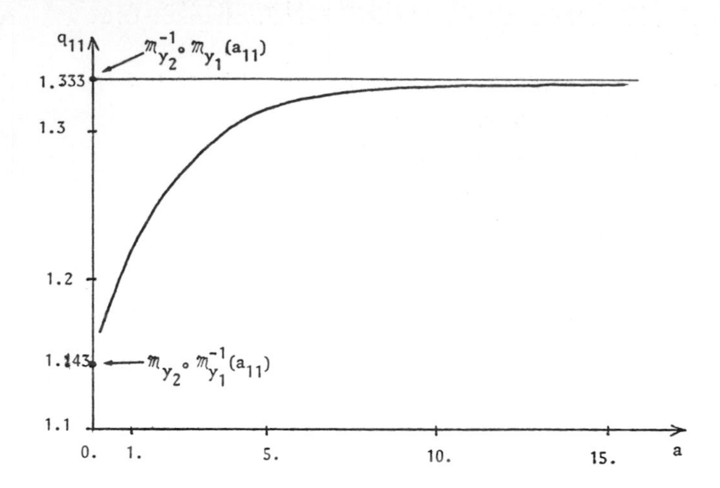

Figure 9 - Variation of q_{11} in function of the side a of Y =]o,a[×]0,1[

Let us introduce

$$\mathcal{m}_{yi}(\phi) = \frac{1}{a_i} \int_o^{a_i} \phi(y_1,y_2)dy_i \quad , \quad a_1 = a, \ a_2 = 1.$$

and

$$\mathcal{m}_{yi}^{-1}(\phi) = \left(\frac{1}{a_i} \int_o^{a_i} \frac{1}{\phi(y_1,y_2)} dy_i\right)^{-1}.$$

We obtain

$$\mathcal{m}_{y_2}^{-1} \circ \mathcal{m}_{y_1}(a_{11}) = 1.333 \ ,$$

$$\mathcal{m}_{y_2} \circ \mathcal{m}_{y_1}^{-1}(a_{11}) = 1.143.$$

If we compare these numbers with the bounds of q_{11} (see. fig. 9) we notice that

$$q_{11} \to \mathcal{m}_{y_2}^{-1} \circ \mathcal{m}_{y_1}^{-1}(a_{11}) \text{ as } a \to 0$$

and

$$q_{11} \to \mathcal{m}_{y_2} \circ \mathcal{m}_{y_1}^{-1}(a_{11}) \text{ as } a \to \infty.$$

Thus, when the ratio of the sides of the representative cell Y is small or large we can approximate q_{11} by two successive homogenizations in one direction. This can be proved if we define the periodicity of a_{ij} by $a_{ij}(\frac{x1}{\varepsilon}, \frac{x2}{\varepsilon^2})$ (see LIONS [1]).

2. - COMPARISON OF THE "TRUE SOLUTION" WITH THE "HOMOGENIZED SOLUTION"

We study here the approximation of u_ε, solution of the equation (4), by the homogenized solution u, solution of equation (9).
Let us consider the unit cell Y as in fig. 5, with

$$a_{ijo} = \delta_{ij} \ , \ a_{ij1} = \lambda\delta_{ij}. \tag{39}$$

We take

$$\Omega =]0,1[\times]0,1[\ ,$$

$$f = 10.$$

$$\varepsilon = 1/2, \ 1/4, \ 1/8.$$

Figure 10 shows the periodicity of the coefficients a_{ij}, in the domain Ω, for $\varepsilon = \frac{1}{8}$.

2.1. Numerical techniques

For solving equations (4) and (9), which give u_ε and u, we use the finite element method with Lagrangian elements of degree one on the triangles.
The homogenized solution which is the solution of a second-order elliptic equation with constant coefficients, is obtained without any difficulty. On the other hand the computation of u_ε needs a triangulation which must be very fine as ε becomes small.
Figure 10 shows a triangulation of Ω for $\varepsilon = 1/8$, which is such that the discontinuities of the coefficients a_{ij} coincide with sides of the triangles. The number of triangles is 4608. In each zone which corresponds to Y_1 (hachured zones), the number of triangles is only 8, this is a minimum to approach u_ε in these regions.
In the case $\lambda \ll 1$, 32 triangles are needed in this zone. However by using the symmetries of the problem, we can reduce it to a problem on a quarter of domain divided in 4608 triangles, which corresponds to a global triangulation with 18432 triangles.
For $\lambda = \frac{1}{114}$ and $\varepsilon = \frac{1}{8}$ we obtain, in the middle M of a zone which corresponds to Y_1, the values of the approximated solution $u_{\varepsilon h}$ given in table 2 (h is the step size).

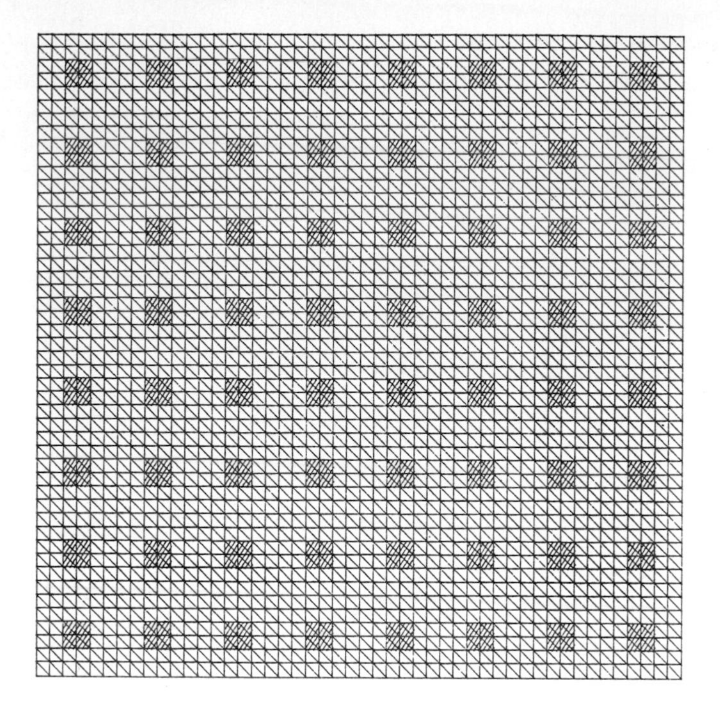

Figure 10 - Triangulation of Ω for $\varepsilon = 1/8$ - 4608 triangles - Hachured zones correspond to Y_1.

Number of triangles	Values of $u_{\varepsilon h}$ at M
1152	0.839
4608	0.996
18432	1.029

Table 2 - Convergence of $u_{\varepsilon h}$, as $h \to 0$, for $\lambda = \frac{1}{114}$, $\varepsilon = \frac{1}{8}$.

One can see that it is very difficult to compute the solution of a two dimensional problem in an heterogeneous medium as the number of heterogeneities is greater than one hundred.
For theoretical results on the convergence of problems with interfaces and singularities see BABUSKA [1].

2.2. Convergence of u_ε towards u.

We have obtained in section 1 the following convergence result

(40) $$u_\varepsilon \to u \text{ in } H_o^1(\Omega) \text{ weakly}$$

and, if coefficients a_{ij} are sufficiently regular (see LIONS [1]), one has

(41) $$||u_\varepsilon - u||_{C^o(\bar{\Omega})} \leq C\varepsilon .$$

In this section we show numerical results on the convergence of u_ε towards u and the figures presented are representations of the solutions u_ε and u restricted along a diagonal of Ω as indicated on figure 11.

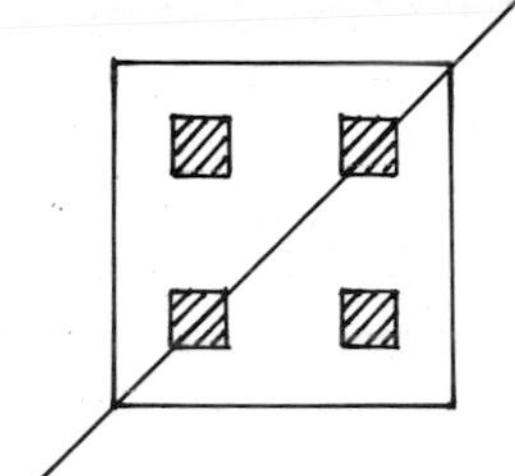

Figure 11

Figures 12, 13, 14 show u_ε and u respectively for $\lambda = 10, \frac{1}{18}, \frac{1}{114}$ and in each one u_ε is represented for ε = 1/2, 1/4, 1/8.
Values λ = 1/18 and λ = 1/114 correspond to the torsion of a fiber-reinforced bar presented in Section 1.3.

One can see in figures 12, 13, 14 that u_ε converges towards u as $\varepsilon \to 0$ but also that the difference between u_ε and u is greater for $\lambda < 1$.
More precisely, we have indicated in table 3, for $\lambda = 10$ and $\lambda = \frac{1}{114}$, the error $u_\varepsilon - u$ in the norms of spaces $L^\infty(\Omega)$, $L^2(\Omega)$, $H_o^1(\Omega)$.

ε	$\lambda = 10$			$\lambda = 1/114$		
	$\text{Sup}\lvert u_\varepsilon - u\rvert$	$\frac{\lVert u_\varepsilon - u\rVert_{L^2}}{\lVert u\rVert_{L^2}}$	$\frac{\lVert u_\varepsilon - u\rVert_{H_o^1}}{\lVert u\rVert_{H_o^1}}$	$\text{Sup}\lvert u_\varepsilon - u\rvert$	$\frac{\lVert u_\varepsilon - u\rVert_{L^2}}{\lVert u\rVert_{L^2}}$	$\frac{\lVert u_\varepsilon - u\rVert_{H_o^1}}{\lVert u\rVert_{H_o^1}}$
1/2	0.128	0.1	0.310	2.24	0.56	0.973
1/4	0.072	0.054	0.357	0.554	0.187	0.92
1/8	0.040	0.028	0.376	0.133	0.049	0.77

Table 3 - Convergence of the error $u_\varepsilon - u$ for $\lambda = 10$, $\lambda = \frac{1}{114}$

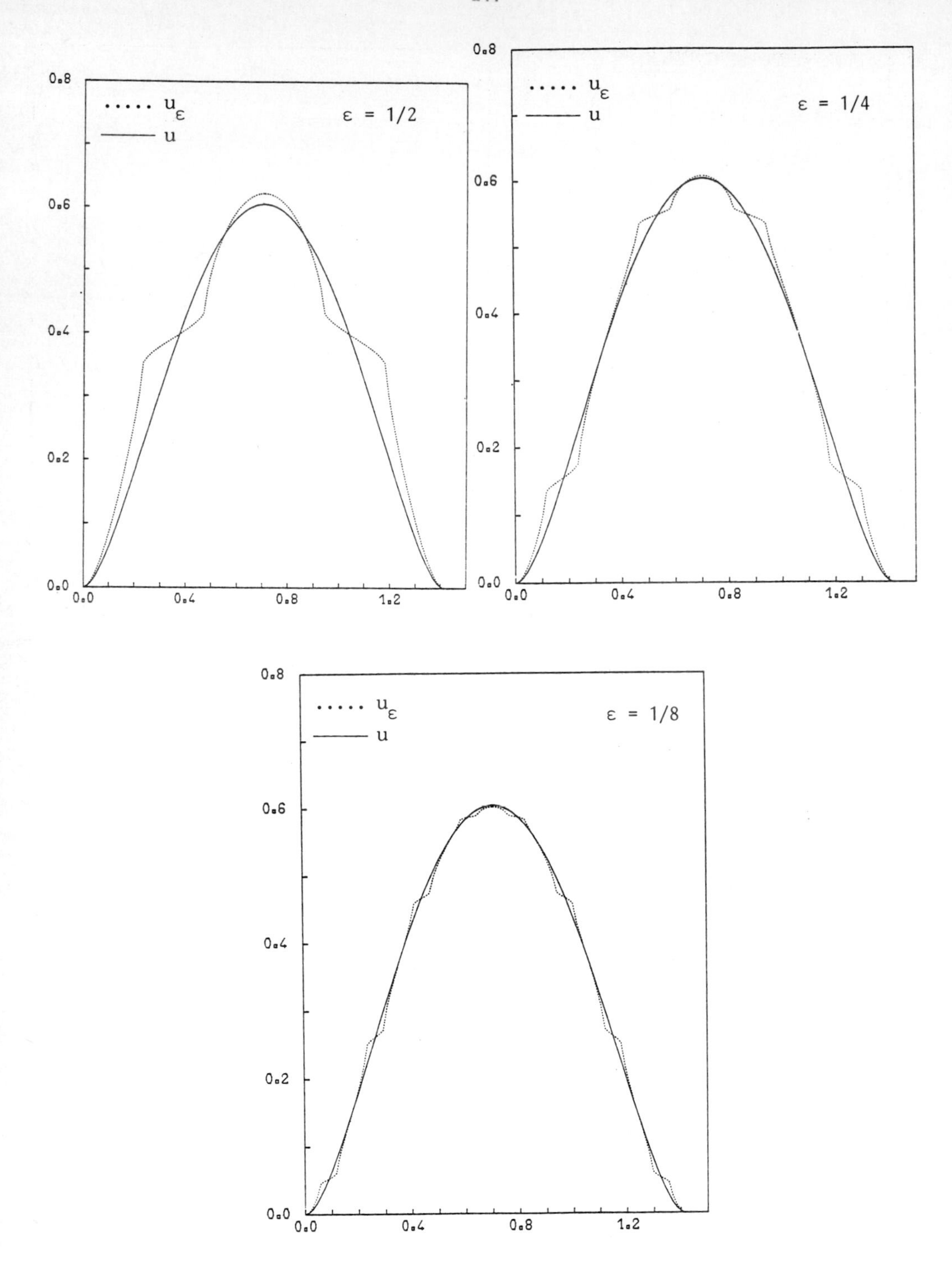

Figure 12 - Comparison of u_ε with u for $\lambda = 10$.

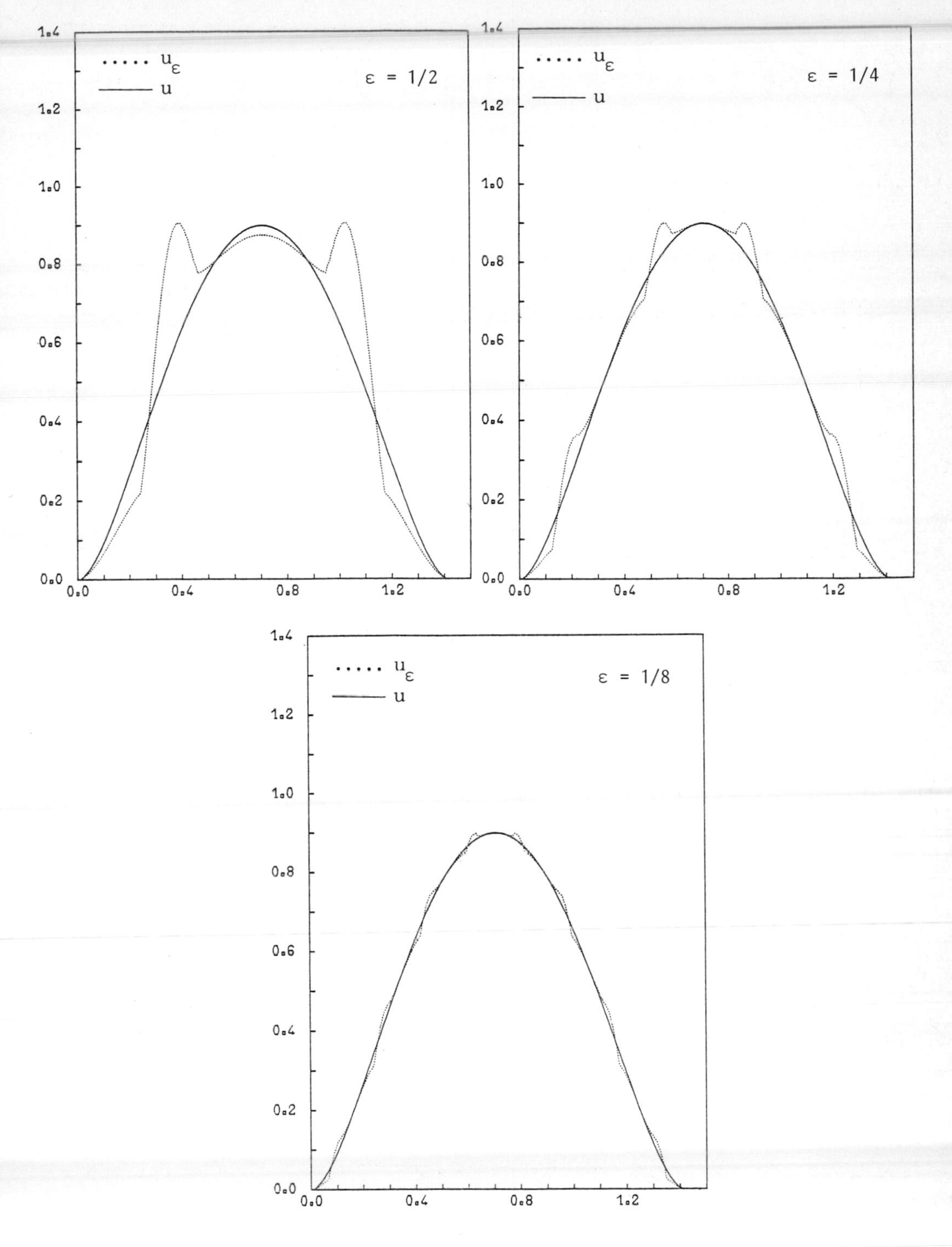

Figure 13 - Comparison of u_ε with u for $\lambda = 1/18$.

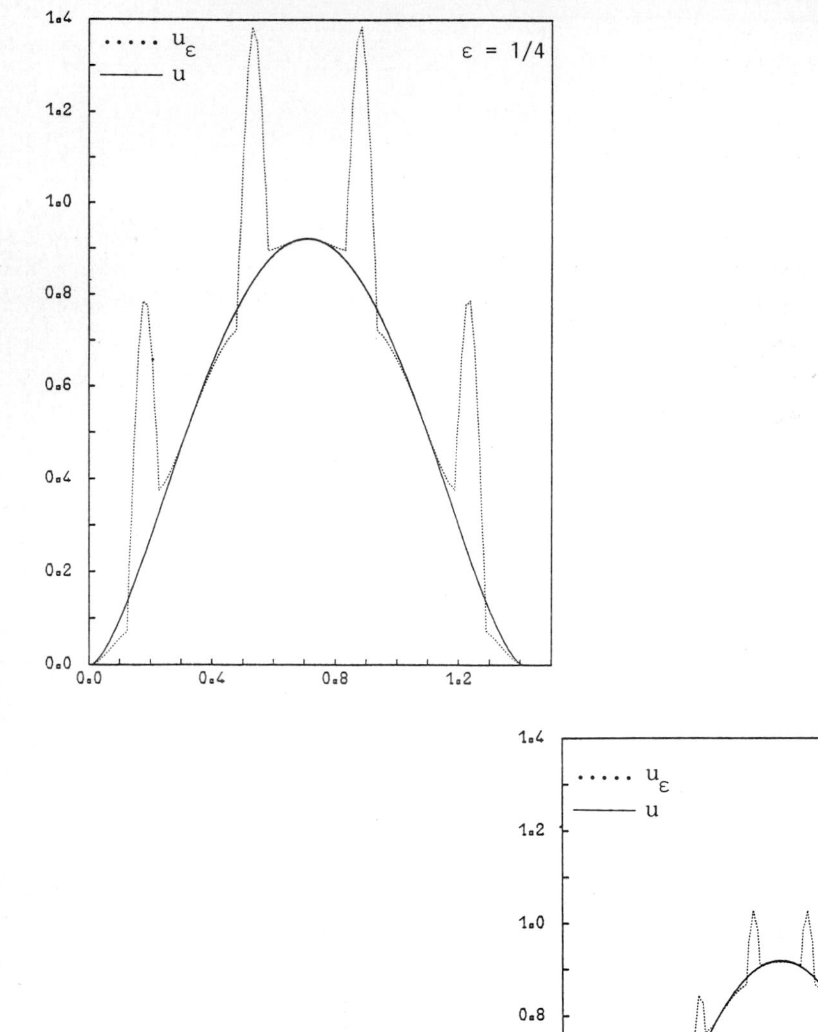

Figure 14 - Comparison of u_ε with u for $\lambda = 1/114$.

We remark in table 3 that $||u_\varepsilon - u||_{H_o^1(\Omega)}$ does not converge but $||u_\varepsilon - u||_{L^2(\Omega)}$ and $||u_\varepsilon - u||_{L^\infty(\Omega)}$ converge as we can see in fig. 12, 13, 14. We also remark that the values of sup $|u_\varepsilon - u|$ have a linear decrease as $\varepsilon \to 0$, though the coefficients a_{ij} be discontinuous.

2.3. An example with a variational inequality

Results of convergence of u_ε towards u have been obtained for simple variational inequalities (see LIONS [1]).
We present here numerical results for the following problem. We introduce the convex set K defined by

$$K = \{v \in H_o^1(\Omega),\ v \leq 0 \text{ in } \Omega\}$$

and we consider u_ε solution of the variational inequality

$$a_\varepsilon(u_\varepsilon, v-u_\varepsilon) \geq (f, v-u_\varepsilon) \quad \forall v \in K,$$

where a_ε is the bilinear form associated to A^ε defined in (5). Then the homogenized problem is : find $u \in K$ solution of

$$a(u, v-u) \geq (f, v-u) \quad \forall v \in K,$$

where a is the bilinear form with constant coefficients associated to the homogenized operator $\mathcal{A}$ defined in section 1.1, and we have (see LIONS [1])

$$u_\varepsilon \to u \text{ in } H_o^1(\Omega) \text{ weakly.}$$

For the computations we consider the unit cell Y as in fig. 5, with

$$a_{ijo} = \delta_{ij}, \quad a_{ij1} = 5\,\delta_{ij}.$$

We take

$$\Omega =]0,1[\times]0,1[$$

$$f = 360\ x_1x_2(1-x_1)(1-x_2) - 5x_1 - 10.$$

$$\varepsilon = 1/2,\ 1/4,\ 1/8.$$

Figure 15 shows sections of the functions u_ε and u as indicated in fig. 11. One can see, in this simple case, the good approximation of u_ε by u for $\varepsilon = 1/4$ or $\varepsilon = 1/8$. One can notice that the free boundary S_ε of the true solutions are very close to the free boundary of the homogenized solution.
On the numerical techniques for solving variational inequalities see GLOWINSKI-LIONS

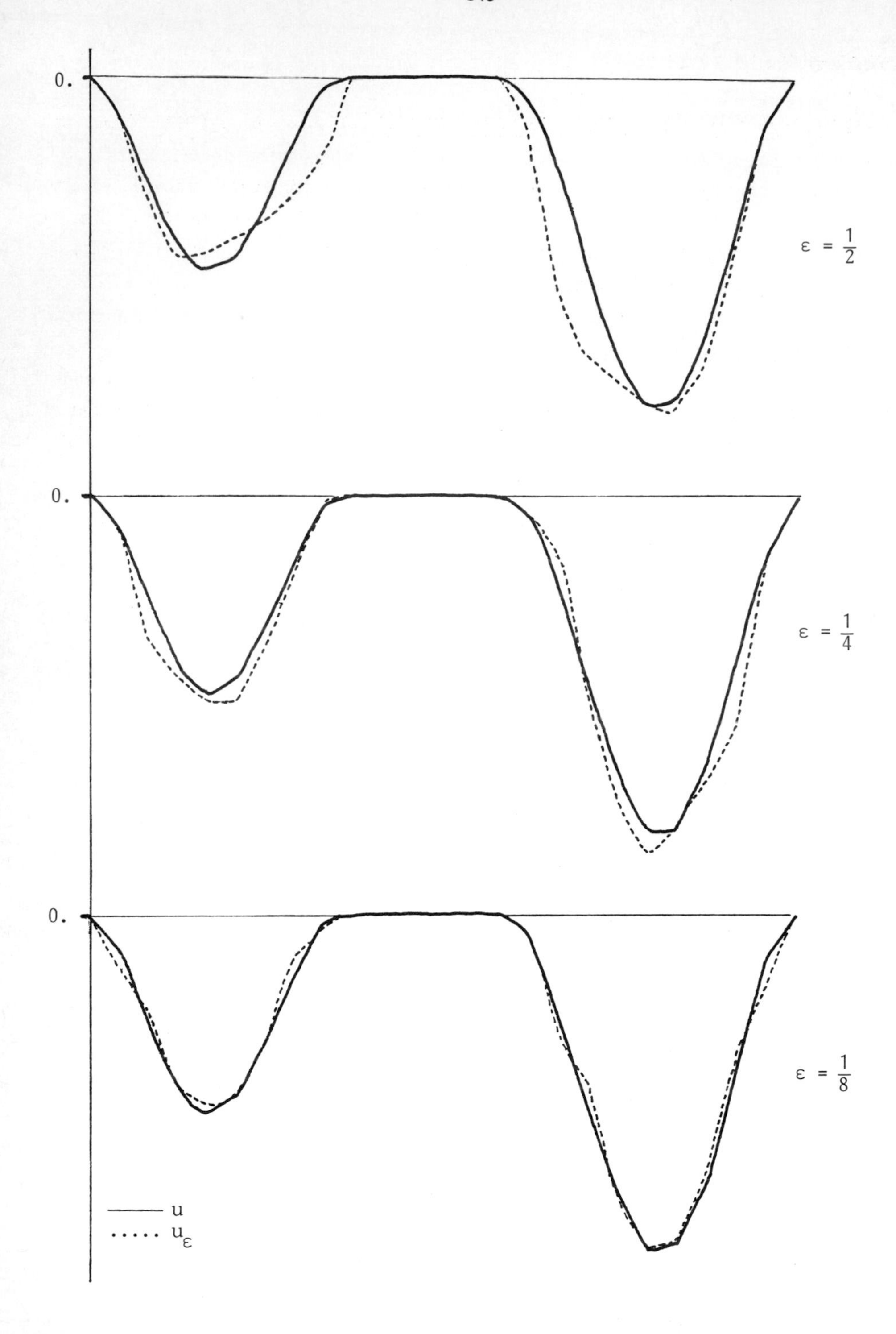

Figure 15 - Comparison of u_ε with u in the case of a variational inequality

TREMOLIERES [1].

3. - APPROXIMATION OF THE PERIODIC BEHAVIOUR OF THE SOLUTIONS u_ε . CORRECTORS.

One can notice in fig. 12, 13, 14 that solutions u_ε have important periodic variations. Thus for $\lambda < 1$ u_ε has "peaks" and for $\lambda > 1$ u_ε has "flat regions". In this section we look for an approximation of this periodic behaviour with terms of order 1 or 2 in ε, provided with the asymptotic expansion.

3.1. Asymptotic expansion - Error estimations.

We consider here the Dirichlet problem (4) introduced in section 1. Let us briefly recall some results presented in BOURGAT-DERVIEUX [1]. The notations are those of section 1. We look for u_ε in the form

$$u_\varepsilon(x) = w_o(x,y) + \varepsilon w_1(x,y) + \varepsilon^2 w_2(x,y) + \dots$$

and we have shown in section 1 that

$$w_o(x,y) = u(x) \tag{42}$$

$$w_1(x,y) = -\chi^i(y)\frac{\partial u}{\partial x_i}(x) + \tilde{w}_1(x). \tag{43}$$

If we continue the computations we obtain

$$w_2(x,y) = -\chi^i(y)\frac{\partial \tilde{w}_1}{\partial x_i}(x) + \eta_{ij}(y)\frac{\partial^2 u}{\partial x_i \partial x_j}(x) + \tilde{w}_2(x) , \tag{44}$$

with $\eta_{ij} \in W$ solution of the equation

$$a_Y(\eta_{ij},\psi) = \int_Y g_{ij}(y)\psi(y)\,dy , \quad \forall\psi\in W, \quad \mathcal{M}(\eta_{ij}) = 0, \tag{45}$$

where g_{ij} is defined by

$$g_{ij}(y) = -\frac{\partial}{\partial y_k}(a_{ki}\chi^j) - a_{ik}\frac{\partial \chi^j}{\partial y_k} + a_{ij} - q_{ij} . \tag{46}$$

Let us introduce $\theta_\varepsilon \in H^1(\Omega)$ solution of the equation

$$\begin{cases} A^\varepsilon\theta_\varepsilon = 0 \\ \theta_\varepsilon = \chi^i(\frac{x}{\varepsilon})\frac{\partial u}{\partial x_i}(x) \text{ on } \Gamma . \end{cases} \tag{47}$$

If $a_{ij} \in C^{1+\alpha}(\bar{\Omega})$, $\alpha > 0$, $f \subset C^{5+\alpha}(\bar{\Omega})$ and the boundary Γ is of class C^5, then one has the following estimations.

(48) $$\|\theta_\varepsilon\|_{C^o(\bar{\Omega})} < K$$

(49) $$\|u_\varepsilon - u - \varepsilon(-\chi^i(\tfrac{x}{\varepsilon})\frac{\partial u}{\partial x_i}(x) + \tilde{w}_1(x) + \theta_\varepsilon)\|_{C^o(\bar{\Omega})} < K\,\varepsilon^2$$

(50) $$\|u_\varepsilon - u - \varepsilon(-\chi^i(\tfrac{x}{\varepsilon})\frac{\partial u}{\partial x_i}(x) + \theta_\varepsilon)\|_{H^1(\Omega)} \leq C\,\varepsilon$$

One can also find results of this type in BENSOUSSAN-LIONS-PAPANICOLAOU [2], BABUSKA [2].

Symmetric context.

We consider here the particuliar case where the coefficients a_{ij} have some properties of symmetry (see section 1.3).
One has the following result (see BOURGAT-DERVIEUX [1]),

PROPOSITION 2 : If the coefficients a_{ij} satisfy the symmetry properties (31) then for the Dirichlet problem (4) one has if $a_{ij}=a_{ji}$

(51) $$\tilde{w}_1 = 0$$

If, in addition, for ε given, Ω is the reunion of entire cells one has

(52) $$\theta_\varepsilon = 0$$

Remark 1 : Assumption $\mathcal{M}(\chi^i) = \frac{1}{|Y|}\int_Y \chi^i(y)dy = 0$ is used in the proof of (51). This justifies the choice made in section 1.1 to determine χ^i.

If $u \in C^{6+\alpha}(\bar{\Omega})$, then using results of proposition 2, we obtain in a symmetric context the following estimations.

(53) $$\|u_\varepsilon - u + \varepsilon\,\chi^i(\tfrac{x}{\varepsilon})\,\frac{\partial u}{\partial x_i}(x)\|_{C^o(\bar{\Omega})} \leq K\,\varepsilon^2$$

(54) $$\|u_\varepsilon - u + \varepsilon\,\chi^i(\tfrac{x}{\varepsilon})\,\frac{\partial u}{\partial x_i}(x)\|_{H^1(\Omega)} \leq C\,\varepsilon$$

(55) $$\|u_\varepsilon - u + \varepsilon\,\chi^i(\tfrac{x}{\varepsilon})\,\frac{\partial u}{\partial x_i}(x) - \varepsilon^2\eta_{ij}(\tfrac{x}{\varepsilon})\,\frac{\partial^2 u}{\partial x_i \partial x_j}(x)\|_{H^1(\Omega)} \leq K\,\varepsilon^{3/2}.$$

One can see that in a symmetric context, only the periodic terms of order 1 and 2 in ε give a better approximation of the solution u_ε. They are respectively called "first order periodic corrector" and "second order periodic corrector".

Remark 2 ; We have shown in BOURGAT DERVIEUX that it is the second order periodic corrector $\eta_{ij}(y)\ \frac{\partial^2 u}{\partial x_i \partial x_j}$, which gives the correction of the peaks of u_ε in the case $\lambda < 1$. J.L. LIONS, using a "stiff expansion" method, obtains in LIONS [2] a similar correction.

3.2. Numerical techniques for the computation of the correctors.

Let us denote by

u_1 the corrected solution obtained with the first order periodic corrector

$$u_1(x) = u(x) - \varepsilon\chi^i(\tfrac{x}{\varepsilon})\ \frac{\partial u}{\partial x_i}\ (x), \tag{56}$$

u_2 the corrected solution obtained with the first and second order periodic correctors,

$$u_2(x) = u(x) - \varepsilon\chi^i(\tfrac{x}{\varepsilon})\ \frac{\partial u}{\partial x_i}\ (x) + \varepsilon^2\eta_{ij}(\tfrac{x}{\varepsilon})\ \frac{\partial^2 u}{\partial x_i \partial x_j}\ (x). \tag{57}$$

We notice that equations (45) and (13) giving respectively η_{ij} and χ^i differ only in their right hand side. So we compute η_{ij} as χ^i. The only difficulty is the discontinuity of coefficients a_{ki} in the term

$$\int_Y - \frac{\partial}{\partial y_k}\ (a_{ki}\chi^j)\psi\ dy$$

which appears in the right hand side of equation (45). Using the Green's formula we obtain :

$$-\int_Y \frac{\partial}{\partial y_k}\ (a_{ki}\chi^j)\psi dy - \int_Y a_{ki}\chi^j\ \frac{\partial\psi}{\partial y_k}\ dy - \int_{\partial Y} a_{ki}\chi^j\psi\ \cos(\vec{n},\vec{y}_k)d\gamma\ .$$

But a_{ki}, χ^j, ψ are Y-periodic functions and $\cos(\vec{n},\vec{y}_k)$ takes opposite values on the opposite sides of Y so

$$-\int_Y \frac{\partial}{\partial y_k}\ (a_{ki}\chi^j)\psi\ dy = \int_Y a_{ki}\chi^j\ \frac{\partial\psi}{\partial y_k}\ dy.$$

For computing the derivatives $\frac{\partial u}{\partial x_i}$ and $\frac{\partial^2 u}{\partial x_i \partial x_j}$, which appear in the calculus of corrected solutions u_1 and u_2, a fine discretization of Ω is needed. We use here 1152 triangles. If the triangles are not locally too different and if the right hand side f is regular we can approximate $\frac{\partial u}{\partial x_i}$ at each node M by

$$\frac{\partial u}{\partial x_i}\ (M) = \frac{1}{N_{V(M)}}\ \sum_{T\in V(M)}\ [\frac{\partial u}{\partial x_i}]_T\ , \tag{58}$$

where

v(M) is the set of triangles of which M is a vertex,

$N_{v(M)}$ is the number of triangles of v(M),

$[\frac{\partial u}{\partial x_i}]_T$ is the value of the derivative $\frac{\partial u}{\partial x_i}$ on the triangle T.

Using formula (58) twice one cas obtain an approximation of $\frac{\partial^2 u}{\partial x_i \partial x_j}$

3.3. First order periodic corrector

Let us consider the case $\lambda > 1$. We take $\lambda = 10$. Figure 16 shows sections of u_ε, u, u_1 along a diagonal of Ω for $\varepsilon = 1/2$.

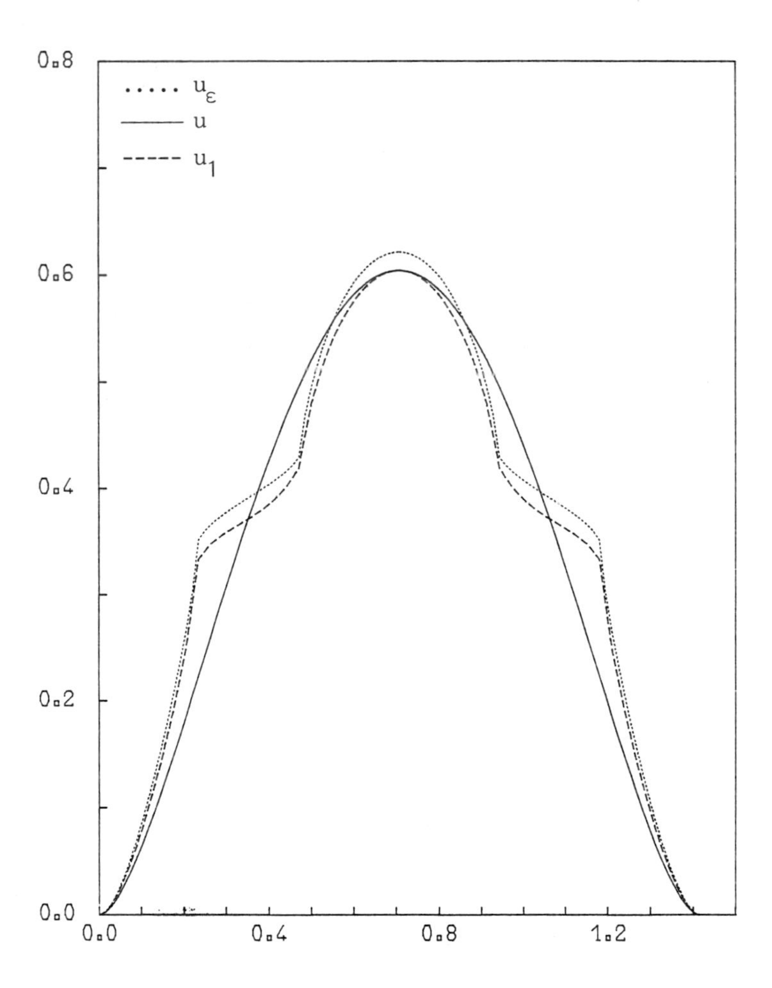

Figure 16 - Diagonal sections of u_ε, u, u_1 for $\lambda=10$, $\varepsilon=1/2$.

One can see in fig. 16 that the first order periodic corrector gives a good approximation of the slopes of u_ε (i.e. of its gradient).
Let us consider now the case $\lambda < 1$. We take $\lambda = 1/114$.
Figure 17 shows diagonal sections of u_ε, u, u_1 for $\varepsilon = 1/4$.

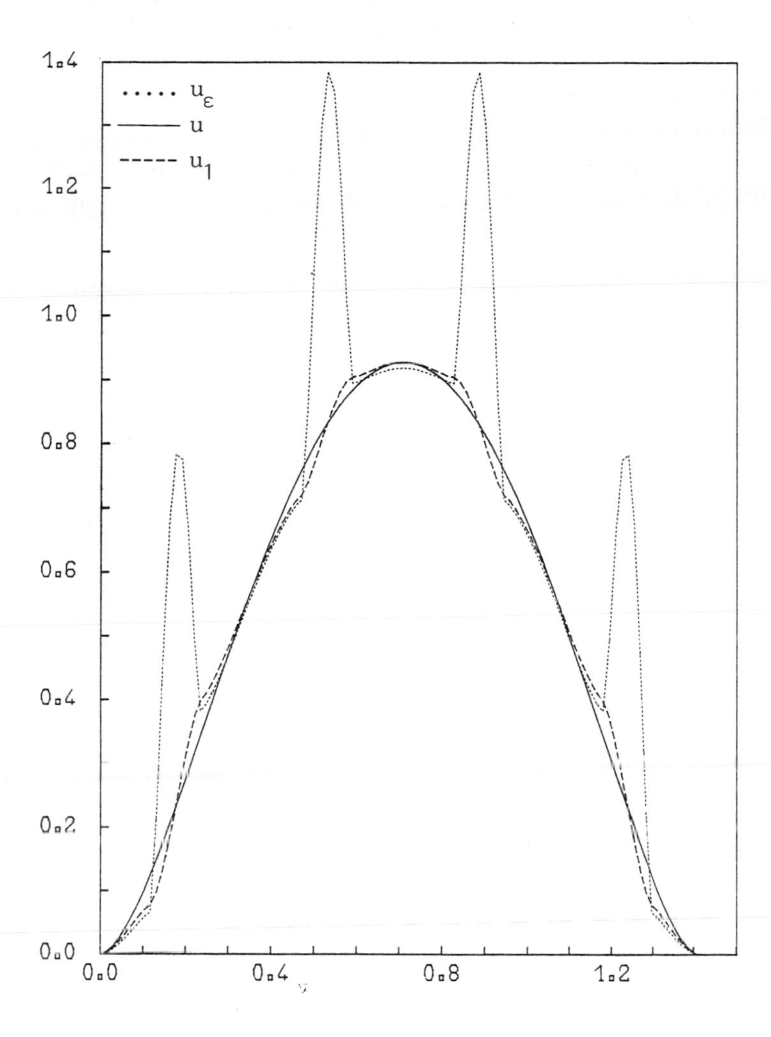

Figure 17 - Diagonal sections of u_ε, u, u_1 for $\lambda = 1/114$, $\varepsilon = 1/4$.

One can see in fig. 17 that the first order periodic corrector does not give any correction of the peaks. More precisely we have indicated in table 4 the error $u_\varepsilon - u_1$ in the norms of spaces $L^\infty(\Omega)$, $L^2(\Omega)$, $H^1(\Omega)$, (compare to table 3).

ε	$\lambda = 10$			$\lambda = 1/114$		
	$\mathrm{Sup}\lvert u_\varepsilon - u_1\rvert$	$\dfrac{\lVert u_\varepsilon - u_1\rVert_{L^2}}{\lVert u_1\rVert_{L^2}}$	$\dfrac{\lVert u_\varepsilon - u_1\rVert_{H^1}}{\lVert u_1\rVert_{H^1}}$	$\mathrm{Sup}\lvert u_\varepsilon - u_1\rvert$	$\dfrac{\lVert u_\varepsilon - u_1\rVert_{L^2}}{\lVert u_1\rVert_{L^2}}$	$\dfrac{\lVert u_\varepsilon - u_1\rVert_{H^1}}{\lVert u_1\rVert_{H^1}}$
1/2	0.020	0.032	0.085	2.24	0.56	0.952
1/4	0.0097	0.012	0.050	0.554	0.183	0.889
1/8	0.0033	0.004	0.033	0.133	0.042	0.707

Table 4 - Convergence of the error $u_\varepsilon - u_1$ for $\lambda = 10$, $\lambda = 1/114$.

3.4. Second order periodic corrector

Figure 18 shows, for $\lambda = 1/114$ and $\varepsilon = 1/4$, diagonal sections of u_ε, u, u_2. If we compare figures 17 and 18 we verify remark 3.2, that is to say that the second order periodic corrector gives the complete correction of the peaks.

We indicate in table 5 the error $u_\varepsilon - u_2$ and one can compare these results to those of table 4.

ε	$\mathrm{Sup}\lvert u - u_2\rvert$	$\dfrac{\lVert u_\varepsilon - u_2\rVert_{L^2}}{\lVert u_2\rVert_{L^2}}$	$\dfrac{\lVert u_\varepsilon - u_2\rVert_{H^1}}{\lVert u_2\rVert_{H^1}}$
1/2	0.181	0.031	0.181
1/4	0.011	0.012	0.012
1/8	0.013	0.012	0.034

Table 5 - Convergence of the error $u_\varepsilon - u_2$ for $\lambda = 1/114$.

We notice in table 5 that the results obtained for $\varepsilon = 1/8$ are worst than those obtained for $\varepsilon = 1/4$. This is due to the error made in the computation of u_ε (see section 2.1). Therefore, if $\lambda < 1$ and ε small, even if the direct computation of u_ε is possible, homogenization gives better results in a symmetric context.

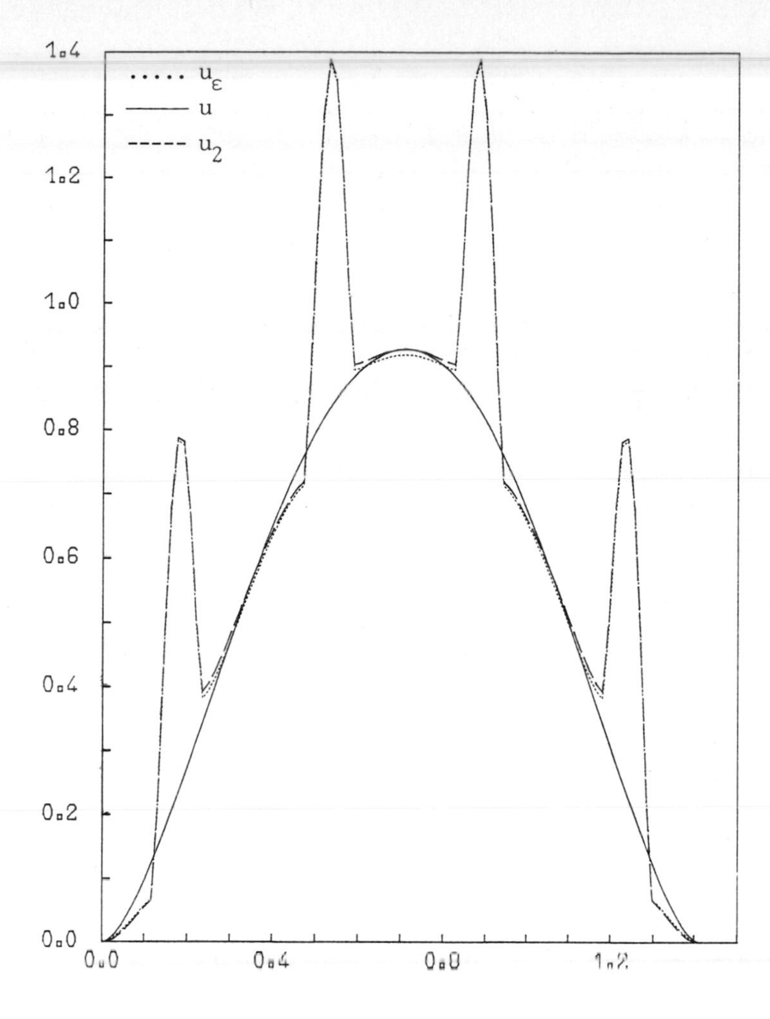

Figure 18 - Diagonal sections of u_ε, u, u_2 for $\lambda = 1/114$, $\varepsilon = 1/4$.

CONCLUSION

It follows from the previous results that the homogenization method is an effective method for solving boundary value problems with periodic coefficients as soon as the number of cells is large. The homogenized solution can be computed easily by solving equations in a representative cell and then by solving an equation with constant coefficients on the whole domain. Furthermore the use of periodic correctors gives, when the coefficients a_{ij} have some properties of symmetry, a good approximation of the periodic behaviour of the solution and this is useful even if the number of cells is not very small.

ACKNOWLEDGEMENTS

I am thankful to A. DERVIEUX and H. LANCHON for their help at various stages of this report.

REFERENCES

I. BABUSKA

[1] "Solution of problems with interfaces and singularities", Tech. Note BN-789 (April 1974), Univ. of Maryland.

[2] "Homogenization and its applications. Mathematical and computational problems". Tech. Note BN-821 (July 1975). Institute for Fluid Dynamics and Applied Mathematics, Univ. fo Maryland.

A. BENSOUSSAN-J.L. LIONS-G. PAPANICOLAOU

[1] "Homogénéisation, correcteurs et problèmes non-linéaires" C.R. Acad. Sc. Paris (1976).

[2] Asympotic analysis of partial derivative equations with highly oscillating coefficients. North Holland- 1978 - Amsterdam.

J.F. BOURGAT-A. DERVIEUX

[1] "Méthode d'homogénéisation des opérateurs à coefficients périodiques. Etude des correcteurs provenant du développement asymptotique". IRIA-LABORIA, Rapport de recherche à paraître.

J.F. BOURGAT-H. LANCHON

[1] "Application of the homogenization method to composite materials with periodic structure". IRIA-LABORIA, Rapport de Recherche N° 208, Dec. 76.

E. DE GIORGI-S. SPAGNOLO

[1] "Sulla convergenza delli integrali dell energia per operatozi ellitici del secondo ordine". Boll. Uni. Math. Ital. 8, 1973, p. 391-411.

S.K. GARG-V. SVATBONAS-G.A. GURTMAN

[1] Analysis of structural composite materials. Marcel Dekker Inc., New-York, 1973.

R. GLOWINSKI-J.L. LIONS-R. TREMOLIERES

[1] Analyse Numérique des Inéquations variationnelles, Paris, Dunod, (1976)

J.L. LIONS

[1] Cours au Collège de France 1975-1976

[2] Remarques sur les aspects numériques de la méthode d'homogénéisation dans les matériaux composites. Colloque IRIA-Novosibirsk, Juin 1976, Dunod (à paraître)

E. SANCHEZ PALENCIA

[1] "Comportement local et macroscopique d'un type de milieux physiques hétérogènes". Int. J. Eng. Sci., 1974, Vol. 12, p. 331-351.

L. TARTAR

[1] Cours Peccot, Collège de France, 1977.

A HOMOGENIZED MULTIGROUP DIFFUSION THEORY FOR THE NEUTRON TRANSPORT EQUATION*

Edward W. Larsen
Theoretical Division
University of California
Los Alamos Scientific Laboratory
Los Alamos, New Mexico 87545/USA

ABSTRACT

We consider the steady multigroup neutron transport equation in a spatially periodic medium, with a spatially periodic source, and vacuum boundary conditions. We require (1) the ratio of a cell diameter to the diameter of the entire medium to be small, and (2) the transport operator to have $N \geq 1$ eigenvalues which are small in magnitude, and simple. Then we show that the transport equation solution is approximated by the solution of an explicit system of N homogenized diffusion equations. We briefly discuss these equations and their properties.

1. INTRODUCTION

The problem of solving the neutron transport equation in a large spatially periodic medium is one which is normally treated by homogenization techniques. This is because for such media, the stated problem is far too complex for direct computational treatment. Therefore various approximate (i.e., homogenized) theories have been developed.

In each of the proposed theories, one performs a cell calculation and uses the results of the cell calculation to determine various homogenized quantities. Sometimes these quantities are homogenized cross sections which are used in transport calculations; more often they are homogenized diffusion coefficients. For this second case (the subject of this paper), there are actually two kinds of approximation which occur: transport theory is simultaneously homogenized *and* replaced by diffusion theory.

Many different prescriptions for homogenized diffusion coefficients have been proposed. Most are defined to exactly preserve certain interaction rates. For example the Benoist[1] coefficients preserve a leakage rate, the Deniz[2] coefficients preserve an asymptotic decay rate as a function of buckling, and the Deniz-Gelbard[3] coefficients preserve a lattice eigenvalue. Somewhat apart from these are the asymptotic diffusion coefficients,[4-8] which do not necessarily preserve any of the above quantities exactly. However, they form an asymptotic solution of the neutron transport equation, and hence they preserve these quantities asymptotically. (The nature of the asymptotic limit is described below in Sec. 2.)

---

*Work performed under the auspices of the U. S. Energy Research and Development Administration.

In each of these theories (at least in their elementary form) the transport equation is approximated by a single diffusion equation. However for a homogeneous medium, where homogenization is unnecessary, it is standard practice to approximate the transport equation by a *system* of coupled diffusion equations.[9] This is because experience has shown that for many problems, a single diffusion equation does not provide a sufficiently accurate approximation.

The construction of systems of homogenized diffusion equations for a spatially periodic medium has been considered previously, the analyses being based - as in the Benoist, Deniz, and Deniz-Gelbard methods mentioned above - on preserving various interaction rates.[9]

However, serious difficulties exist in the use of some of the homogenized diffusion equations which are based on the preservation of interaction rates. For one thing, the theoretical foundations are weak.[9] For another, the diffusion coefficients can become negative, infinite, complex[9] - or even multivalued.[3]

In this paper we construct an asymptotic solution of the multigroup neutron transport equation for a large, spatially periodic medium. Our analysis yields a system of homogenized diffusion equations whose solution is used in the leading term of the asymptotic solution. Our method both extends the earlier asymptotic method, and it improves the earlier results in several respects. We also distinguish between the kinds of problems for which a system of homogenized diffusion equations is necessary to approximate the transport equation, and those for which a single equation will suffice.

2. DEFINITIONS AND RESULTS

Since the multigroup form of the transport equation is the one most often used in reactor analysis, we shall use it here:

$$(\underset{\sim}{\Omega}\cdot\underset{\sim}{\nabla}_x)\underline{\psi}(\underset{\sim}{x},\underset{\sim}{\Omega}) + \underline{\underline{\Sigma}}(\underset{\sim}{x})\cdot\underline{\psi}(\underset{\sim}{x},\underset{\sim}{\Omega}) - \int \underline{\underline{S}}(\underset{\sim}{x},\underset{\sim}{\Omega}\cdot\underset{\sim}{\Omega}')\cdot\underline{\psi}(\underset{\sim}{x},\underset{\sim}{\Omega}')d^2\Omega' = \underline{Q}(\underset{\sim}{x},\underset{\sim}{\Omega}),$$

$$\underset{\sim}{x} \in R, \quad |\underset{\sim}{\Omega}| = 1. \tag{2.1}$$

In (2.1), $\underline{\psi}(\underset{\sim}{x},\underset{\sim}{\Omega})$ is an n x 1 vector whose j-th component is the neutron flux for the j-th energy group at position $\underset{\sim}{x}$ and traveling in the direction ψ; the functions $\underline{\underline{\Sigma}}$ and $\underline{\underline{S}}$ are n x n matrices which are periodic in $\underset{\sim}{x}$ across a cell C; $\underline{Q}$ is an n x 1 vector representing a source, which is periodic in $\underset{\sim}{x}$ across a cell, and R is a region in physical space.

For simplicity, we impose vacuum boundary conditions:

$$\underline{\psi}(\underset{\sim}{x},\underset{\sim}{\Omega}) = \underline{0} \quad \text{for } \underset{\sim}{x} \in \partial R, \quad \underset{\sim}{\Omega} \text{ pointing into R.} \tag{2.2}$$

Then (2.1), (2.2) is a well-posed problem for $\underline{\psi}$.

We denote the operator on the left side of (2.1) by $\underline{\underline{T}}$. Since $\underline{\underline{T}}$ has spatially periodic coefficients, we define a point eigenvalue λ_i and an eigenvector $\underline{\phi}_i(\underset{\sim}{x},\underset{\sim}{\Omega})$ of $\underline{\underline{T}}$ by requiring $\underline{\underline{T}}\cdot\underline{\phi}_i = \lambda_i\underline{\phi}_i$ and $\underline{\phi}_i$ to be periodic in $\underset{\sim}{x}$ across a cell.

We also require the following of $\underline{\underline{T}}$: there exists a neighborhood of the origin in the complex plane ($|\lambda| < \ell$), with order one radius [$\ell = O(1)$], within which the spectrum of $\underline{\underline{T}}$ consists of a finite number of simple point eigenvalues (λ_1, λ_2, ..., λ_m with $|\lambda_i| \le |\lambda_{i+1}|$). (One could easily allow the eigenvalues to have finite multiplicity, but for simplicity we do not consider this here.)

If, in (2.1), $R = \mathbb{R}^3$, then $\underline{\psi}$ can be expressed exactly as

$$\underline{\psi}(\underset{\sim}{x},\underset{\sim}{\Omega}) = \sum_{i=1}^{m} A_i\underline{\phi}_i(\underset{\sim}{x},\underset{\sim}{\Omega}) + \ldots, \tag{2.3}$$

where the dots refer to the contribution to $\underline{\psi}$ from the spectrum of $\underline{\underline{T}}$ in $|\lambda| \ge \ell$, and

$$\lambda_i A_i = Q_i, \tag{2.4}$$

$$Q_i = \iint \underline{\phi}_i^*(\underset{\sim}{x},\underset{\sim}{\Omega})\cdot\underline{Q}(\underset{\sim}{x},\underset{\sim}{\Omega})\,d^2\Omega\, d^3x. \tag{2.5}$$

In (2.5) and the rest of this paper, an integral with respect to $\underset{\sim}{x}$ is always taken over a cell, and we write d^3x as the increment of volume in a cell even if the cell is one or two-dimensional.

Also in (2.5), the 1 x n vector $\underline{\phi}_i^*$ represents the normalized eigenfunction corresponding to the eigenvalue λ_i of the adjoint operator $\underline{\underline{T}}^*$:

$$\underline{\underline{T}}^*\cdot\underline{\phi}^*(\underset{\sim}{x},\underset{\sim}{\Omega}) = -(\underset{\sim}{\Omega}\cdot\underset{\sim}{\nabla}_x)\underline{\phi}^*(\underset{\sim}{x},\underset{\sim}{\Omega}) + \underline{\phi}^*(\underset{\sim}{x},\underset{\sim}{\Omega})\cdot\underline{\underline{\Sigma}}(\underset{\sim}{x}) - \int \underline{\phi}^*(\underset{\sim}{x},\underset{\sim}{\Omega}')\cdot\underline{\underline{S}}(\underset{\sim}{x},\underset{\sim}{\Omega}\cdot\underset{\sim}{\Omega}')\,d^2\Omega', \tag{2.6}$$

$$\iint \underline{\phi}_i^*(\underset{\sim}{x},\underset{\sim}{\Omega})\cdot\underline{\phi}_j(\underset{\sim}{x},\underset{\sim}{\Omega})\,d^2\Omega\, d^3x = \delta_{ij}. \tag{2.7}$$

We note from (2.3), (2.4) that if one or more of the eigenvalues of $\underline{\underline{T}}$ are small in magnitude, then the contribution to the expansion (2.3) from just these eigenvalues dominates, and the terms denoted by "+ ..." in (2.3) can be neglected. For this case the infinite medium problem described above is "almost critical," and an asymptotic solution exists for the "real" problem (2.1), (2.2) which generalizes the infinite medium solution, and which is determined by the solution of a system of homogenized diffusion equations.

This asymptotic solution has the form

$$\underline{\psi}(\underset{\sim}{x},\underset{\sim}{\Omega}) = \sum_{i=1}^{m} A_i(\underset{\sim}{x})\underline{\phi}_i(\underset{\sim}{x},\underset{\sim}{\Omega}) + O(\varepsilon). \tag{2.8}$$

Here $A_i(\underset{\sim}{x})$, $1 \le i \le m$, are slowly varying functions of $\underset{\sim}{x}$ $[\nabla_{\underset{\sim}{x}} A_i = O(\varepsilon)]$ which are determined by

$$\sum_{j=1}^{m} [-\nabla_{\underset{\sim}{x}} \cdot \underset{\approx}{D}_{ij} \cdot \nabla_{\underset{\sim}{x}} + \underset{\sim}{\nu}_{ij} \cdot \nabla_{\underset{\sim}{x}}] A_j(\underset{\sim}{x}) + \lambda_i A_i(\underset{\sim}{x}) = Q_i, \tag{2.9}$$

$$A_i(\underset{\sim}{x}) = 0, \qquad \underset{\sim}{x} \varepsilon \partial R, \tag{2.10}$$

with the homogenized coefficients $\underset{\approx}{D}_{ij}$ and $\underset{\sim}{\nu}_{ij}$ defined by:

$$\underset{\sim}{\nu}_{ij} = \iint [\underset{\sim}{\phi}_i^*(\underset{\sim}{x},\underset{\sim}{\Omega}) \cdot \underset{\sim}{\phi}_j(\underset{\sim}{x},\underset{\sim}{\Omega})] \underset{\sim}{\Omega} \, d^2\Omega d^3x, \tag{2.11}$$

$$\underset{\approx}{D}_{ij} = \iint \underset{\sim}{\Omega} \left\{ \underset{\sim}{\phi}_i^*(\underset{\sim}{x},\underset{\sim}{\Omega}) \cdot \underset{\equiv}{T}^{-1} \cdot \left[\underset{\sim}{\phi}_j(\underset{\sim}{x},\underset{\sim}{\Omega}) \underset{\sim}{\Omega} - \sum_{k=1}^{m} \underset{\sim}{\phi}_k(\underset{\sim}{x},\underset{\sim}{\Omega}) \underset{\sim}{\nu}_{kj} \right] \right\} d^2\Omega d^3x. \tag{2.12}$$

Also,

$$\varepsilon = \frac{\text{typical diameter of C}}{\text{typical diameter of R}} . \tag{2.13}$$

The boundary conditions (2.10) have been chosen so that the asymptotic solution (2.8) satisfies the condition (2.1) to leading order. For the remainder of this paper, we shall derive and discuss the system of homogenized diffusion equations (2.9), (2.11), (2.12).

3. ASYMPTOTIC ANALYSIS

To derive the asymptotic solution, we require the dimensionless parameter ε defined in (2.13) to be small. In other words, we require a typical diameter of R to be large compared to a typical diameter of a cell. Then, away from ∂R, the solution of (2.1), (2.2) will vary almost periodically in the spatial variable and will locally have the form (2.3), (2.4). Our ansatz is therefore:

$$\underset{\sim}{\psi}(\underset{\sim}{x},\underset{\sim}{\Omega}) \simeq \sum_{i=1}^{m} a_i(\underset{\sim}{r}) \underset{\sim}{\phi}_i(\underset{\sim}{x},\underset{\sim}{\Omega}) + \sum_{j=1}^{\infty} \varepsilon^j \underset{\sim}{\psi}_j(\underset{\sim}{r},\underset{\sim}{x},\underset{\sim}{\Omega}), \tag{3.1}$$

$$\underset{\sim}{r} = \varepsilon \underset{\sim}{x}. \tag{3.2}$$

We shall require each $\underset{\sim}{\psi}_j$, $j \ge 1$, to be periodic in $\underset{\sim}{x}$ across a cell and to satisfy

$$\iint \underset{\sim}{\phi}_i^*(\underset{\sim}{x},\underset{\sim}{\Omega}) \cdot \underset{\sim}{\psi}_j(\underset{\sim}{r},\underset{\sim}{x},\underset{\sim}{\Omega}) d^2\Omega d^3x = 0, \qquad 1 \le i \le m. \tag{3.3}$$

This requirement has the physical virtue of keeping the most important parts of $\underline{\psi}$ - the contributions from the smallest eigenvalues - in the leading order term. There is however a more important reason for imposing (3.3), which will shortly become apparent.

We also require one or more of the eigenvalues of $\underline{\underline{T}}$ to be small and we neglect the projection of $\underline{Q}$ onto the eigenspaces of $\underline{\underline{T}}$ corresponding to $|\lambda| \geq \ell$, for the reasons outlined in Sec. 2. Therefore for the remainder of this paper, we take

$$\underline{Q}(\underset{\sim}{x},\underset{\sim}{\Omega}) = \sum_{i=1}^{m} Q_i \underline{\phi}_i(\underset{\sim}{x},\underset{\sim}{\Omega}). \tag{3.4}$$

Now we combine (2.1) and (3.1)-(3.4), to obtain

$$\sum_{i=1}^{m} [\lambda_i a_i(\underset{\sim}{r}) + \varepsilon \underset{\sim}{\Omega} \cdot \underset{\sim}{\nabla}_r a_i(\underset{\sim}{r})] \underline{\phi}_i(\underset{\sim}{x},\underset{\sim}{\Omega}) + \varepsilon \underline{\underline{T}} \cdot \underline{\psi}_1(\underset{\sim}{r},\underset{\sim}{x},\underset{\sim}{\Omega}) + \varepsilon^2 [\underset{\sim}{\Omega} \cdot \underset{\sim}{\nabla}_r \underline{\psi}_1(\underset{\sim}{r},\underset{\sim}{x},\underset{\sim}{\Omega}) + \underline{\underline{T}} \cdot \underset{\sim}{\psi}_2(\underset{\sim}{r},\underset{\sim}{x},\underset{\sim}{\Omega})]$$

$$= \sum_{i=1}^{m} Q_i \underline{\phi}_i(\underset{\sim}{x},\underset{\sim}{\Omega}) + O(\varepsilon^3). \tag{3.5}$$

Multiplying (3.5) on the left by $\underline{\phi}_j^*$, integrating Ω over the unit sphere and $\underset{\sim}{x}$ over a cell, and imposing (2.7) and (3.3) yields

$$\lambda_j a_j(\underset{\sim}{r}) + \varepsilon \sum_{i=1}^{m} (\underset{\sim}{\nu}_{ji} \cdot \underset{\sim}{\nabla}_r) a_i(\underset{\sim}{r}) + \varepsilon^2 \iint \underline{\phi}_j^*(\underset{\sim}{x},\underset{\sim}{\Omega}) \cdot [(\underset{\sim}{\Omega} \cdot \underset{\sim}{\nabla}_r) \underline{\psi}_1(\underset{\sim}{r},\underset{\sim}{x},\underset{\sim}{\Omega})] d^2\Omega d^3x = Q_j + O(\varepsilon^3), \tag{3.6}$$

where ν_{ij} is defined by (2.11). Thus, to obtain a closed system of equations for $a_i(\underset{\sim}{r})$ with $O(\varepsilon^3)$ error, we must determine $\underline{\psi}_1$ with at most $O(\varepsilon)$ error.

To do this we multiply (3.6) by $\underline{\phi}_j$, sum over j, and subtract the resulting equation from (3.5). After rearranging, we get

$$\underline{\underline{T}} \cdot \underline{\psi}_1(\underset{\sim}{r},\underset{\sim}{x},\underset{\sim}{\Omega}) = - \sum_{i=1}^{m} \left[\underline{\phi}_i(\underset{\sim}{x},\underset{\sim}{\Omega}) \underset{\sim}{\Omega} - \sum_{j=1}^{m} \underline{\phi}_j(\underset{\sim}{x},\underset{\sim}{\Omega}) \underset{\sim}{\nu}_{ji} \right] \cdot \underset{\sim}{\nabla}_r a_i(r) + O(\underset{\sim}{\varepsilon}). \tag{3.7}$$

By definition of the $\underset{\sim}{\nu}_{ij}$, $\underline{\psi}_1$ can be constructed satisfying (3.3) with j=1. (This is true even if one of the eigenvalues of $\underline{\underline{T}}$ is zero.) Also, since the projection of the right side of (3.7) onto the eigenspaces of $\underline{\underline{T}}$ corresponding to the spectrum in $|\lambda| < \ell = O(1)$ is zero, $\underline{\underline{T}}^{-1}$ acting on this function will be $O(1)$. (This is the second reason for imposing (3.3); it leads to the result that $\underline{\psi}_1$ is $O(1)$, regardless of the number of small eigenvalues of $\underline{\underline{T}}$.)

Solving (3.7) for ψ_1 and substituting into (3.6) gives

$$\lambda_j a_j(\underset{\sim}{r}) + \sum_{i=1}^{m} [\varepsilon \underset{\sim}{\nu}_{ji} \cdot \underset{\sim}{\nabla}_r - \varepsilon^2 \underset{\sim}{\nabla}_r \cdot \underset{\approx}{D}_{ji} \cdot \underset{\sim}{\nabla}_r] a_i(\underset{\sim}{r}) = Q_j + O(\varepsilon^3) \tag{3.8}$$

where D_{ij} is defined by (2.12).

The final step is to express the results in terms of the original spatial variable $\underset{\sim}{x}$. To do this, we define $A_j(\underset{\sim}{x}) = a_j(\varepsilon \underset{\sim}{x})$. Then by (3.2), $\underset{\sim}{\nabla}_x = \varepsilon \underset{\sim}{\nabla}_r$, and so Eqs. (3.1), (3.2) and (3.8) reduce to (2.8) and (2.9).

4. DISCUSSION

If none of the eigenvalues of $\underline{\underline{T}}$ are small in magnitude, than the assumption (3.4) is not justified; in spite of this, (3.8) gives

$$A_j(\underset{\sim}{x}) = a_j(\varepsilon \underset{\sim}{x}) \simeq \frac{Q_j}{\lambda_j}, \tag{4.1}$$

which is the infinite medium solution, correct a few mean-free paths from ∂R.

However, if one or more of the eigenvalues λ_j of $\underline{\underline{T}}$ are small, i.e., comparable in size to $\varepsilon|\underset{\sim}{\nu}_{ji}|$ or $\varepsilon^2|\underset{\approx}{D}_{ji}|$ for some i, then for these j the derivative terms in (3.8) become comparable in size to the undifferentiated terms. Thus the full diffusion equation (3.8) must be retained, and the corresponding solutions a_i are large (of order Q_i/λ_i) in the central part of R.

The solutions a_i for "large" λ_i are still approximately given by (4.1), and derivatives of these a_i are approximately zero. Thus these a_i basically uncouple from the rest of the equations (3.8).

Therefore, only those a_i which correspond to small λ_i are significantly coupled by (3.8), since only these will have O(1) derivatives away from ∂R.

This shows that if one or more of the λ_i are small, then increasing ℓ (hence including more eigenfunctions - which will correspond to eigenvalues greater in magnitude than ℓ) will not substantially affect the asymptotic solution.

The previous asymptotic analyses which led to a homogenized diffusion equation required one and only one eigenvalue of $\underline{\underline{T}}$ to be small.[4-8] For this case, the system (3.8) - or (2.9) - reduces to a single equation which coincides asymptotically with the earlier results. However, this situation does not typically occur in practice, especially for problems with several energy groups which are weakly coupled.

In addition, the previous asymptotic analyses required $\underline{\underline{T}}$ to be split into two parts and written as $\underline{\underline{T}} = \underline{\underline{T}}_o + \varepsilon^2 \underline{\underline{T}}_2$, where $\underline{\underline{T}}_o$ has zero as a simple point eigenvalue. In general, there is no unique or "best" way to express $\underline{\underline{T}}$ in this manner, and different choices of $\underline{\underline{T}}_o$ and $\underline{\underline{T}}_2$ will give slightly different asymptotic results. The present analysis avoids this ambiguity because it does not require $\underline{\underline{T}}$ to either be split or rewritten in terms of ε.

Finally, we mention that the determination of the homogenized diffusion coefficients $\underset{\sim}{\nu}_{ij}$ and $\underset{\approx}{D}_{ij}$ (in that order) does not involve solving a system of equations. It is only *after* these coefficients are evaluated that one must contend with the system of equations (2.9), for the functions $A_j(\underset{\sim}{x})$ in (2.8).

REFERENCES

1. P. Benoist, "Theorie du Coefficient de Diffusion dans un Reseau Comportant des Cavites," CEA-R-2278 Centre d'Etudes Nucleaires - Saclay (1964).

2. V. Deniz, "Study of the Kinetics of Thermalized Neutron Populations in Multiplying or Nonmultiplying Heterogeneous Media," Nucl. Sci. Eng. 28, 397 (1967).

3. E. M. Gelbard, "Anisotropic Neutron Diffusion in Lattices of Zero-Power Plutonium Reactor Experiments," Nucl. Sci. Eng. 54, 327 (1974).

4. E. W. Larsen, "Neutron Transport and Diffusion in Inhomogeneous Media. I," J. Math. Phys. 16, 1421 (1975).

5. E. W. Larsen, "Neutron Transport and Diffusion in Inhomogeneous Media. II," Nucl. Sci. Eng. 60, 357 (1976).

6. M. Williams, "Homogenization of Linear Transport Problems," Thesis Dissertation, New York University (1976).

7. A Bensoussan, J. L. Lions, and G. C. Papanicolaou, "Boundary Layers and Homogenization of Transport Processes," lecture notes available from the Dept. of Mathematics, University of Utah, Salt Lake City, Utah (1976).

8. E. W. Larsen and M. Williams, "Neutron Drift in Heterogeneous Media," Nucl. Sci. Eng., to appear.

9. A. F. Henry, Nuclear Reactor Analysis, MIT Press, Cambridge, Mass. (1975).

ESTIMATION DE COEFFICIENTS HOMOGENEISES

L. TARTAR

Analyse Numérique et Fonctionnelle
C.N.R.S. et Université Paris-Sud
Bâtiment 425, 91405 Orsay

I - Introduction

L'homogénéisation permet de trouver les équations satisfaites par des grandeurs macroscopiques à partir des équations vérifiées par les grandeurs physiques et de la composition microscopique (ou des informations qu'on possède sur la structure microscopique).

Le modèle mathématique consiste à étudier la solution d'une équation (ou d'un système d'équations) aux dérivées partielles dont les coefficients ont un comportement oscillatoire marqué ; en général la solution sera très proche, pour une topologie faible, de la solution d'une équation homogénéisée (pouvant être de nature très différente de l'équation originale) qui a des coefficients assez réguliers appelés coefficients homogénéisés ou effectifs.

Quand on a une information complète sur la structure microscopique (si les coefficients sont périodiques de période très petite par exemple) on peut trouver des formules explicites (nécessitant des calculs plus ou moins classiques) donnant la valeur des coefficients effectifs ; dans le cas général l'existence d'une formule explicite est conjecturée.

Quand on ne possède sur la structure microscopique que des renseignements partiels (d'ordre statistique par exemple) on ne sait démontrer que quelques inégalités satisfaites par les coefficients homogénéisés ; c'est à des résultats de ce type qu'on va s'intéresser ici.

II - Le problème modèle

Considérons la solution d'une équation du type

$$- \sum_{ij} \frac{\partial}{\partial x_i} \left(a^{\varepsilon}_{ij}(x) \frac{\partial u^{\varepsilon}}{\partial x_j}\right) = f^{\varepsilon} \quad \text{dans} \quad \Omega \subset \mathbb{R}^N \tag{1}$$

où les coefficients a^ε_{ij} (qui sont des fonctions mesurables) vérifient la condition d'ellipticité uniforme

$$
(2) \qquad \begin{cases} \exists\ \alpha>0 \quad \forall \lambda \in \mathbf{R}^N & \sum_{ij} a^\varepsilon_{ij}(x)\ \lambda_i\ \lambda_j \geqslant \alpha|\lambda|^2 \quad \text{presque partout} \\ \exists\ M>0 & |a^\varepsilon_{ij}(x)| \leqslant M \quad \text{presque partout.} \end{cases}
$$

Si f^ε converge vers f^0 et u^ε vers u^0 (dans des topologies qu'on précisera plus loin) alors u_0 satisfait une équation du type

$$
(3) \qquad - \sum_{ij} \frac{\partial}{\partial x_i} \left(q_{ij}(x) \frac{\partial u^0}{\partial x_j}\right) = f^0
$$

où les coefficients q_{ij} sont appelés coefficients homogénéisés.

[Dans le cas où les coefficients a^ε_{ij} sont de la forme $a_{ij}(\frac{x}{\varepsilon})$, avec a_{ij} périodique, on peut montrer qu'il existe un développement

$$
(4) \qquad u^\varepsilon(x) = u^0(x) + \varepsilon\, u^1(x,\tfrac{x}{\varepsilon}) + \ldots] \ .
$$

On définit ainsi une convergence d'un type nouveau, baptisée H-convergence (ou convergence au sens de l'homogénéisation)et notée $a^\varepsilon_{ij} \xrightarrow{H} q_{ij}$, caractérisée par la

Propriété caractéristique : Soit $a^\varepsilon_{ij} \xrightarrow{H} q_{ij}$. Si u^ε vérifie

$$
(5) \qquad \frac{\partial u^\varepsilon}{\partial x_j} \longrightarrow \frac{\partial u^0}{\partial x_j} \quad \text{dans } L^2(\Omega) \text{ faible pour } j = 1, \ldots , N
$$

$$
(6) \qquad - \sum_{ij} \frac{\partial}{\partial x_i} \left(a^\varepsilon_{ij} \frac{\partial u^\varepsilon}{\partial x_j}\right) \longrightarrow f^0 \quad \text{dans } H^{-1}(\Omega) \text{ fort}
$$

alors on a

$$
(7) \qquad \sum_j a^\varepsilon_{ij} \frac{\partial u^\varepsilon}{\partial x_j} \longrightarrow \sum_j q_{ij} \frac{\partial u^0}{\partial x_j} \quad \text{dans } L^2(\Omega) \text{ faible pour } i = 1, \ldots , N .
$$

∎

Remarque 1 : En particulier u^0 vérifie (3).

∎

Le résultat fondamental de la théorie est le théorème de compacité suivant :

Théorème 1. *Soit a^ε_{ij} une suite de fonctions vérifiant (2) ; il existe une sous suite H-convergeant vers q_{ij}, où les q_{ij} vérifient (2) avec le même α (mais en général avec une constante M' différente de M) .*

∎

Dans deux cas importants (admettant quelques variantes) on sait expliciter les q_{ij} :

<u>Théorème 2</u>. *Si les coefficients* a^{ε}_{ij} *ont la forme* $a_{ij}(\frac{x}{\varepsilon})$ *avec* a_{ij} *périodique, toute la suite H-converge (quand* ε *tend vers 0) vers des* q_{ij} *constants déterminés par la résolution de* N *équations aux dérivées partielles de la manière suivante :*

i) pour $k = 1, \dots, N$ *on calcule la solution* w_k *(déterminée à une constante additive près) de l'équation*

$$\begin{cases} - \sum_{ij} \frac{\partial}{\partial y_i} (a_{ij}(y) \frac{\partial w_k}{\partial y_j}) = 0 \\ w_k - y_k \text{ périodique} \end{cases}$$

ii) on calcule q_{ik} = *moyenne sur une période de* $\sum_j a_{ij}(y) \frac{\partial w_k}{\partial y_j}$

■

<u>Théorème 3</u>. *Si les* a^{ε}_{ij} *sont des fonctions de* x_1 *seulement, alors* $a^{\varepsilon}_{ij} \xrightarrow{H} q_{ij}$ *(les* q_{ij} *sont aussi des fonctions de* x_1 *seulement) équivaut à :*

i) $\frac{1}{a^{\varepsilon}_{11}} \longrightarrow \frac{1}{q_{11}}$ *dans* L^{∞} *faible* $*$

ii) Pour $i \neq 1$ $\begin{cases} \frac{a^{\varepsilon}_{i1}}{a^{\varepsilon}_{11}} \longrightarrow \frac{q_{i1}}{q_{11}} \text{ dans } L^{\infty} \text{ faible } * \\ \frac{a^{\varepsilon}_{1i}}{a^{\varepsilon}_{11}} \longrightarrow \frac{q_{1i}}{q_{11}} \text{ dans } L^{\infty} \text{ faible } * \end{cases}$

iii) Pour $i \neq 1$, $j \neq 1$ $a^{\varepsilon}_{ij} - \frac{a^{\varepsilon}_{i1}\, a^{\varepsilon}_{1j}}{a^{\varepsilon}_{11}} \longrightarrow q_{ij} - \frac{q_{i1}\, q_{1j}}{q_{11}}$ *dans* L^{∞} *faible* $*$

■

<u>Remarques 2</u> : Si $a^{\varepsilon}_{ij} = a^{\varepsilon}_{ji}$ $\forall\, i,j$, alors $q_{ij} = q_{ji}$ $\forall\, i,j$.
Dans ce cas si $\exists\, \beta>0, \forall \lambda, |\sum a^{\varepsilon}_{ij}(x)\lambda_i\lambda_j| \leqslant \beta|\lambda|^2$ presque partout, alors q_{ij} vérifie cette inégalité pour la même valeur de β .

Dans ce cas la H-convergence coïncide avec la G-convergence (voir [1]) .

3 : La H-convergence est locale . les q_{ij} sont déterminés de manière unique (à un ensemble de mesure nulle près) et la connaissance des a^{ε}_{ij} sur un ouvert ω (ou même un ensemble de mesure positive) suffit à caractériser les q_{ij} sur ω .

Cette convergence est cependant différente des convergences faibles usuelles : on peut fabriquer aisément deux suites a^{ε}_{ij} , b^{ε}_{ij} telles que $F(a^{\varepsilon}_{ij})$ et $F(b^{\varepsilon}_{ij})$ aient même limite faible pour toute fonction continue F mais H-convergent vers des limites différentes.

4 : On peut montrer que (5), (6), (7) entraînent

$$\sum_{ij} a^{\varepsilon}_{ij} \frac{\partial u^{\varepsilon}}{\partial x_j} \frac{\partial u^{\varepsilon}}{\partial x_i} \longrightarrow \sum_{ij} q_{ij} \frac{\partial u^{o}}{\partial x_j} \frac{\partial u^{o}}{\partial x_i} \text{ dans } L^1_{loc} \text{ faible .} \tag{8}$$

5 : Si dans (1) u_{ε} est le potentiel électrostatique alors $E_{\varepsilon} = -\text{grad } u_{\varepsilon}$ est le champ électrique et $D_{\varepsilon} = -A_{\varepsilon} \text{ grad } u_{\varepsilon}$ est le champ d'induction électrique ; u_o désigne le potentiel moyen (macroscopique), E_o , D_o les champs moyens qui sont alors reliés par $D_o = qE_o$ où q est le tenseur de conductivité effectif (qui n'est pas une moyenne de a_{ε}). La densité d'énergie éléctrostatique est $e_{\varepsilon} = (E_{\varepsilon}, D_{\varepsilon})$ et (8) exprime que la densité moyenne est $e_o = (E_o, D_o)$.

C'est là un fait remarquable car pour une fonction G la valeur moyenne de $G(E_{\varepsilon}, D_{\varepsilon})$ n'est pas en général $G(E_o, D_o)$.

6 : Les méthodes utilisées sont de type variationnel et n'utilisent aucune propriété spéciale à l'ordre deux .

∎

III - Estimations

On va maintenant se restreindre au cas des coefficients symétriques.

La H-convergence a des propriétés intéressantes par rapport à l'ordre (on note $M \leq N$ pour deux matrices si $(M\lambda, \lambda) \leq (N\lambda, \lambda)$ $\forall \lambda \in \mathbb{R}^N$). On note A_{ε} la matrice de coefficients a^{ε}_{ij} , Q celle de coefficients q_{ij} .

Théorème 4. *Si* $A_{\varepsilon} \leq B_{\varepsilon}$ *(avec* $A^*_{\varepsilon} = A_{\varepsilon}$*) avec* $A_{\varepsilon} \xrightarrow{H} Q$ *et* $B_{\varepsilon} \xrightarrow{H} R$ *alors* $Q \leq R$.

∎

Théorème 5. *Soit* $A^*_{\varepsilon} = A_{\varepsilon}$ *avec* $A_{\varepsilon} \xrightarrow{H} Q$; *si* $A_{\varepsilon} \longrightarrow A_o$ *dans* $(L^{\infty})^{N^2}$ *faible* * *et* $A^{-1}_{\varepsilon} \longrightarrow B^{-1}_o$ *dans* $(L^{\infty})^{N^2}$ *faible* * *alors* $B_o \leq Q \leq A_o$. ∎

Remarques 7 : Dans le cas des formules explicites des théorèmes 2 et 3 ces inégalités ne se démontrent pas plus aisément que dans le cas général.

<u>8</u> : En utilisant les formules explicites on peut trouver des cas où $Q - B_o$ et $A_o - Q$ ont tous deux une valeur propre nulle.

<u>9</u> : Avec les interprétations de la remarque <u>5</u> le Théorème 5 s'écrit :

conductivité effective $\leqslant$ *conductivité moyenne*

résistivité effective $\leqslant$ *résistivité moyenne*

<u>10</u> : Appliquons le Théorème 5 au cas $a^{\varepsilon}_{ij} = \delta_{ij}\, a(\frac{x}{\varepsilon}, \frac{y}{\varepsilon})$ où a a la période 1 en x et y et est défini sur le carré unité par

$$a(x,y) = \begin{Bmatrix} \alpha & \text{si} \quad 1/3 \leqslant x,y \leqslant 2/3 \\ 1 & \text{sinon} \end{Bmatrix}.$$

Alors $q_{ij} = q\, \delta_{ij}$ (pour des raisons évidentes de symétrie) et vérifie

$$\frac{9\alpha}{8\alpha+1} \leqslant q \leqslant \frac{8+\alpha}{9} \;.$$

Comme on le voit cette formule n'est pas très précise quand $\alpha \longrightarrow 0$ ou ∞

Si on considère la fonction $b(x,y) = \begin{Bmatrix} \alpha & \text{si} \quad 1/3 \leqslant x \leqslant 2/3 \\ 1 & \text{sinon} \end{Bmatrix}$ (et qu'on applique le Théorème 3 pour calculer la H-limite de $b(\frac{x}{\varepsilon}, \frac{y}{\varepsilon})$) , on trouve grâce au Théorème 4 d'autres inégalités sur q :

pour $\alpha \leqslant 1$ $\quad \frac{2+\alpha}{3} \leqslant q$ ce qui améliore pour $0 \leqslant \alpha \leqslant 1/4$

pour $\alpha \geqslant 1$ $\quad q \leqslant \frac{3\alpha}{2\alpha+1}$ ce qui améliore pour $4 \leqslant \alpha$.

Ces nouvelles inégalités sont beaucoup plus précises pour $\alpha \longrightarrow 0$ ou ∞ .

∎

Les inégalités précédentes utilisent la limite faible de A_ε (ou A_ε^{-1}) qui se calcule en intégrant A_ε sur des ensembles de mesure positive puis en prenant la limite. On va obtenir des inégalités plus précises en intégrant d'abord sur des ensembles de dimension $N-1$ ou 1 . Pour chaque direction d'hyperplan on va obtenir deux inégalités ; si on choisit la direction $x_1 = 0$ on devra utiliser la convergence suivante :

<u>Notation</u> : On écrira $M_\varepsilon \xrightarrow{*1} M$ pour dire que $\frac{1}{m^{\varepsilon}_{11}} \rightharpoonup \frac{1}{m_{11}}$, $\frac{m^{\varepsilon}_{i1}}{m^{\varepsilon}_{11}} \rightharpoonup \frac{m_{i1}}{m_{11}}$,

$$\frac{m^{\varepsilon}_{1j}}{m^{\varepsilon}_{11}} \rightharpoonup \frac{m_{1j}}{m_{11}} \;,\; m^{\varepsilon}_{ij} - \frac{m^{\varepsilon}_{i1}\, m^{\varepsilon}_{1j}}{m^{\varepsilon}_{11}} \rightharpoonup m_{ij} - \frac{m_{i1}\, m_{1j}}{m_{11}} \text{ dans } L^\infty \text{ faible } * .$$

Théorème 6. *Soit* $\omega(x_2,\dots,x_N)$ *une fonction positive d'intégrale 1 et soit* ρ *la mesure* $\delta_{x_1} \otimes \omega$.

Si on a $A_\varepsilon * \rho \xrightarrow{*1} B_o$, *alors* $Q * \rho \leqslant B_o$. ∎

Théorème 7. *Soit* $\omega(x_1)$ *une fonction positive d'intégrale 1 et soit* ρ *la mesure* $\omega \otimes \delta_{x_2 = \dots = x_N = 0}$.

Si on a $A_\varepsilon^{-1} * \rho \xrightarrow{*1} C_o^{-1}$, *alors* $Q * \rho \geqslant C_o$. ∎

Remarques 11 : L'algorithme du Théorème 6 consiste à faire une moyenne arithmétique en $x_2,\dots,x_N$ puis à effectuer l'analogue d'une moyenne harmonique en x_1 (dans le cas $A_\varepsilon = a_\varepsilon I$) ; l'algorithme du Théorème 7 commence par une moyenne harmonique en x_1 .

12 : La convergence $*1$, qui est déjà apparue au Théorème 3, est en fait très naturelle. L'objet mathématique A est en fait une application qui transforme une 1-forme différentielle $E = \sum_i E_i \, dx_i$ en une (N-1)-forme $D = \sum_i D_i \, dx_1 \dots \widehat{dx_i} \dots dx_N$; le produit $E \wedge D = \sum_i E_i D_i \, dx$, quand on l'écrit en utilisant $D_1, E_2, \dots, E_N$ (qui sont les seules composantes ayant un sens sur l'hyperplan $x_1 = 0$) fait apparaître les quantités $\frac{1}{a_{11}}$, $\frac{a_{i1}}{a_{11}}$ et $a_{ij} - \frac{a_{i1}\,a_{ij}}{a_{11}}$.

13 : Si on applique les Théorèmes 6 et 7 à l'exemple de la Remarque 10 on obtient les inégalités plus précises :

$$\frac{2 + 7\alpha}{3(2\alpha+1)} \leqslant q \leqslant \frac{3(2+\alpha)}{7 + 2\alpha} .$$

Quand $\alpha \to 0$ (ou ∞) on peut voir (sur la formule du Théorème 2 car on peut alors exhiber les solutions) que q tend vers $\frac{2}{3}$ (ou $\frac{3}{2}$) . ∎

Démonstrations : a) Pour chaque $\lambda \in \mathbb{R}^N$ on sait fabriquer une suite de fonctions (vectorielles) v_ε telle que $v_\varepsilon \rightharpoonup \lambda$ dans $L^2(\Omega)^N$ faible, rot $v_\varepsilon = 0$, div$(A_\varepsilon v_\varepsilon)$ converge fortement dans $H^{-1}(\Omega)$ et donc $A_\varepsilon v_\varepsilon \rightharpoonup Q\lambda$ dans $L^2(\Omega)^N$ faible et $(A_\varepsilon v_\varepsilon, v_\varepsilon) \rightharpoonup (Q\lambda,\lambda)$ dans $L^1_{loc}(\Omega)$ faible. [C'est l'essentiel de l'information donnée par la définition de la H-convergence et le Théorème 1 (que l'on admettra ici) ; en gros il suffit de résoudre div$(A_\varepsilon$ grad $u_\varepsilon - Q\lambda) = 0$ avec $u_\varepsilon - \lambda.x \in H^1_0(\Omega)$ et de poser v_ε = grad u_ε].

b) (Théorème 4) Soit v_ε une suite comme ci-dessus associée à

A_ε et w_ε associée à B_ε, toutes deux pour le vecteur λ. On passe à la limite dans l'inégalité $(A_\varepsilon(v_\varepsilon-w_\varepsilon),(v_\varepsilon-w_\varepsilon)) \geqslant 0$: $(A_\varepsilon v_\varepsilon,v_\varepsilon)$ et $(A_\varepsilon v_\varepsilon,w_\varepsilon)$ convergent vers $(Q\lambda,\lambda)$ dans $L^1_{loc}(\Omega)$ faible ; $(A_\varepsilon w_\varepsilon,w_\varepsilon) \leqslant (B_\varepsilon w_\varepsilon,w_\varepsilon)$ qui converge vers $(R\lambda,\lambda)$ dans $L^1_{loc}(\Omega)$ faible. Ceci donne $(R\lambda,\lambda) \geqslant (Q\lambda,\lambda)$.

c) (Théorème 5) Soit v_ε associée à A_ε pour le vecteur λ. On passe à la limite dans $(A_\varepsilon(v_\varepsilon+\mu),(v_\varepsilon+\mu)) \geqslant 0$ ce qui donne

$$(Q\lambda,\lambda) + 2(Q\lambda,\mu) + (A_o\mu,\mu) \geqslant 0 \text{ d'où } A_o \geqslant Q .$$

Puis on passe à la limite dans $(A_\varepsilon^{-1}(A_\varepsilon v_\varepsilon+\mu)\ ,\ (A_\varepsilon v_\varepsilon+\mu)) \geqslant 0$ ce qui donne $(Q\lambda,\lambda) + 2(\lambda,\mu) + (B_o^{-1}\mu,\mu) \geqslant 0$ d'où $B_o \leqslant Q$.

d) (Théorème 6) On utilise l'inégalité $\rho*(A_\varepsilon(v_\varepsilon+w_\varepsilon),v_\varepsilon+w_\varepsilon) \geqslant 0$ et on choisit $w_\varepsilon(x_1)$ avec $\text{rot}\, w_\varepsilon = 0$ (donc $w_{\varepsilon 2},\ldots, w_{\varepsilon N}$ sont constants) et $w_\varepsilon \longrightarrow \mu$ dans $L^\infty(\Omega)$ faible $*$; $\rho*(A_\varepsilon v_\varepsilon,v_\varepsilon) \longrightarrow \rho*(Q\lambda,\lambda)$ dans $L^1_{loc}(\Omega)$ faible et $\rho*(A_\varepsilon v_\varepsilon,w_\varepsilon) \longrightarrow \rho*(Q\lambda,\mu)$ dans $L^2(\Omega)$ faible. On choisit ensuite w_ε pour minimiser la limite de $\rho*(A_\varepsilon w_\varepsilon,w_\varepsilon) = ((\rho * A_\varepsilon)w_\varepsilon,w_\varepsilon)$ (puisque w_ε ne dépend que de x_1) ; la plus petite limite est $(B_o\mu,\mu)$ d'après le lemme :

<u>Lemme</u> <u>1</u>. *Soit* $M_\varepsilon \xrightarrow{*\,1} M_o$ *(avec* $M_\varepsilon \geqslant 0$*). Alors si* $w_\varepsilon \longrightarrow \mu$ *dans* $L^\infty(\Omega)$ *faible* $*$ *avec* $w_{\varepsilon 2},\ldots,w_{\varepsilon N}$ *constants et* $(M_\varepsilon w_\varepsilon,w_\varepsilon) \longrightarrow \ell$ *dans* $L^\infty(\Omega)$ *faible* $*$ *on a nécessairement* $\ell \geqslant (M_o\mu,\mu)$ *et il existe une suite telle que* $\ell = (M_o\mu,\mu)$. *[La meilleure suite s'obtient en écrivant* $(M_\varepsilon w_\varepsilon)_1 = \mu_1$, $w_{\varepsilon j} = \mu_j$ *pour* $j > 1$*].*

On a donc $((\rho * Q)\lambda,\lambda) + 2((\rho * Q)\lambda,\mu) + (B_o\mu,\mu) \geqslant 0$, $\forall\, \lambda,\mu$; donc $\rho * Q \leqslant B_o$.

e) (Théorème 7) On utilise l'inégalité $\rho * (A_\varepsilon^{-1}(A_\varepsilon v_\varepsilon+w_\varepsilon),A_\varepsilon v_\varepsilon+w_\varepsilon) \geqslant 0$ et on choisit $w_\varepsilon(x_2,\ldots,x_N)$ avec $\text{div}\, w_\varepsilon = 0$ (donc $w_{\varepsilon 2},\ldots,w_{\varepsilon N}$ sont constants) et $w_\varepsilon \longrightarrow \mu$ dans $L^\infty(\Omega)$ faible $*$;
$\rho * (A_\varepsilon v_\varepsilon,v_\varepsilon) \longrightarrow \rho * (Q\lambda,\lambda)$ dans $L^1_{loc}(\Omega)$ faible et
$\rho * (v_\varepsilon,w_\varepsilon) \longrightarrow \rho * (\lambda,\mu) = (\lambda,\mu)$ dans $L^2(\Omega)$ faible. On choisit ensuite w_ε pour minimiser la limite de $\rho * (A_\varepsilon^{-1}w_\varepsilon,w_\varepsilon) = ((\rho * A_\varepsilon^{-1})w_\varepsilon,w_\varepsilon)$ (puisque w_ε ne dépend pas de x_1) ; la plus petite limite est $(C_o^{-1}\mu,\mu)$.

On a donc $((\rho * Q)\lambda,\lambda) + 2(\lambda,\mu) + (C_o^{-1}\mu,\mu) \geqslant 0$, $\forall\, \lambda,\mu$; donc $\rho * Q \geqslant C_o$.

∎

Dans les démonstrations précédentes on a utilisé le fait que si $v_\varepsilon \rightharpoonup v$, $w_\varepsilon \rightharpoonup w$ dans $L^2(\Omega)^N$ faible avec $\text{rot}\, v_\varepsilon = 0$ et $\text{div}\, w_\varepsilon$ convergeant fortement dans $H^{-1}(\Omega)$ alors $(v_\varepsilon, w_\varepsilon) \rightharpoonup (v,w)$ dans $\mathcal{D}'(\Omega)$. Ce résultat (qui s'obtient facilement en écrivant $v_\varepsilon = \text{grad}\, u_\varepsilon$ et en intégrant par parties) est le premier exemple d'applications d'une méthode nouvelle : la compacité par compensation ; en utilisant d'autres exemples on peut obtenir d'autres inégalités sur Q qui malheureusement ne sont pas très explicites.

Pour fabriquer d'autres inégalités on considère d'abord une forme quadratique B définie sur l'espace de dimension $2N^2$ formé des couples de matrices (V,W) vérifiant la condition

$$\text{i)} \quad \begin{cases} B(V,W) \geqslant 0 \text{ si } \exists\, \xi \in \mathbb{R}^N,\ \xi \neq 0,\ \exists\, \lambda \in \mathbb{R}^N \text{ tels que} \\[2ex] \qquad V_{ij} = \lambda_j \xi_i \quad \forall\, i,j \quad \text{et} \quad \sum_i w_{ij} \xi_i = 0 \quad \forall\, j \end{cases}$$

Si ensuite $L(V,W)$ est une forme linéaire en V,W on définit

$$\text{ii)} \qquad F(A) = \sup_V \left(B(V,AV) + L(V,AV)\right)$$

Théorème 8. *Soit F définie par ii) avec B vérifiant i). Alors si $A_\varepsilon \xrightarrow{H} Q$ et $F(A_\varepsilon) \rightharpoonup \bar{F}$ dans $L^\infty(\Omega)$ faible *, on a $F(Q) \leqslant \bar{F}$.* ∎

Démonstration : Etant donnée une matrice V on peut trouver une suite V_ε convergeant vers V dans $L^2(\Omega)^{N^2}$ faible, telle que $\text{rot}(V_\varepsilon \lambda) = 0 \quad \forall\, \lambda$, avec $\text{div}(A\, V_\varepsilon \lambda)$ convergeant dans $H^{-1}(\Omega)$ fort $\forall\, \lambda$ (et donc $A_\varepsilon V_\varepsilon \rightharpoonup QV$ dans $L^2(\Omega)^{N^2}$ faible). Par compacité par compensation on en déduit que si $B(V_\varepsilon, A_\varepsilon V_\varepsilon)$ converge vaguement vers une mesure μ_0 on a $\mu_0 \geqslant B(V,QV)$. Donc $F(A_\varepsilon)$ est supérieur à $B(V_\varepsilon, A_\varepsilon V_\varepsilon) + L(V_\varepsilon, A_\varepsilon V_\varepsilon)$ qui converge vaguement vers

$\mu_0 + L(V,QV) \geqslant B(V,QV) + L(V,QV)$ d'où $\bar{F} \geqslant B(V,QV) + L(V,QV)$ pour tout V et donc $\bar{F} \geqslant F(Q)$. ∎

Remarques 14 : Le Théorème 5 correspond au cas où B a la forme $(V\lambda, W\lambda)$ avec $\lambda \in \mathbb{R}^N$ et L a la forme $a(V\lambda,\mu) + b(W\lambda,\mu)$.

15 : Le Théorème 8 est valable sans hypothèse de symétrie sur A_ε ; mais dans le cas non symétrique même l'analogue du Théorème 5 donne un résultat compliqué car on trouve les inégalités suivantes :

si $A_\varepsilon \left(\frac{A_\varepsilon+A_\varepsilon^*}{2}\right)^{-1} A_\varepsilon^* \longrightarrow B_0$, $\left(\frac{A_\varepsilon+A_\varepsilon^*}{2}\right)^{-1} A_\varepsilon^* \longrightarrow C_0$ et $\left(\frac{A_\varepsilon+A_\varepsilon^*}{2}\right)^{-1} \longrightarrow D_0$

alors $\forall w,z \in \mathbb{R}^N \quad (B_0w,w) + (C_0w,z) + (w,C_0z) + (D_0z,z) \geq \left(\left(\frac{Q+Q^*}{2}\right)^{-1} (Q^*w+z), Q^*w+z\right)$.

16 : Un problème important est de savoir caractériser les valeurs de Q possibles quand on connaît toutes les limites faibles de fonctions $G(A_\varepsilon)$. Par exemple si on mélange deux matériaux isotropes de conductivités α et β avec les proportions $1 - \theta$ et θ, quelles sont les différentes conductivités possibles pour les matériaux homogénéisés? Plus généralement, connaissant certaines propriétés statistiques de la répartition des matériaux que peut-on déduire des coefficients homogénéisés?

17 : Tout ce qui précède visait à exposer les techniques et la forme des résultats et en particulier est valable pour d'autres situations de type variationnel; tout lecteur courageux saura démontrer de même des inégalités pour le cas du système de l'élasticité. La méthode doit plutôt être affinée d'abord pour obtenir une caractérisation dans le cas modèle.

■

IV - Commentaires

L'homogénéisation dans un cadre général, c'est-à-dire non restreint au cas de coefficients périodiques, a été étudiée d'une part par Spagnolo et De Giorgi dans le cas d'opérateurs symétriques d'ordre 2 sous le nom de G-convergence (on trouvera un exposé de cette approche dans [1]) et d'autre part par Murat et Tartar dans le cas général (on trouvera un exposé dans [2]-[3] et dans les notes [4] quand elles auront paru!). La compacité par compensation a été développée par Murat et Tartar [5]-[6].

Les autres aspects de l'homogénéisation (cas périodique) pourront être consultés (en plus des exposés contenus dans cet ouvrage) dans Babuska [7] et Bensoussan-Lions-Papanicolaou [8].

V - Bibliographie

[1] Spagnolo - Convergence in energy for elliptic operators dans "Numerical Solution of Partial Differential Equations III" ed. B. Hubbard. Academic Press, 1976, p.469-498.

[2] Tartar - Quelques remarques sur l'homogénéisation Colloque Franco-Japonais, Tokyo Sept. 1976 - A paraître.

[3] Murat - Thèse, Paris 1977.

[4] Tartar - Cours Peccot, Collège de France, Paris 1977 - A paraître.

[5] Murat - Compacité par compensation - A paraître dans Ann. Scu. Norm. Pisa.

[6] Tartar - Une nouvelle méthode de résolution d'équations aux dérivées partielles non linéaires
Colloque de Besançon 1977 - A paraître.

[7] Babuska - Homogèneization and its applications
dans "Numerical Solution of Partial Differential Equations III"
ed. B. Hubbard. Academic Press , 1976 , p.89-116.

[8] Bensoussan-Lions-Papanicolaou - Asymptotic Methods in Periodic Structures
North Holland, Amsterdam 1978.

MEDICAL APPLICATIONS
APPLICATIONS MEDICALES

SOME APPLICATIONS OF COMPUTATIONAL MATHEMATICS TO MEDICAL PROBLEMS

M. Bestehorn, U. Ebert, D. Steinhausen, H. Werner
Rechenzentrum der Universität Münster
D 44 Münster/Westf., Germany

If data processing is mentioned in connection with medicine one usually thinks of the administrative applications, e.g. registration of the patients, bookkeeping of the medical services and preparing of the bill at the end of the treatment, in short the rationalisation of all the administrative work in the hospital.

One step further with a closer look we would find medical records (anamnesis, diagnosis, therapy) in addition to those data required and mentioned before. Some medical facts are stored for statistical and epidemiological studies, some for preventive care. Thus the computer might be used to provide regularly invitations for threatened patients. Such a data base also enables the researcher in the medical field to quickly extract cases he may use for special investigations, thus saving him the time he usually would have spent in running over hugh medical files.

In this paper we are investigating a third aspect of computer applications to medicine. We think of medicine as part of natural science. In natural science it has become more and more usual to work quantitatively. In physics and chemistry this is common practice for decades if not centuries. In biological science statistical support of investigations is nowadays obvious. Step by step methods of medicine become also accessable to quantitative investigations if this term is interpreted liberally. This might be related to the fact that more and more technical equipment is used for diagnosis and therapy.

Today it is observed that the physicist-in particular the man working in theoretical physics-will produce his own set of equations, also called his model, and then he will do his own programming. He has been trained to use the mathematical machinery and it is without any criticism to remark that he might not be able to keep up with the most recent developments of numerical methods and computer science for his work but he will resort to approved methods and may also use the software packages available at computing centers.

On the other hand the physician will usually not devise his own quantitative model and he will not write his own programs. He will rely on the support of mathematicians, computer scientists, and engineers. The computer scientist or mathematician, however, cannot attack medical problems without appropriate advice of a physician. This is the reason why we usually have a close cooperation between representatives of the subjects mentioned.

The a priori condition for the quantitative treatment of medical problems is the possibility to formulate a model i.e. to set up quantitative relations which will describe the process we are out to investigate. Of course in setting up a model not all facts that might influence the result will be taken into account. A careful examination of the model, however, will force us to keep in mind which facts have been considered and which are deliberately neglected. Just as in other natural sciences we will establish a functional relation between two categories of quantities, those that can or have been measured and others that are of medical relevance and must be evaluated.

Today we have a well defined portion of mathematics, that is called mathematical physics, but we are far away from a similar state with respect to medicine. There are, however, certain fields beyond statistics such as approximation theory and systems identification, which will definitely belong to those branches of mathematics relevant to medicine. Compare also the work of Peskin [8 , 9].

In this expository talk I like to indicate in some comparatively simple models typical situations the mathematician will face in the field of medicine. It should be emphasized that in these cases not mathematics is the center of our research and not the brilliance of mathematical theorems is what really counts, but the medical result is the main concern and we like to get it with as simple a mathematical tool as possible.

There are some real cases where after possibly lenghty investigations and calculations a relatively simple set of relations emerges that can be numerically solved in order to find the needed medical quantities.

A second very general set will lead to inverse problems of differential equations or problems of systems identification. They are characterized by the fact that we have to estimate parameters that are frequently derived from rather ill conditioned problems and that are not obtained by the standard procedures taught in the mathematics courses.

To give an example suppose that we know the requested results can be obtained theoretically as the solution of a linear differential equation with constant coefficients. It might be necessary to find the coefficients of the differential equation rather than the solution. If we know the coefficients and the initial conditions it is no problem to reproduce the curve but here we have to specify coefficients so that the calculated curve matches the measured data.

In the theory of differential equations we are used to deal with existance, uniqueness and continuous dependence (as specifying a well posed problem) but we are not acquainted with that type of problem described above. It will lead to approximation theory. Looking at the applications we have done, I feel that parameter estimation is usually related to the diagnosis and we are asking how precise we can estimate the parameters so that there is apparently a close relation to the uniqueness problem.

In the third category we could collect problems in which we are able to mathematically support the therapy. Again we like to come out with parameters which will usually be used to set up an instrument or to describe certain doses of medications to serve a therapeutic purpose. In these cases it is not so important how precise the parameters are specified as long as the pursued aim is achieved without unnecessary inconvenience to the patient. In my opinion this relates more to establishing the existence of a solution.

To be more specific we will look into three typical examples that have been treated by us at the computing center of the University of Münster in the last years and we will mention other problems falling into the same categories.

1) An application from ophtalmology showing that finally simple formulas may lead to striking medical results:

The treatment of one-sided aphacia by combination of contact lens and spectacle glasses.

The standard way of adaptation of spectacles by an ophtalmologist is well known to many of us - the physician lets the patient try several glasses until he is satisfied with the gained vision. Numerical approaches became interesting when ophtalmologists started to implant plastic lenses in aphacic eyes, i.e. eyes from which the lens had been removed either because of a cataract or an accident. In this case implanting the optimal lens at the first trial seems absolutely indispensable. Ultrasonic measurements provide the geometric data of the living eye, nowadays. Building a mathematical model is reduced to some calculations of geometrical optics, lenghty and tedious but elementary in character. Basis assumptions are that the thickness of all used lenses might be neglected and that only the neighbourhood of the optical axis is considered, thus no distortion is introduced.

If we analyse an optical system consisting of a phacic eye, a contact lens, and a spectacle glass (compare figure 1 for notations) denoting the refractive powers of the lenses and the cornea of the eye as indicated by D_B, D_H, D_C, D_L (in diopters, say), the only quantity not directly accessable to measurement is the refractive power D_L of the eye lens. One gets the relations [5 ,11]

$$(1) \quad \bar{D}_B = \frac{D_B}{1-\ell \cdot D_B}\ , \quad C = \frac{n}{D_C+D_H+\bar{D}_B}\ , \quad \delta = \frac{d}{n}\ ,$$

$$(2) \quad D_L = \frac{n}{L-d} - \frac{n}{C-d}\ ;$$

n = refractive index of aqueous humor.

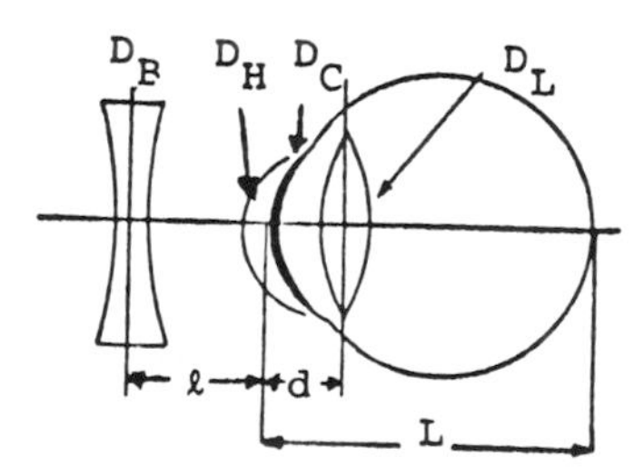

figure 1

and these formulae may either be used to give the focal power of a living eye's lens or to calculate the lens to be implanted in the distance d from the cornea.

As a byproduct the focal length is seen to be equal to

$$(3) \quad F = \frac{L-d}{(1-\ell\cdot D_B)\cdot(1-\delta\cdot D_C-\delta D_H)-\delta\cdot D_B}\ , \qquad \delta = \frac{d}{n}\ .$$

With this length F one gets also a statement on the size of the image perceived on the retina of the eye.

Professor Dr. H. Gernet, the ophtalmologist of our team, knew about the difficulties almost every patient has after one sided surgical removal of the eye lens. The patient no longer has a three dimensional perception, he has trouble to coordinate the two pictures of both eyes associated to one object. The medical answer so far: Cover up one eye by frosted glass or remove the second eye lens if there is any reason to justify it.

Comparison of the two quantities F(normal eye) and F(aphacic eye) immediately gave a clue - they differ considerably and so do the sizes of the two pictures.

After this observation the solution was at hand:

Try to provide

1. a sharp mapping onto the retina in each eye and
2. equality of the focal lengths of the two optical systems (associated with the right and left eye).

This leads to three conditions, hence three parameters are needed. They can be found (principally) in the spectacle glasses <u>and</u> one contact lens.

The procedure now is obvious, neglecting certain medical details and physical limitations. Remember $\delta=0$, $D_L=0$ for aphacic eye. Determine F of phacic eye, find spectacle glass of aphacic eye from

$$(4) \quad F(\text{aphacic}) = \frac{L(a)}{1-\ell\cdot D_B(a)} \overset{!}{=} F\ ,$$

use $D_H(a)$ to satisfy (2) with $D_L=O$ i.e. by

$$(5)\quad \frac{n}{L(a)} = \frac{n}{C(a)} = D_C(a) + \left[D_H(a)+\overline{D}_B(a)\right] .$$

The result looks strange to insiders. We might find a negative glass for the spectacle to the aphacic eye. But this is compensated by the strong (positive) contact lens.

In fact it suffices to satisfy (4) only within a certain liberal error of about 5%. This may be used to gain other additional advantages as e.g. slightly enlarged images of the objects.

All patients treated so far (almost 200) are absolutely satisfied with this type of prescription freeing them from headache and saving them extra eye surgery, giving back the threedimensional perception.

The formulas are so simple that they may be used by an ophtalmologist without further assistance of a mathematician.

2) As second category of problems we consider an example of fitting measured data:

The determination of the albumen fractions of serum.

If the albumen-composition of the serum is needed for the diagnosis, the serum may be decomposed into its albumen portions by the so called electrophoresis. The separation is based on the different quotients of surface charge and the molecular weight of the macromolecules and the different velocities resulting from this in an electrical field.

By the electrophoresis and subsequent chemical procedures the different fractions are separated and their density is represented by the blackening of a strip of special paper. Its optical analysis leads to the curve E(t) of extinction (fig. 2), and the areas of the peaks associated with the different albumen fractions are proportional to the number of particles of the fractions.

We make the hypothesis that the length of the paths, which the molecules of each fraction covered during a certain time, is normally distributed about a mean distance. Thus we are led to approximate the extinction curve E(t) by a sum of weighted Gauss curves $N(\mu,\sigma,t)=\frac{1}{\sigma\sqrt{2\pi}}\exp\left(-\left(\frac{t-\mu}{\sigma}\right)^2\cdot\frac{1}{2}\right)$

that means $E(t) \approx \sum_{i=1}^{k} p_i \, N(\mu_i, \sigma_i, t)$,

if k different albumen fractions are expected. Matching is done with respect to an appropriate norm. In cases of numerical measurements use of the L_2-norm seems to be advisible, i.e., we have to find a parameter vector $p = (p_1, \mu_1, \sigma_1, \ldots, p_k, \mu_k, \sigma_k)$ so that

$$\Phi(p) = \| E(t) - \sum_{i=1}^{n} p_i \cdot N(\mu_i, \sigma_i, t) \| = \min.$$

Since only a finite number of measurements is available the discrete norm will be used. We assume y_k to be the measured data at the point t_k, (k=1,...,n) and define

$$f(p) := \left[(y_1 - \sum_{i=1}^{k} p_i \; N(\mu_i, \sigma_i, t_1)), \ldots, (y_n - \sum_{i=1}^{k} p_i \; N(\mu_i, \sigma_i, t_n)) \right].$$

We assume that a first guess p_o for the parameter vector is available. If f(p) is Frechet differentiable and A is the functional matrix at the point p_o, then

$$\| f(\tilde{p}) - f(p_o) - A(\tilde{p} - p_o) \| = o(\| p_o - \tilde{p} \|) ,$$

so that we can linearize the problem

$$\| f(p_o) + A(\tilde{p} - p_o) + o(\| \tilde{p} - p_o \|) \| = \| f(\tilde{p}) \| = \min$$

to get

$$\| f(p_o) + Ah \| = \min$$

and calculate $\tilde{p} = p_o + h$

This leads directly to the Gauss-Newton algorithm:

i) Determine h_i so that $\| f_i + A_i h_i \| = \min_h \| f_i + A_i h \|$.

ii) Examine for $p_{i+1} = p_i + h_i$, if p_{i+1} is a minimal point or if the error $\| f_{i+1} \|$ is small enough. If this is not true return to i).

This algorithm has the following disadvantages:

a) If A_i is singular, then i) cannot be satisfied.

b) Even if the start vector is "good", it might happen that the norm of the error vector increases.

Otherwise the iteration converges to a stationary point of $\Phi(p)$.

Unfortunately the cases a) and b) indeed arise in practical applications. In nonlinear optimization it is suggested to replace i) by

$$\left\| \begin{pmatrix} f_i \\ 0 \\ \vdots \\ 0 \end{pmatrix} + \begin{pmatrix} A_i \\ B_i \end{pmatrix} h \right\| = \min .$$

Marquardt, Morrison, Levenberg introduced $B_i = \nu_i I$, where ν_i^2 is greater than the greatest Eigenvalue of $A_i^T A_i$ and I is the identity matrix. This procedure slows down the rate of convergence. Following Osborne [2] we try to overcome this draw back by using ν as a control parameter that will go to 0, if convergence takes place.

With this algorithm we solved several discrete non-linear least square problems successfully. For one case of an extinction curve the result is shown in figure 2.

In (mathematically) similar cases the physicians ask for <u>the course of a medication</u> in the blood or in some region of the patient's body in dependence of the time [3].

So we handled a case in which the vitamin-A-level in the blood of the patient had to be determined as a function of time. This reduces mathematically to matching measured values of concentrations by a "Bateman function"

$$y(t) = y_o + \frac{a \cdot k_1}{k_1 - k_2} \left[e^{-k_2 (t-b)} - e^{-k_1 (t-b)} \right]$$

were y_o is the initial concentration of the vitamin-A-level at time of application,

- b is the time between oral application and commencement of resorption,
- k_1 is the rate of resorption,
- k_2 is the rate of reduction, and
- a is a factor proportional to the dosage.

Hence we again face a problem of exponential approximation (figure 3).

3) In this example we describe a model for the situation that seems typical for applications of mathematics in therapeutics [4].

<u>Optimal radiation-treatment planning</u>

We have to specify parameters that control the radiation treatment of

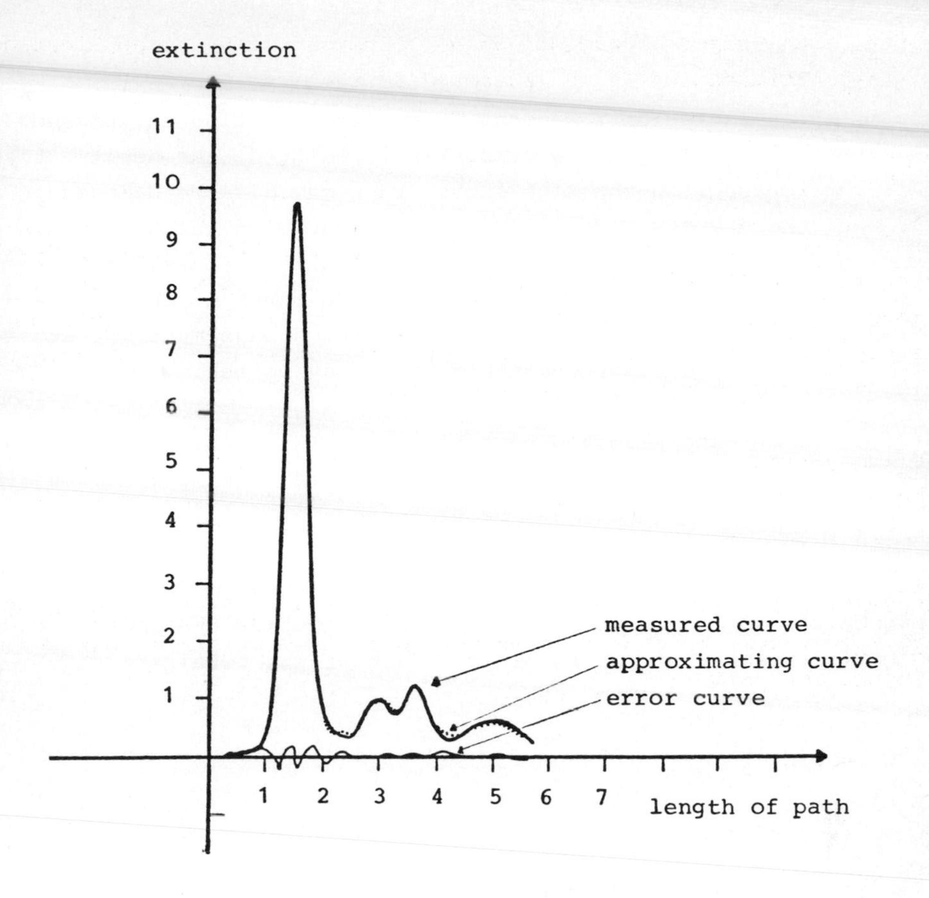

figure 2

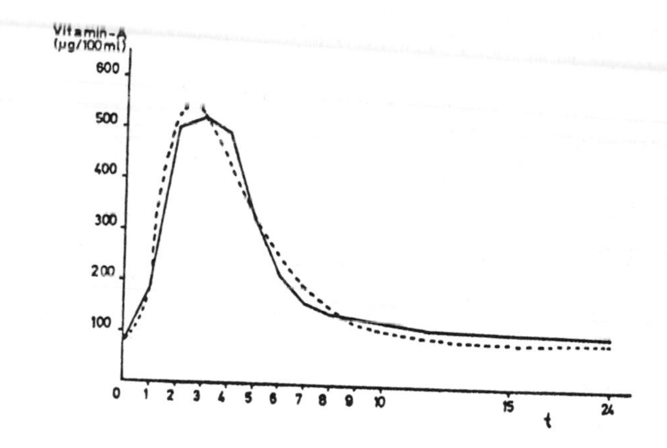

figure 3 —— = measured curve
---- = approximating curve

cancerous tissue.

The aim of every radiation treatment is to destroy malignant tissue within a patient's body. When a tumour is irradiated, the healthy tissue which is confronted with the high-energy rays may also be damaged. The task of radiotherapy is to determine a treatment plan which destroys the tumor and injures the healthy tissue as little as possible. Moreover, one has to take into account that in some parts of the body and some organs (e.g. the liver or spinal cord) the dosage must not exceed a certain amount since otherwise these parts may perish.

Therefore, a treatment plan has to satisfy the following conditions: In some parts of the patient's body minimal doses must be achieved, in other parts maximal doses may not be exceeded.

These restrictions are reflected in a mathematical model by constraints on the dosage in the body. We try to realize the goal of minimal injury of healthy tissue by minimization of the total irradiation time.

The dosage at a point of the body depends on the source of radiation, the condition of the patient, and the time of irradiation. Essential points are the type of the source of radiation, its position, its shape and its alignment. In the patient's body different tissues absorb radiation in different manners. Thus the dosage at a point P is described by

$$D(P,p_1,\ldots,p_s)\cdot t ,$$

where t denotes the irradiation time and $p_1,\ldots,p_s$ designate the remaining parameters. We may assume that D is continuous. The central ray of the cone of radiation is named the <u>central axis</u>.

The physician has two typical ways to arrange the radiation treatment. If the source irradiates the patient for a fixed time from a fixed location with constant parameters, one calls it a <u>fixed field treatment</u>.

In the following, a second model with a moving source of radiation is developed. Hence we have to find a space curve, a parametrization of this curve as function of the time and the alignment to specify the location of the radiation source.

The model for radiation treatment which shall be examined here tries to distribute the directions in space from which the patient is irradiated in as uniform a manner as possible. So it is called the Distribution-model (model for spacial distribution).

The underlying idea is easily explained. If one tries to destroy a tumour with one field then certainly the tissue lying between the tumour and the treatment unit will perish, too. Because of the absorption of radiation by the patient's body the dose there has to be even higher than in the tumour. As the doses of separate fields add in every place it is more advantageous to use more than one field. These fields should come from very different directions and overlap only in the tumour. This concept is generalized in this model: it is assumed that the source of radiation moves along one or more orbits around the patient. The orbits are usually taken from a family of closed space curves, e.g. circles, that may be characterized by a finite number of data. For each orbit the central axis of the field has to pass through a fixed point within the patient, the target point, which is situated in the tumour. The remaining parameters of the source are also constant and are prescribed in advance.

Then the Distribution-model determines the velocity v [cm/sec], with which the source moves along the orbit, or its inverse w=1/v [sec/cm], the marginal "staying" rate at a point of the orbit. The objective function of the optimization problem is the total irradiation time, and it is to be minimized. As a result, advantageous parts of the orbits are used, i.e. the source of radiation will move slowly there. Disadvantageous parts will be avoided, i.e. the source will move very rapidly or not all over these parts.

Considering at first one fixed orbit O we get the following mathematical problem:

For each point P of the body B and for each weight function w=1/v we have the total dose

$$D_S(P,w) = \int_O D(P,s)\cdot w(s)\cdot ds$$

and the total time of irradiation

$$F_S(w) = \int_O w(s)\cdot ds \ .$$

There exist constraints that have to be fulfilled, the given minimal dose $M_\ell(P)$ and maximal dose $M_u(P)$ at every point P of B. For medical reasons it is convenient to specify a maximal marginal staying rate M, i.e. 1/M is the minimum velocity of the source; there should be no quasi-fixed fields.

Thus the Distribution-model can be expressed in the form

$$M_\ell(P) \leq D_s(P,w) \leq M_u(P), \qquad P \in B$$

$$0 \leq |w| \leq M,$$

$$\text{minimize } F_s(w).$$

With appropriate assumptions on the geometric configuration of body B, the tumour, the sensitive organs, the physical parameters, and the space of function w one can establish the existence of an solution to this optimization problem.

In practice we do not have complete information about the patient's body, but only an approximation. Moreover, the weight function w should be easy to realize technically. This leads to a discretization of the problem by which the model is reduced to a problem in linear programming.

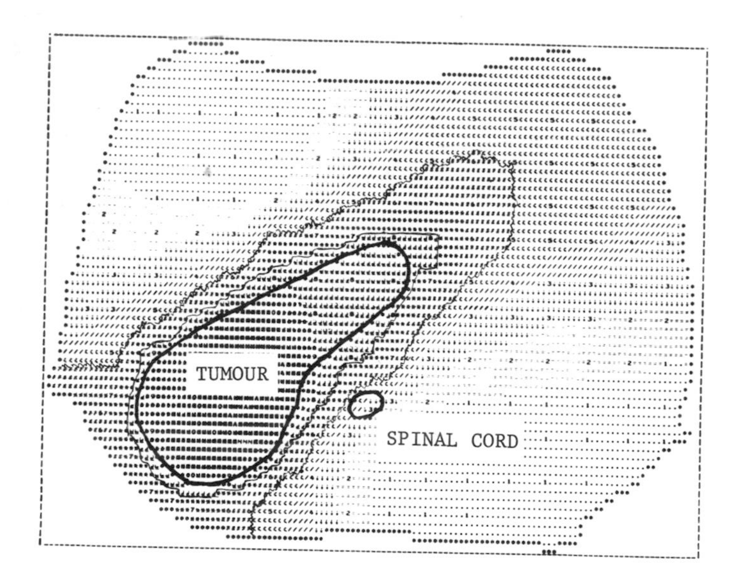

figure 4 Result of the Distribution-model
—— 5000 rd isodose ∿∿ 4000 rd isodose

We have implemented the model explained above. The figure 4 describes the irradiation of a kidney tumour. It has already grown very extensively and the lymphatic ganglions situated in front of the spinal cord must also be irradiated.

In addition to the tumour the spinal cord has been drawn because it contains relatively sensitive nerve fibres. It should receive not more than 4000 rd (radiation absorbed dose) because otherwise it could be irreparably damaged. The tumour itself must get as a minimum a dosage of 5000 rd to be destroyed. Moreover, there are restrictions on the dosage at the skin (4500 rd).

In figure 4 there are two admissible orbits for the source of radiation; each has a fixed target point. The obtained intensity can be seen from the different shaded zones.

The physician prefers the fixed field treatment against the distribution treatment because the adjustment is easier for a fixed field than for a moving source. By means of the automatic treatment planning it is possible to decide whether fixed fields can be used or whether moving fields are indispensable.

In figure 5 we illustrate a case in which the tumour is similarly located as before but smaller. It is here possible to use a fixed field treatment for the irradiation.

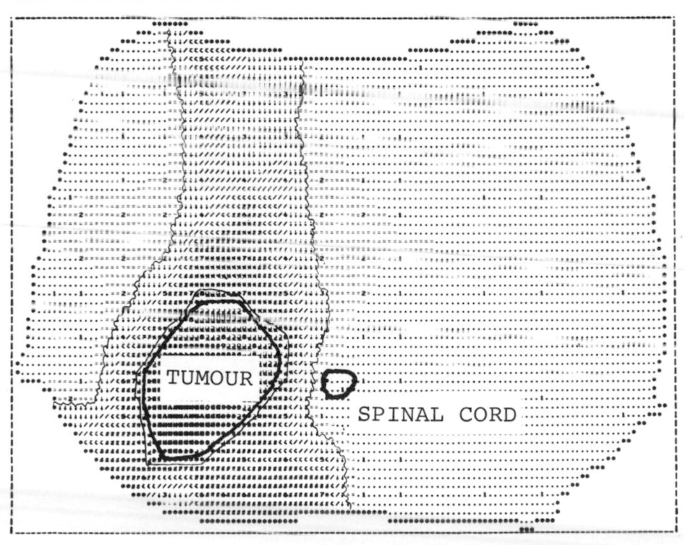

figure 5 Result of a fixed field treatment
—— 5000 rd isodose ∿∿ 2000 rd isodose

Dialysis treatment evaluation and control

Nowadays in a relatively high number of cases the function of the kidney has to be substituted by an dialyzer used as artificial kidney. In [1] some biochemical and mathematical relations are derived to establish a so called dialysis index D_I, which is defined as the quotient of the calculated weekly amount of a marker solute removed and the minimum weekly amount of the marker solute to be removed. Thus the dialysis index gives a measure of success of a dialysis treatment. As shown in [1] it is a function of the weekly dialysis time T, some equipment parameters $\vec{P}$ of the dialyzer used, and of some measurements $\vec{B}$ of the individual patient:

$$D_I = f(T,\vec{P},\vec{B})$$

The problem of optimal treatment evaluation (similar to the radiation treatment planning) is to guarantee or enhance a desired minimal $D_I \geqslant 1$ while minimizing the dialysis time T and choosing the appropriate dialyzer (i.e. $\vec{P}$).

Mathematically this leads to a mixed nonlinear-integer programming problem.

In closing we should mention that the mathematician or computer scientist may encounter problems in the region of rehabilitation, that seem totally unrelated to his field. One example is the translation (conversion) of (machine readable) inkprint into Braille print which is composed of characters of 6 dots read by blind persons with their fingers. The rules for the use of Braille print (different from language to language) demand frequent letter combinations to be contracted. The contractions are governed by sophisticated rules that are partly even based on the semantics of the word or morphem in which they appear. Therefore the said translation can only approximately be achieved by means of computers and presents a great challenge to the mathematician. It is comparable to the design of a language translater. For the German Braille and English Braille converters some fairly good solutions are described in [6,10].

<u>Literatur</u>

[1] Babb, A.L., Strand, M.J., Uvelli, D.A., Milutinove, J., Scriber, B.H., Quantitative Description of Dialysis Treatment: A Dialysis Index
Kidney International, Vol. 7, No 1, Supplement No 2 (1975)

[2] Bestehorn, M., Implementierung eines Verfahrens zur Approximation empirischer Daten mit einer Summe von gewichteten Gaußverteilungen, Rechenzentrum Universität Münster, Schriftenreihe Nr. 5 (1974)

[3] Dost, F.H., Grundlagen der Pharmakokinetik, Georg Thieme Verlag, Stuttgart (1968)

[4] Ebert, U., Planning of Radiation Treatment, in: C.A. Micchelli and T.J. Rivlin (eds.), Optimal Estimation in Approximation Theory, New York (1977), extended version: Optimale Auslegung von Bestrahlungsplänen, Rechenzentrum Universität Münster, Schriftenreihe Nr. 19 (1976)

[5] Gernet, H., Ostholt, H., Werner, H., Neue klinische Grundlagen zur Binkhorst-Linseneinpflanzung bei Altersstar, Sitz. Ber. der 123. Vers. des Vereins Rhein-Westf. Augenärzte (1971)

[6] Gildea, R.A.J., Hübner, G., Werner, H. (Ed.), Computerized Braille Production, Rechenzentrum Universität Münster, Schriftenreihe Nr. 9 (1974)

[7] Osborne, M.R., A class of Methods for Minimizing a Sum of Squares, The Australian Computer Journal 4 (1972), S.164-169

[8] Peskin, C.S., Mathematical Aspects of Heart Physiology, Lecture Notes, Courant Institute, NYU (1973-74)

[9] Peskin, C.S., Partial Differential Equations in Biology, Lecture Notes, Courant Institute, NYU (1975-76)

[10] Werner, H., Dost, W., Automatisierte Herstellung von Blindenschrift mit Hilfe einer Datenverarbeitungsanlage, IBM-Nachrichten, Nr. 194 (1969), S.594-599

[11] Werner, H., Gernet, H., Neuser, G., Rechenschemata, Nomogramme und Tabellen zur Berechnung der Brennweiten linsenhaltiger ametroper brillen- und haftschalenkorrigierter Augen sowie der Brechkräfte der Augenlinsen, GMD Bonn, Nr. 86 (1974)

[12] Werner, H., Gernet, H., Ostholt, H., Neuser, G., Formeln zur Berechnung der optischen Größen, die bei den mit Brillen und Haftschalen korrigierten Augen auftreten, Rechenzentrum Universität Münster, Schriftenreihe Nr. 4 (1974)

[13] Werner, H., Ostholt, H., Gernet, H., Beitrag zur augenseitigen Optik, v. Graefes Arch. klin. exp. Ophtal. 199, S.281-291 (1976)